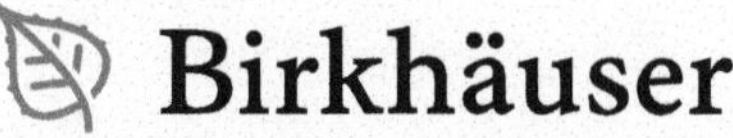

Uluğ Çapar

A Guide to Generalized Functions

Linear, Nonlinear, Random, and Infinite Dimensional Distributions

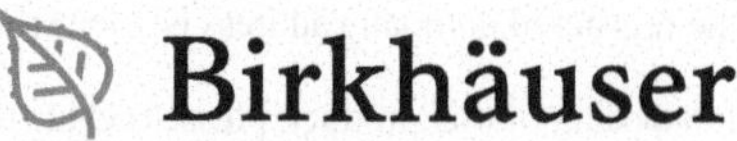

Uluğ Çapar
Faculty of Engineering and Natural
Sciences
Sabancı University
Istanbul, Türkiye

ISBN 978-3-032-09183-3   ISBN 978-3-032-09184-0   (eBook)
https://doi.org/10.1007/978-3-032-09184-0

Mathematics Subject Classification: 28C20, 46A08, 46A11, 46A13, 46E35, 46F05, 46F10, 46F20, 46F25, 46F30, 60B11, 60G57, 60H07

This book is published under the imprint Birkhäuser, www.birkhauser-science.com by the registered company Springer Nature Switzerland AG
The registered company address is: Gewerbestrasse 11, 6330 Cham, Switzerland

If disposing of this product, please recycle the paper.

Acknowledgments The author is indebted to his deceased spouse Sezer Çapar for her constant support during the initial stage of this book project. His thanks are also due to Profs. Hüsnü Ata Erbay of Özyegin University in Istanbul, Mustafa Özdemir of the Mediterranean University in Antalya and Devin Sezer of M.E.T.U. for their guidance in certain directions including the usage of LATEX. He is also thankful to Prof. Kazım İlhan Ikeda of University of Bosphorus and Asst. Prof. Mete Han Karakaş of UCLA for providing the reprints of some rare articles.

Competing Interests The author has no competing interests to declare that are relevant to the content of this manuscript.

Introduction

The theory of generalized functions has four main faces: linear, nonlinear, random and infinite dimensional distributions.

The aim of the present work is to treat these four faces at a fairly rigorous level in a unifying format displaying their interfaces within the confines of a single manageable size volume. There are excellent and classical texts, treatises available in each category. However a single book dealing with each of these categories with a comparable and interlinked format appears to be notably absent. In this way the readership that the present text addresses to is the young researchers, graduate students or mathematicians, applied mathematicians, theoretical physicists who have felt the need to acquire an immediate background, even to specialize in the theory of distributions at some stage of their academic life. Thus before embarking upon a galore of classical books and studying research papers they can get a fairly good idea of different facets of the theory by reading a single text of a few hundred pages.

(I) The theory of generalized functions was initiated by Sobolev and Laurent Schwartz in the mid-fifties of the last century in an attempt to give a rigorous mathematical meaning to such singular objects as the Dirac delta function and its derivatives. Physicists had been dealing with such objects and were ocassionally coming up with erroneous or contradictory results. The new theory was redefining these objects belonging to a class wider than the regular classical functions called 'distributions'. It was basically a theory of duality between topological vector spaces. Hence a distribution was in fact a continuous linear functional assigning a number to the elements of a specific linear space $\mathcal{D}$ called the space of test functions in contrast to the classical conception of a function assigning a number to each point in $\mathbb{R}^n$. The distribution was also infinitely differentiable with a new rule of derivative, thus freeing analysis from the difficulties in presence of non-differentiable functions. Besides this rule of differentiation called the 'distributional derivative' was coinciding with the usual derivative when restricted to classical differentiable functions. This theory of (linear) generalized functions was also suitable for the study of the linear partial differential equations with constant coefficients which may have singularities on their right hand sides and/or in the boundary conditions.

The impressive success of the linear theory in the domain of the differential equations was due to the well-known Malgrange-Ehrenpreis theorem which stated that every linear partial differential operator with constant coefficients had a fundamental solution which was basically a distribution.

(II) Encouraged by the initial successes of the linear theory, researchers set to extend the result of the Malgrange-Ehrenpreis theorem to variable coefficient partial differential operators and possibly to non-linear differential operators. But then Lewy produced a counter-example in which a very simple linear partial differential operator with variable coefficients failed to possess any distribution solution. Thus it was realized that the Schwartz theory of linear distributions was not capable of tackling even the linear phenomena, let the non-linear ones aside.

Further investigations revealed other weaknesses, inadequacies even pathologies of the linear theory of distributions. As a striking example to illustrate this point we can consider the so-called Stability Paradoxes: the systems of equations

$$\begin{cases} u = 0 \\ u^2 = 1 \end{cases}$$

and

$$\begin{cases} u = 0 \\ u^2 = \delta \end{cases}$$

have weak solutions! Actually such anomalies were resulting from the fact that there was not a suitable rule of product in the space of distributions $\mathcal{D}'$. There were different attempts to introduce such a product rule in $\mathcal{D}'$, but all had serious shortcomings. To remedy this deficiency one might consider imbedding $\mathcal{D}'$ into a differential algebra.

But this was also hindered by the Laurent Schwartz's celebrated Impossibility Result. This result roughly states that under very mild and reasonable conditions, the imbedding of $\mathcal{D}'$ into an associative, commutative algebra has to give some concessions regarding the natural properties of the derivative and the product rule (such as the Leibniz rule). For instance such an algebra can not extend the pointwise multiplication of continuous functions, also it can not possess any element δ satisfying $x\delta = 0$. But $\mathcal{D}'$ does possess such a delta, namely the Dirac delta function itself. However such an imbedding is not altogether impossible provided that some properties are abandoned.

In 1984 J. F. Colombeau achieved to imbed $\mathcal{D}'$ into an associative, commutative differential algebra warding off the consequences of the Schwartz's Impossibility Result in an optimal manner, i.e. the concessions given are optimal in many respects. Colombeau's algebra, usually denoted by $\mathcal{G}$, preserved the usual algebraic and differential structure of smooth functions and was stable under a large class of non-linear operations. Also equipped with a peculiar coupled calculus it was capable of dealing with some non-linear partial differential operators. If a differential equation or a boundary value problem failed to have a classical or distribution solution, a solution could be sought in the larger body of Colombeau's non-linear distributions. Colombeau's theory was applied to diversified areas of the non-linear phenomena ranging from shock-wave equations to fluid mechanics, from elastic-plastic deformation in continuum mechanics to certain problems of quantum field theory.

However in Colombeau's and similarly Rosinger's theory of non-linear distributions, the requirement of polynomial growth of non-linear maps was an essential feature. As a consequence many singular problems with nonlinearities growing faster than any polynomial were bound to lay outside the scope of these theories. Egorov proposed an exponential of generalized functions not based on the asymptotic considerations. Delcroix and Scarpalezos constructed spaces of generalized functions associated to a given asymptotic scale and stable under non-linear maps growing faster than polynomials.

(III) One of the standard models of a stochastic process is a family of measurable functions on a probability space $(\Omega, \mathcal{F}, P)$ indexed by a parameter usually interpreted as time. An equivalent view point is to visualize a stochastic process as a random function. For a fixed chance element $\omega \in \Omega$, the random function $f(\omega, t)$ is a trajectory of the process traced in the course of time. One of the most important stochastic processes is the Brownian motion (or the Wiener process) popping up in a large class of natural and social phenomena. Equally important is the so-called white-noise which usually described as the formal derivative of the Brownian motion. However it is well-known that the trajectories of Brownian motion are nowhere differentiable as they are modeled after the very irregular motions of pollen particles submerged in a fluid medium bombarded by the surrounding molecules. Gelfand, one of the greatest mathematicians of the 20th century overcame this difficulty by defining a stochastic process not a random function, but rather a random generalized function (in Schwartz sense) so that the trajectories would be infinitely differentiable in the distribution sense. This was the starting point of the theory of generalized stochastic processes.

The probabilistic behavior of a stochastic process is determined by the probability measure induced on the space of trajectories which is usually a function space. This probability measure is obtained by a certain extension process starting by the finite dimensional probability distributions. In the case of generalized processes the trajectory space consists of linear generalized functions, hence belong to the continuous dual of the test function space. Then in order to obtain the probability measure of the generalized process one has to extend the so-called cylindrical measures. When this is possible, the resulting measure is a special Gaussian measure supported in the dual space.

Gelfand and Badrikian carried out generalizations in this regard when they considered o more abstract set-up to define Gaussian measures in the duals of countably Hilbertian nuclear spaces or general topological vector spaces. The resulting theory is named the theory of cylindrical measures. Here the crucial matter is to investigate the conditions under which the Gaussian cylinder set measures are countably additive.

Oberguggenberger-Russo ([Ober-Rus]) and Çapar-Aktuğlu [Çap-Ak 1] attempted to construct non-linear version of the random generalized functions, in particular the random Colombeau distributions.

(IV) Letting $\mathcal{C}_0([0, 1])$ to be the B-space of real-valued continuous functions on $[0, 1]$ vanishing at $t = 0$ under the uniform norm, define the coordinate functional W_t on this space by $W_t(\omega) = \omega(t)$. It was shown by Wiener that there exists a unique probability measure μ on $\mathcal{C}_0([0, 1])$ such that the map $(t, \omega) \to W_t(\omega)$ is a Wiener process. If H denotes the Hilbert sub-space of $\mathcal{C}_0([0, 1])$ consisting of absolutely continuous functions with square integrable derivatives which is dense in $\mathcal{C}_0([0, 1])$ and called Cameron-Martin space, the triplet $(\mathcal{C}_0([0, 1]), H, \mu)$ is an example of an abstract Wiener space.

A general abstract Wiener space is also a triplet (W, H, μ), W being a B-space obtained by the completion of the Hilbert space H (the Cameron-Martin space) under a weaker norm and μ is the Wiener measure which is the extension of Gaussian cylinder measure on H. Since the underlying space W is in general infinite dimensional, the analysis carried out on the abstract Wiener space is referred to as an infinite dimensional differential analysis. However such an analysis of the Wiener functionals faces a difficulty since most of the usual functionals such as Ito integrals and solutions of many stochastic differential equations are not differentiable in the sense of Fréchet, even not continuous. In the seventies of the last century Malliavin managed to introduce a kind of weak differential calculus utilizing the quasi-invariance of the Wiener measure in the direction of the Cameron-Martin space vectors. Under his sense of Sobolev differentiation important functionals mentioned above became smooth. We can define the gradient, divergence and the Ornstein-Uhlenbeck operators and by means of them we can construct the Sobolev space $\mathbb{D}^\infty(\mathbb{R}) = \bigcap_{k \in \mathbb{N}} \bigcap_{1 < p < \infty} \mathbb{D}_k^p(\mathbb{R})$ of functionals and its dual $\mathbb{D}^{-\infty}(\mathbb{R})$. These spaces are the counterparts of test function space $\mathcal{D}(\mathbb{R})$ and its dual $\mathcal{D}'(\mathbb{R})$ respectively of the linear theory. $\mathbb{D}^{-\infty}$ is the Meyer-Watanabe generalized functionals space.

An alternative approach to infinite dimensional analysis is based on the white-noise space. Consider a Brownian motion $W_t(\omega) = \omega(t)$, $\omega \in \Omega$ vanishing at $t = 0$ a.s. that has its paths in $\Omega = \mathcal{C}_0$ equipped with the topology of uniform convergence on bounded sets. Thus a Wiener functional $f(\omega)$ on Ω can be regarded as a functional of the sample paths. If Ω_0^- denotes rapidly decreasing members of Ω, we know from the sample path properties of Brownian motion that $P(\Omega_0^-) = 1$. As $\Omega_0^- \subset \mathcal{S}'(\mathbb{R})$, (tempered distributions), both the Brownian process and its distribution derivative $\dot{W}$ of its sample paths can be regarded as tempered distribution-valued random elements. Letting $J : \Omega_0^- \to \mathcal{S}'(\mathbb{R})$, it induces , by using the Wiener measure concentrated on Ω_0^- a measure ν on $(\mathcal{S}'(\mathbb{R}), \mathcal{B}(\mathcal{S}'(\mathbb{R})))$; $(\mathcal{B}$: the Borel σ-algebra) which is called the white noise measure. A measurable function on $(\mathcal{S}'(\mathbb{R}), \mathcal{B}(\mathcal{S}'(\mathbb{R})), \nu)$ will be then a white noise functional. A Wiener functional f and a white noise functional F are related by $F = f \circ J^{-1}$. The advantage of this approach is that we can utilize nice properties of $\mathcal{S}'$ as the dual of a nuclear space. Also by means of the second quantization operator (the counterpart of the Ornstein-Uhlenbeck operator in Wiener spaces) we can construct the Sobolev spaces $(\mathcal{S})_p$, $p \in \mathbb{N}$. We let $(\mathcal{S})$ and $(\mathcal{S})^*$ as the projective and inductive limits of $(\mathcal{S})_p$ respectively which are in duality. $(\mathcal{S})$ is called the Hida testing functionals and $(\mathcal{S})^*$ the space of Hida distributions. The Hida's testing functional space $(\mathcal{S})$ is smaller than the Meyer-Watanabe testing functional space $\mathbb{D}^\infty$ although it can be continuously and densely imbedded into the latter. Accordingly the Hida distributions space $(\mathcal{S})^*$ is larger than the Meyer-Watanabe distribution space $\mathbb{D}^{-\infty}$. The theory of Hida distributions was favorably applied to the rigorous formulation of certain operators in quantum physics.

The last Sect. 5.12 based on a recent research of the author ([Çap 1]) is an attempt of a synthesis bringing different sides of the theory together. This is done by imbedding Meyer-Watanabe and Hida distribution into a Colombeau type algebra, thus reconciling two faces of the theory of generalized functions. In this way a product rule, other than indirect and somewhat artificial Wick product is offered for infinite dimensional distributions. As a straightforward application, the Feynman integrand is easily interpreted as a generalized function in the proposed new sense.

The present book is intended to be a guide or handbook for mathematicians and physicists who have intended to specialize in one or more categories of the theory of generalized functions along with their interactions. The author has tried to maintain as rigorous a level as possible. However the text can not be claimed to be comprehensive since there is a vast body of knowledge accumulated in the field and a large number of publication, collected research work etc.

Lengthy proofs, especially those of theorems which have become a classical acquisition of the theory, such as Meyer's inequalities and the Gross theorem, have been omitted. Otherwise the objective of creating a text of manageable size could not have been accomplished. However this is also the trend of famous books as [Wat-1], [Üst-Zak].

For the preparation of the manuscript, the author has used his class notes accumulated during his long teaching career in different universities extending over 5 decades. He has also consulted and made use of the classical texts [Bad], [Bia], [Chow], [Col-1], [Gel-Shil], [Hi-Kuo], [Hor], [Hörm-1], [Hu-Yan], [Kuo-1], [Kuo-2], [Nua], [Ri-Youn], [Rod], [Ros], [Rud], [Schf], [Schw-1], [Shi], [Stak], [Oba], [Üst-Zak], [Wat-1], [Yos] at various levels of utilization by properly referencing.

Sections 4.4, 5.12 include the original research of the author.

The subject matter extend over quite a number of mathematical areas such as functional analysis, theory of functions, measure theory, operator theory, differentiable manifolds, probability theory, stochastic processes and stochastic analysis. The standard concepts and terminology, like those usually covered in the first year graduate courses on these areas are assumed as known by the readers. Those concepts, the definitions and basic properties of which are sufficiently covered in the text, such as Hermite polynomials, countably Hilbertian nuclear spaces are only indicated in the Index. Concepts and terminology falling outside of these two categories such as Fock spaces, tensor products of linear operators etc. are covered in the GLOSSARY in alphabetic order. This was preferred to giving many complements, appendices etc.

In order not to stifle the text into a large number of formula numbers, the special symbols like *, **, †, ‡, ♮, * have been used for intermediate computations, auxiliary formulae if pertinent.

Contents

Chapter 1
Linear Distributions

1.1 Introduction

The concept of generalized functions resulted from the attempts to free analysis from the difficulties in presence of non-differentiable functions or functions with even stronger irregularities. One possible way to get around this difficulty could be to extend the class of functions into a much larger body of objects called 'generalized functions' or simply 'distributions' such that the above shortcomings would disappear.

In order that such an extension be meaningful and useful it should possess the following properties :

(a) All functions from a certain level of regularity on should be included in the new class.
(b) There should be a new rule of differentiation called 'distributional derivative' which coincides with the standard differentiation when the objects are classical differentiable functions.
(c) Every object in this class should have distributional derivative which is again a distribution. Thus the distributions are infinitely differentiable.
(d) There should be enough machinery (limits, transforms etc.) and adequate supply of convergence theorems to carry out analysis in the class.

Curious enough this abstract mathematical requirements are backed up by the physical phenomena in that certain physical entities can not be described satisfactorily by the classical notion of a function as $y = f(x)$. To be more specific: point electrostatic charge, mass concentrated at a single point in space, unit impulse imparted to a system over an infinitely short time interval or a unit vertical force applied at a single point of a one dimensional space are all idealizations. In reality these physical entities are not located in such sets of zero measure. Consider for instance the last example. In fact by any possible device the force can not be exerted at an isolated point a of zero length. There is a force density distribution like in the following figure which can be represented by a function (not necessarily smooth) with a support in a

© The Author(s), under exclusive license to Springer Nature Switzerland AG 2026
U. Çapar, *A Guide to Generalized Functions*,
https://doi.org/10.1007/978-3-032-09184-0_1

Fig. 1.1 Sequence of curves having unit area underneath leading to the idea of a Dirac delta function

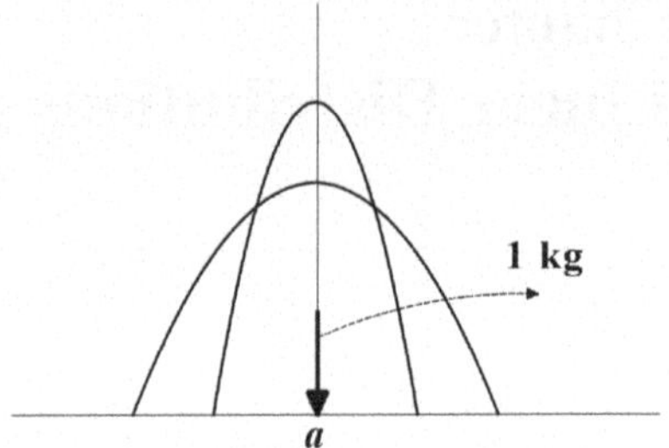

short interval around the point a. To get closer to the idealization we may visualize a sequence of peaking-up functions with narrower and narrower supports shrinking to the single point $\{a\}$ while the area under their graphs is kept as unity. The resulting object reached as a limit is not a conventional function: has a singleton support $\{a\}$ and $f(a) = +\infty$. Furthermore $\int f\,dx = 1$ would be absurd since $f(x) = 0$ for all $x \neq a$, i.e. it is almost everywhere zero (except on the set $\{a\}$ which has measure (length) zero), therefore its integral will be zero by the Lebesgue integration theory (Fig. 1.1).

The limiting object is called the <u>Dirac delta function</u>. Physicist were dealing with such singular objects and their formal derivatives and occasionally they were coming up with erroneous or contradictory results. To straighten such abnormalities these type of objects should be redefined within a well established mathematical theory and this is exactly the <u>theory of generalized functions</u>. This theory is due to S.L. Sobolev (1945) but mainly to Schwartz [Schw-1]. Unifying the needs of mathematics and physics one has to interpret the notion of a function. To simplify take the case of $n = 1$, instead of looking at f as an object assigning the number $f(x)$ to every $x \in \mathbb{R}$, reinterpret as assigning the number $\int f\phi\,dx$ to any suitably chosen <u>test function</u> ϕ. This would be well-defined for instance when f is <u>locally integrable</u>, i.e. measurable and $\int_K |f|\,dx < \infty$ for any compact $K \subset \mathbb{R}$ and ϕ is sufficiently regular. Again this is well in accordance with the physical considerations since measured quantities are almost every time averages.

The set of all test functions will form a space which can be suitably denoted by $\mathcal{D}$. Then $\int f\,\phi\,dx$ will be a special case of a functional on the space $\mathcal{D}$ and with an appropriate topology of $\mathcal{D}$ this functional can be assumed to be continuous.

The above description yields an intuitive framework for what should be a generalized function, i.e. a continuous functional on the space of test functions.

1.2 Test Functions Space $\mathcal{D}(\Omega)$ and the Distributions $\mathcal{D}'(\Omega)$

$\mathcal{D}(\Omega)$: For Ω (open) $\subset \mathbb{R}^n$, this is the space of all complex valued functions $f \in \mathcal{C}^\infty(\Omega)$ with a compact subset of Ω as the support.

$\mathcal{D}_K$: For K a compact subset of $\mathbb{R}^n$, this is the space of all $f \in \mathcal{C}^\infty(\Omega)$ whose support lies in K. Choose compact sets K_N, $(N = 1.2, \ldots)$ such that K_N lies in the interior of K_{N+1} and $\Omega = \bigcup_N K_N$. Then $\mathcal{D}(\Omega) = \bigcup_{K \subset\subset \Omega} \mathcal{D}_K$. ($\subset\subset$ denotes compact subsets). Consider the non-decreasing norms for $f \in \mathcal{D}(\Omega)$:

$$\|f\|_N = \sup\{|D^\alpha f(x)| : x \in \Omega, |\alpha| \leq N\}, \ N = 0, 1, 2, \ldots \qquad (*)$$

where for each multi-index $\alpha = (\alpha_1, \ldots, \alpha_n)$, the differential operator D^α is $D^\alpha = (\frac{\partial}{\partial x_1})^{\alpha_1}, \ldots, (\frac{\partial}{\partial x_n})^{\alpha_n}$ whose order is $|\alpha| = \alpha_1 + \cdots \alpha_n$. If $|\alpha| = 0$, $D^\alpha f = f$.

The restriction of these norms to any fixed K induces the same topology on $\mathcal{D}_K$ as the one generated by the seminorms $p_N(f) = \sup\{|D^\alpha f(x)| : x \in N_K, |\alpha| \leq N\}$, $N = 1, 2, 3, \ldots$.

To see this it is sufficient to notice that to each K compact corresponds an integer N_0 such that $K \subset K_N$ for all $N \geq N_0$. For these N, $\|\phi\|_N = p_N(\phi)$, , $\phi \in \mathcal{D}_K$ and both p_N and $\| . \|_N$ are non-decreasing.

The locally convex topology generated by the countable set $(*)$ of seminorms in $\mathcal{D}(\Omega)$ is metrizable , however it is not complete, (for locally convex topology see the GLOSSARY). To show this let us consider $n = 1$, $\Omega = \mathbb{R}$ and $\phi \in \mathcal{D}(\mathbb{R})$ with $supp\ \Phi \subset [0, 1]$ and $\phi > 0$ in $(0, 1)$. Also let

$$\alpha_m(x) = \phi(x - 1) + \frac{1}{2}\phi(x - 2) + \cdots + \frac{1}{m}\phi(x - m) \quad \text{having support in} \quad [1, m + 1].$$

Notice that $\|\phi(x - j)\|_N = \|\phi(x)\|_N$, $(j = 1, 2, \ldots, m)$. For $n > m$: $\alpha_n(x) - \alpha_m(x) = \frac{1}{m+1}\phi(x - m - 1) + \frac{1}{m+2}\phi(x - m - 2) + \cdots + \frac{1}{n}\phi(x - n)$ and as the supports of the terms on the right-hand side are non-overlapping we can state that $\|\alpha_n(x) - \alpha_m(x)\|_N \leq (\frac{1}{m+1} + \cdots + \frac{1}{n})\|\phi\|_N$, $N \in \mathbb{N}$.

A local base of neighborhoods of the topology induced by the norms given in $(*)$ can be formed by the sets

$$U_N = \{\phi \in \mathcal{D}(\mathbb{R}) : \|\phi\|_N < \frac{1}{N}\}, \quad (N = 1, 2, \ldots).$$

For given ϵ there exists some positive integer k, such that m, n greater than k $\|\alpha_n(x) - \alpha_m(x)\|_N < \epsilon$. Thus $\{\alpha_m\}$ is a Cauchy sequence, but $\lim \alpha_m$ does not have compact support so that it is not in $\mathcal{D}(\mathbb{R})$ hence the suggested topology is not complete.

We introduce now another locally convex topology in $\mathcal{D}(\Omega)$ in which all Cauchy sequences converge. This is the inductive limit topology τ, namely the finest locally convex topology under which all of the canonical injections $\mathcal{D}_K \hookrightarrow \mathcal{D}(\Omega)$ are continuous, ($K \subset\subset \Omega$, $\mathcal{D}(\Omega) = \bigcup_{K \subset\subset \Omega} \mathcal{D}_K$) when $\mathcal{D}_K$ are equipped with the Fréchet topology τ_K as described above. Equivalently the collection of all absolutely convex subsets $W \in \mathcal{D}(\Omega)$ such that $W \cap \mathcal{D}_K \in \mathcal{U}_K$, is a base of neighborhoods , ($\mathcal{U}_K$:

collection of all neighborhoods of the topology τ_K). The topology τ is then the collection of all unions of sets of the form $\phi + W$, $\phi \in \mathcal{D}(\Omega)$ and W a set in the base of neighborhoods.

Let us select a sequence of compact sets K_n as $K_n \subset K_{n+1}^{\circ}$ and $\bigcup_{n \in \mathbb{N}} K_n = \Omega$. Then $\mathcal{D}_{K_{n+1}}$ induces the same topology on $\mathcal{D}_{K_n}$ and the inductive limit topology above is also the strict inductive limit of $\mathcal{D}_{K_n}$, that is the finest convex topology on $\mathcal{D}(\Omega)$ for which the canonical imbeddings are continuous according to the properties of inductive topologies, (c.f.the GLOSSARY).

Some Properties

(I) The topology of any $\mathcal{D}_K$ coincides with the subspace topology that $\mathcal{D}_K$ inherits from $(\mathcal{D}(\Omega), \tau)$:

(i) By the construction of the topology in $\mathcal{D}(\Omega)$, any neighborhood in $(\mathcal{D}(\Omega), \tau)$ restricted to K is a neighborhood in τ_K. Hence the subspace topology is included in τ_K.

(ii) Conversely let V be an open set in τ_K, then the definition of the topology τ_K implies that to every $\phi \in V$ there correspond N and $\delta > 0$ such that

$$\{\psi \in \mathcal{D}_K : \|\psi - \phi\|_N < \delta\} \subset V \qquad (\sharp).$$

Define an absolutely convex neighborhood in $(\mathcal{D}(\Omega), \tau)$ as: $W_\phi = \{\psi \in \mathcal{D}(\Omega) : \|\psi\|_N < \delta\}$. Notice that by $(\sharp)$, $\mathcal{D}_K \cap (\phi + W_\phi) = \phi + (\mathcal{D}_K \cap W_\phi) \in V$. Calling $G = \bigcup_{\phi \in V}(\phi + W_\phi)$ we have $G \cap \mathcal{D}_K = \bigcup_{\phi \in V} \mathcal{D}_K \cap (\phi + W_\phi) = \bigcup_{\phi \in V} \{\phi + \mathcal{D}_K \cap W_\phi\} = V$ (the last equality by the fact that inside the brace is a neighborhood of ϕ included in V).

G is obviously an open set in τ and showing that τ_K is included in the subspace topology.

(II) Let B be a bounded subset of $\mathcal{D}(\Omega)$, i.e. it is absorbed by every neighborhood of the origin, (see the GLOSSARY for the 'absorbent set'). Then $B \subset \mathcal{D}_K$ for some $K \subset\subset \Omega$. Besides B is bounded in $\mathcal{D}_K$, i.e. there are constants M_j such that $\sup_{\psi \in B} \|\psi\|_j \leq M_j$ for every $j \in \mathbb{N}$.

(III) $\mathcal{D}(\Omega)$ has also the Heine-Borel property since a bounded set B is included in some $\mathcal{D}_K$ by II) and $\mathcal{D}_K$ has the Heine-Borel property, (i.e. every open cover has a finite subcover).

(IV) $\mathcal{D}(\Omega)$ as a strict inductive limit of complete $\mathcal{D}_K$ spaces is complete.

Theorem 1 (Convergence in $\mathcal{D}(\Omega)$)

(a) *If $\{\phi_j\}$ is a Cauchy sequence in $\mathcal{D}(\Omega)$, then there is a compact subset $K \subset\subset \Omega$ such that $\{\phi_j\} \subset \mathcal{D}_K$ and $\lim_{j,k \to \infty} \|\phi_j - \phi_k\|_N = 0$ $(N = 0, 1, 2, \ldots)$.*

(b) *Suppose $\phi_j \to 0$ in the topology of $\mathcal{D}(\Omega)$. Then there is $K \subset\subset \Omega$ which contains the support of every test function ϕ_j and $D^\alpha \phi_j \longrightarrow 0$ uniformly as $j \to \infty$ for every multi-index α.*

(Similarly we say that $\phi_j \longrightarrow \phi$ if the sequence $\phi_j - \phi \longrightarrow 0$ as $j \to \infty$ in the topology of $\mathcal{D}(\Omega)$).

Proof (a) Cauchy sequences are bounded . Then the property (II) above implies that a Cauchy sequence $\{\phi_j\}$ lies in some $\mathcal{D}_K$. But then by the property (I) $\{\phi_j\}$ is also a Cauchy sequence relative to τ_K.

(b) As $\{\phi_j\}$ is convergent it is bounded in $\mathcal{D}(\Omega)$. Then according to the property (II) , $\{\phi_j\}$ lies in $\mathcal{D}_K$ for some $K \subset\subset \Omega$. Since the topology of $\mathcal{D}_K$ is the subspace topology inherited from $\mathcal{D}(\Omega)$ (property (I)), ϕ_j also converges to 0 in $\mathcal{D}_K$. But then $\|\phi_j\|_N = \sup\{|D^\alpha\phi_j(x)| : x \in K, \ |\alpha| \le N\} \to 0$, $N = 0, 1, 2, \ldots$ and this means uniform convergence. $\square$

Remark The result is also true when we have a family $\{\phi_\epsilon\}_{0<\epsilon<1}$ of test functions. Then this means there is a compact $K \subset \Omega$ containing the supports of all ϕ_ϵ and $D^\alpha\phi_\epsilon \to 0$ uniformly as $\epsilon \to 0$ for every multi-index α. (In this case the meaning of convergence: given $\delta > 0$ and a neighborhood (of the origin) U, $\exists \epsilon_0 > 0$ such that $0 < \epsilon < \epsilon_0 \Rightarrow \phi_\epsilon \in U$).

Proof $\{\phi_\epsilon\}_{0<\epsilon<1}$ is bounded, thus contained in a compact set K. If the assertion is wrong , then given a neighborhood U in $\mathcal{D}(\Omega)$, $\exists$ a sequence $\{\epsilon_k\}$ converging to 0 such that ϕ_{ϵ_k} does not converge to 0 which contradicts the theorem.

Occassionally another test functions space is used in the theory of distributions named 'the space of rapidly decreasing functions'. First we give the definition of a special differential operator replacing D^α which simplifies some of the formalism.

Definition 1 If α is a multi-index, then $D_\alpha = (i)^{-|\alpha|} D^\alpha = (\frac{1}{i}\frac{\partial}{\partial x_1})^{\alpha_1} \cdots (\frac{1}{i}\frac{\partial}{\partial x_n}^{\alpha_n})$.
Consider all functions $f \in \mathcal{C}^\infty(\mathbb{R}^n)$ for which

$$\sup_{|\alpha|\le N} \ \sup_{x\in\mathbb{R}^n} (1 + |x|^2)^N |(D_\alpha f)(x)| < \infty, \ N = 0, 1, \ldots (\vee)$$

The requirement is that $P.D_\alpha$ is a bounded function on $\mathbb{R}^n$ for every complex polynomial P and for every multi-index α. Since this is true with $(1 + |x|^2)^N P(x)$ in place of $P(x)$, it follows that every $P.D_\alpha$ lies in $L^1(\mathbb{R}^n)$.

Definition 2 All functions in $\mathcal{C}^\infty(\mathbb{R}^n)$ satisfying $(\vee)$ form a vector space denoted by S_n and is called <u>rapidly decreasing functions</u>. On S_n the countable collection of norms $(\vee)$ defines a locally convex topology (c.f [Rud], Theorem 7.4).

Under this topology S_n is complete (i.e. a Fréchet space). This is because if $\{f_i\}$ is a Cauchy sequence in S_n, then for every pair of multi-indices α and β the functions $x^\beta D^\alpha f_i(x)$ converge then uniformly on $\mathbb{R}^n$ to a bounded function $g_{\alpha\beta}$ as $i \to \infty$. It follows that $g_{\alpha\beta}(x) = x^\beta D^\alpha g_{00}(x)$ and hence that $f_i \to g_{00}$ in S_n. Thus S_n is complete, ([Rud], Theorem 7.4 a)).

Definition 3 (*Linear distributions-Schwartz distributions*) Let the space $\mathcal{D}(\Omega)$ of the test functions be equipped with the inductive topology as described above. A continuous linear functional on $\mathcal{D}(\Omega)$ is called a <u>linear distribution or a Schwartz distribution</u> or a linear <u>generalized function</u>. Throughout the text it will also be referred as <u>distributions</u> if there is no danger of confusion with the other types of distributions.

The dual $\mathcal{D}'(\Omega)$ is therefore the space of all distributions defined on Ω. A distribution $T \in \mathcal{D}'(\Omega)$ takes on the value $< T, \phi >$ at $\phi \in \mathcal{D}(\Omega)$.

Proposition 1 *A linear functional on $\mathcal{D}(\Omega)$ is continuous (i.e. it is a linear distribution) $\Longleftrightarrow$ to every compact $K \subset\subset \Omega$ corresponds a positive number M and a non-negative integer N such that for all $\phi \in \mathcal{D}_K$*

$$| < T, \phi > | \leq M \|\phi\|_N \equiv M \max_{|p| \leq N} \sup_{x \in K} |D^p \phi(x)|.$$

<u>Note</u>. If given a distribution T, there is N satisfying this inequality for all compact K (M may vary), then the smallest such N is called the <u>order of T</u>. If no N will do for all K, then T is said to be of <u>infinite order</u>.

Proof A linear functional T on $\mathcal{D}(\Omega)$ is continuous iff its restriction $T \circ 1_K$ is continuous in $\mathcal{D}_K$ for every compact $K \subset\subset \Omega$. But iff condition for this is that there should be a seminorm p_i on $\mathcal{D}_K$ and a positive constant M such that $| < T, \phi > | \leq M p_i(x)$, $x \in K$, $\phi \in \mathcal{D}_K$. If we consider the seminorms given before, we arrive at the conclusion. $\qquad\qquad\square$

<u>SOME EXAMPLES</u> (Linear Distributions)

(1) Delta functions (Dirac Measures)

  (a) Let Ω be an open subset of $\mathbb{R}^n$ and x a point of Ω. The linear form $\delta_x(\phi)$: $\phi \longrightarrow \phi(x)$ is continuous on $\mathcal{D}(\Omega)$. This is by Proposition 1: If $x \notin K$, then $\phi(x) = 0$ and the inequality is trivially satisfied. Otherwise $|\phi(x)| \leq \sup_{y \in K} |\phi(y)| = \|\phi\|_0$ and δ_x becomes a linear distribution of order 0. In relation to the physical phenomena discussed in the introduction , δ_x may be interpreted as the 'unit mass or charge' placed at the point x. If $x = 0$, then $\delta_0 \doteq \delta$ is usually called the <u>Dirac measure</u> on $\mathbb{R}^n$, also called the <u>Dirac delta function</u>. For $n = 1$, $\delta_0(\phi) = \phi(0)$.

  (b) Surface delta function: For $n = 3$ let ξ be a (sufficiently regular) surface in $\mathbb{R}^3$. Define a distribution δ_s by $< \delta_s, \phi >= \int_\xi \phi\, ds$, (a surface integral), $\phi \in \mathcal{D}(\mathbb{R}^3)$.

  (c) Curve delta function: Let l be a sufficiently regular curve in $\mathbb{R}^3$. Define $< \delta_l, \phi >= \int_l \phi\, dl$ (a line integral), $\phi \in \mathcal{D}(\mathbb{R}^3)$.

(2) Let f be a locally integrable function in Ω, an element of L^1_{loc}, i.e. f is Lebesgue measurable and $\int_K |f(x)|dx < \infty$ for every compact $K \subset\subset \Omega$ and dx being the Lebesgue measure in $\mathbb{R}^n$. Define:
$< T_f, \phi >= \int_\Omega f(x)\phi(x)dx$, $\phi \in \mathcal{D}(\Omega)$.   Since $| < T_f, \phi > | \leq (\int_K |f|dx)\|\phi\|_0$, by Proposition 1 $T_f : \mathcal{D}(\Omega) \to \mathbb{C}$ is continuous , thus $T_f \in \mathcal{D}'(\Omega)$. T_f is called a <u>regular distribution</u>. The inclusion $T_f \in \mathcal{D}'(\Omega)$ is also injective:
In the case that f is continuous, let $f \neq 0$. There exists a point $x_0 \in \Omega$ such that $f(x_0) \neq 0$. We can write $f = f_1 + if_2$ where f_1 and f_2 are real valued continuous functions. There exists $\alpha > 0$ and a ball $B_r(x_0)$ such

that the inequalities $f_i(x) > \alpha$ or $f_i(x) < -\alpha$ holds for $x \in B_r(x_0)$ at least for one of the indices $1, 2$. By one of the standard results of functional analysis, ([Hor], Proposition 2.12.5), there exists $\psi \in \mathcal{D}(\Omega)$ such that $\psi(x) \geq 0, \psi(x) = 1$ for $x \in B_{r/2}(x_0)$ and $\psi(x) = 0$ for $x \in [B_r(x_0)]^c$. Then $< T_f, \psi(x) >= \int_{B_r(x_0)} f_1(x_0)\psi(x)dx + i \int_{B_r(\psi(x))} dx \neq 0$ since if for instance $f_1(x) > \alpha \in B_r(x_0)$, then $\int_{B_r(x_0)} f_1(x)\psi(x)dx \geq B_{r/2}(x_0)f_1(x)dx \geq \alpha|B_{r/2}(x_0)|$ where $|.|$ is the volume of $B_{r/2}(x_0)$. Hence $T_f \neq 0$. Thus if $f, g \in \mathcal{C}(\Omega)$ such tat if $f - g \neq 0$, then $T_{f-g} \neq 0$ showing the injectivity.In the case of locally integrable functions two regular distributions T_f and T_g are equal $\Leftrightarrow f = g$ a.e..

Functions more regular than the locally integrable ones, e.g. continuous functions, smooth functions, rapidly decreasing functions, L^p functions can also be used to define distributions. This is why the distributions are called generalized functions.

In particular $\mathcal{C}^\infty(\Omega) \subset \mathcal{D}'(\Omega)$. Algebraically the multiplication on $\mathcal{C}^\infty(\Omega)$ extends to a $\mathcal{C}^\infty$-module structure on $\mathcal{D}'(\Omega)$. For $T \in \mathcal{D}'(\Omega)$ and $\psi \in \mathcal{C}^\infty(\Omega)$ we have $\mathcal{C}^\infty \times \mathcal{D}'(\Omega) \longrightarrow \mathcal{D}'(\Omega), (\psi, T) \to \psi T$ where $(\psi T)(\phi) = T(\psi\phi), \phi \in \mathcal{D}(\Omega)$, (c.f. also 1.6, operation V).

(3) Suppose μ is a complex Borel measure or a positive measure on Ω with $\mu(K) < \infty$ for every compact $K \subset \Omega$. Then the equation $< T_\mu, \phi >= \int_\Omega \phi d\mu$ ($\phi \in \mathcal{D}(\Omega)$) defines a distribution.

(4) Heaviside distribution: Consider $H(x) = \begin{cases} 1 & \text{if } x > 0 \\ 0 & \text{if } x < 0 \end{cases}$. This function defines a regular distribution T_H called the <u>Heaviside distribution</u> by the integral $< T_H, \phi >= \int_0^\infty \phi(x)dx, \ (\phi \in \mathcal{D}(\mathbb{R}))$.

<u>Note</u>. Distributions in (2) and (4) are genuine functions while the one in (3) is a measure. The Dirac functions of (1) are not conventional functions, however they can be re-interpreted as degenerate measures.

More properties of the topology of $\mathcal{D}(\Omega)$

Theorem 2 $\mathcal{D}(\Omega)$ *in its inductive limit topology is barreled, bornological and a Montel space. For these terminology from the theory of locally convex topological vector spaces and their properties see the GLOSSARY.*

Proof $\mathcal{D}_K$'s are Fréchet spaces, therefore they are barreled (GLOSSARY) . Then $\mathcal{D}(\Omega)$ as the inductive limit of $\mathcal{D}_K$'s is also barreled . For the same reason it is bornological. Finally $\mathcal{D}(\Omega)$ is a Montel space since it is both barreled and has the Heine-Borel property (property (III)) preceding Theorem 1. $\square$

Topology in the space $\mathcal{D}'(\Omega)$ of distributions

Calling $E = \mathcal{D}(\Omega)$, we usually consider $E' = \mathcal{D}'(\Omega)$ equipped with the strong topology $\beta(E', E)$ (c.f. the GLOSSARY). Then

Theorem 3 $\mathcal{D}'(\Omega)$ *is barreled, bornological and complete. Besides it is also a Montel space.*

Sketch of the Proof. $E' = \mathcal{D}'(\Omega)$ being the dual of a strictly inductive limit of the spaces $\mathcal{D}_K$, is barreled and bornological according to a theorem of Grothendieck, (c.f. [Hor] Theorem 3.16.2). As $\mathcal{D}(\Omega)$ is bornological (Theorem 2), $\mathcal{D}'(\Omega)$ is complete, (c.f. [Rob-R], Chap. vi, pp. 104, Corollary 1). Furthermore as $\mathcal{D}(\Omega)$ is a Montel space, $\mathcal{D}'(\Omega)$ is also a Montel space.

As a consequence of these topological properties, the weak convergence in $\mathcal{D}'(\Omega)$ implies the strong convergence as it is proved in the following proposition, (the weak convergence is the convergence in weak topology, c.f. GLOSSARY). This result also enables us to check the closure of the space of distributions under certain operations.

Proposition 2 *Let $\{T_j\}_{j \in \mathbb{N}}$ be a sequence of distributions and suppose that for every $\phi \in \mathcal{D}(\Omega)$ $\lim_{j \to \infty} < T_j, \phi >= T(\phi)$ exists. Then $T : \phi \longrightarrow T(\phi) \equiv < T, \phi >$ is a distribution and $\{T_j\}$ converges to T also in the strong topology of $\mathcal{D}'(\Omega)$.*

Proof The elementary filter (c.f. [Rob-R], III.5) associated with the sequence $\{T_j\}$ is $\sigma(E', E)$-bounded $(E = \mathcal{D}(\Omega))$, i.e. it is pointwise bounded: for each $\phi \in \mathcal{D}(\Omega)$, $\{T_j(\phi)\}$ is a bounded set of numbers, (σ denotes the weak topology). We first show that $\{T_j\}$ is equicontinuous.

Let $I = [-\epsilon, \epsilon]$, then $B = \bigcap_{j=1}^{\infty} T_j^{-1}(I)$ is absolutely convex and closed, (closedness is trivial, convexity and balancedness are also easily verified). It is also absorbent as $\{T_j(\phi)\}$ is bounded there exists λ, such that $\{T_j(\phi)\} \subset \lambda I$ or $\phi \in \lambda B$. Hence B is a barrel in the barreled space $\mathcal{D}(\Omega)$ (Theorem 2) and is therefore a neighborhood. But $T_j(B) \subset I$ for all $j \in \mathbb{N}$ and so $\{T_j\}$ is equicontinuous.

For every $\phi \in B$ there exists a positive integer j_0 such that for $j \geq j_0$ $|T(\phi) - T_j(\phi)| \leq \epsilon$. Thus

$$|T(\phi)| \leq |T(\phi) - T_j(\phi)| + |T_j(\phi)| \leq 2\epsilon,$$

showing continuity at the origin which is sufficient for the continuity of T. Clearly T is linear too.

As for the strong convergence of T_j.

Since the same sets are bounded in every topology of a dual pair (c.f. the GLOSSARY) $D = \{T_j\}$ is also bounded in the strong topology $\beta(E', E)$, $(E' = \mathcal{D}'(\Omega))$.

Assertion: $\beta(E', E)$ induces the same topology on D as $\sigma(E', E)$ does. To prove this assertion we may assume that D is closed. Otherwise we prove the assertion for the β-closure $\bar{D}$, if β and σ induce the same topology on $\bar{D}$, then they also do so on D. As $\mathcal{D}'(\Omega)$ is a Montel space under β, it has the Heine-Borel property (c.f. [Rob-R], IV.3 Supplement 2)., therefore D is compact. Let D_σ and D_β be the set D equipped with the topologies $\sigma(E', E)$ and $\beta(E', E)$ respectively, $(\sigma(E', E)$ is the weak topology). Consider the identity bijection $f : D_\beta \to D_\sigma$ which is continuous since $\sigma(E', E)$ is a Hausdorff topology coarser than $\beta(E', E)$. f^{-1} is also continuous. This is because if F is a closed subset of the compact D, then it is compact, thus $f(F)$ is also compact hence closed. This makes f a homeomorphism, hence on $D = \{T_j\}_{j=1}^{\infty}$

the topologies induced by β and σ are identical, meaning that T_j converges strongly too. $\qquad\square$

Remarks (1) The statement also holds when there is a family $\{T_\epsilon\}_{0<\epsilon<1}$ of distributions satisfying $T(\phi) = \lim_{\epsilon\to 0} < T_\epsilon, \phi >$. Then T is a distribution and $\{T_\epsilon\}$ converges to T also strongly in $\mathcal{D}'(\Omega)$ as $\epsilon \to 0$. If the assertion is wrong , then there exists a sequence $\{T_{\epsilon_k}\}$, ϵ_k a sequence of positive numbers tending to zero, such that the sequence $\{T_{\epsilon_k}\}$ does not tend to T strongly. But this contradicts the Proposition 2 since $\{T_{\epsilon_k}\}$ converges to T in the topology $\sigma(\mathcal{D}', \mathcal{D})$, therefore converges strongly to T.

<u>Note</u>. By the convergence of a family $\{T_\epsilon\}$ as $\epsilon \to 0$ strongly to T we understand: given $\delta > 0$ and a strong neighborhood V of T there exists ϵ_0 such that T_ϵ belongs to V when $0 < \epsilon < \epsilon_0$.

(2) The statement of the Proposition 2 can also be obtained as a corollary to the Banach-Steinhaus theorem for locally convex spaces, (c.f. [Rob-R]).

<u>EXAMPLES</u> (Applications of Proposition 2)

(I) $f(x) = \frac{1}{x}$, $x \in \mathbb{R}$ is not locally integrable since it is not integrable in any interval containing the origin. However for each $\phi \in \mathcal{D}(\mathbb{R})$, $\lim_{\epsilon\to 0}\int_{|x|>\epsilon} \frac{\phi(x)}{x}dx$ (†) exists. To see this:

Define ξ by $\xi(x) = \frac{\phi(x) - \phi(0)}{x}$ if $x \neq 0$ and $\xi(0) = \lim_{\epsilon\to 0}\frac{\phi(x) - \phi(0)}{x} = \phi'(0)$. Then $\xi(x)$ is a continuous function on $\mathbb{R}$. Suppose compactly supported ϕ vanishes out when $|x| \geq a$. Then

$$\int_{|x|>\epsilon} \frac{\phi(x)}{x}dx = \phi(0)\int_{\epsilon<|x||<a} \frac{dx}{x} + \int_{\epsilon<|x|<a} \xi(x)dx = \phi(0)(\ln\frac{a}{\epsilon} - \ln\frac{a}{\epsilon})$$

$+ \int_{\epsilon<|x|<a} \xi(x)dx = \int_{\epsilon<|x|<a} \xi(x)dx.$ Then $\lim_{\epsilon\to 0}\int_{|x|>\epsilon} \frac{\phi(x)}{x}dx = \int_{-a}^{a} \xi(x)dx$, so that (†) exists.

On the other hand for fixed ϵ, , $\phi \to \int_{|x|>\epsilon} \frac{\phi(x)}{x}dx$ defines a regular distribution T_ϵ on $\mathbb{R}$. This we can see by Proposition 1 considering $|\int_{|x|>\epsilon} \frac{\phi(x)}{x}dx| \leq 2\ln\frac{a}{\epsilon} \sup_x |\phi(x)|$. Thus by Remark 1 to the Proposition 2 the limit (†) defines a distribution. Since this limit is named the Cauchy principal value (valeur principal) of the integral $\int_{-\infty}^{\infty} \frac{\phi(x)}{x}dx$ in analysis we denote this distribution by $v.p.\frac{1}{x}$, so that $< v.p.\frac{1}{x}, \phi >= \lim_{\epsilon\to 0}\int_{|x|>\epsilon} \frac{\phi(x)}{x}dx$.

(II) <u>Delta Sequences , delta filters</u> Let $\rho(x)$ $(x \in \mathbb{R}^n)$ be a non-negative integrable function with support in a closed ball centered at the origin. Further assume that $\int_{\mathbb{R}^n} \rho(x)dx = 1$ holds. For a sequence $\epsilon_j \downarrow 0$ define $\rho_{\epsilon_j}(x) = \frac{1}{(\epsilon_j)^n}\rho(\frac{x}{\epsilon_j})$. Then $\rho_{\epsilon_j}(x)$ as a sequence of regular distributions converges in $\mathcal{D}'(\Omega)$ weakly also strongly by the Proposition 2. In the present case the limit distribution is the Dirac measure.

Proof First we notice

(i) $\int_{\mathbb{R}^n} \rho_{\epsilon_j}(x)dx = 1$,
(ii) $\lim_{j\to\infty} \int_{\|x\|>A} \rho_{\epsilon_j}(x)dx = 0$ for every $A > 0$,
(iii) $\lim_{j\to\infty} \int_{\|x\|<A} \rho_{\epsilon_j}(x)dx = 1$ for every $A > 0$.

We have to show that for every $\phi \in \mathcal{D}(\Omega)$, $\lim_{j\to\infty} \int_{\mathbb{R}^n} \rho_{\epsilon_j}(x)\phi(x)dx = \delta(\phi) = \phi(0)$. In fact

$$|\int_{\mathbb{R}^n} \rho_{\epsilon_j}(x)\phi(x)dx - \phi(0)| = |\int_{\mathbb{R}^n} \rho_{\epsilon_j}(x)(\phi(x) - \phi(0))dx| \leq |\int_{\|x\|<A} \rho_{\epsilon_j}(x)(\phi(x) - \phi(0))dx|$$

$$+ |\int_{\|x\|>A} \rho_{\epsilon_j}(x)(\phi(x) - \phi(0))dx| \leq \eta(A) + M(\int_{\|x\|>A} \rho_{\epsilon_j}(x)dx \quad (*))$$

where $\eta(A) = \max_{\|x\|\leq A} |\phi(x) - \phi(0)|$ and $|\phi(x) - \phi(0)| \leq M$, $\forall x \in \mathbb{R}^n$. The last inequality in $(*)$ follows from i), the nonnegativity of ρ_{ϵ_j}. We have by the continuity of $\phi(x)$: $\lim_{A\to 0} \eta(A) = 0$. Given $\epsilon_j > 0$ we can therefore choose A such that $\eta(A) < \epsilon/2$. With A so chosen we use (ii) to select j_0 such that for $j \geq j_0$, $\int_{\|x\|>A} \rho_{\epsilon_j}(x)dx < \frac{\epsilon_j}{2M}$. Thus for such choises the right side of $(*)$ will be less than ϵ proving the result. $\qquad\square$

A sequence like $\{\rho_{\epsilon_j}\}$ is called a 'delta sequence'. A simple example for $n = 1$ would be obtained by $\rho(x) = \Pi(x) = \begin{cases} 0 & |x| > 1/2 \\ 1 & |x| \leq 1/2 \\ 1/2 & |x| = 1/2 \end{cases}$ and $\rho_{\epsilon_j}(x) = \frac{1}{\epsilon_j}$ $\Pi(\frac{x}{\epsilon_j})$, ($\Pi(x)$ is called "the gate function").

In accordance with Remark 1 , if $\phi(x)$ is an integrable function with compact support, then $\{\phi_\epsilon(x) = \frac{1}{\epsilon^n}\phi(\frac{x}{\epsilon})\}$ constitutes a delta sequence (filter) converging weakly (also strongly) to $\delta(x)$ as $\epsilon \to 0$. It will be a fortiori so if $\phi \in \mathcal{D}(\Omega)$ and such delta filters will play an important role in the nonlinear distribution theory as discussed in Chap. 3. More generally $\phi_{\epsilon,x}(\lambda) = \frac{1}{\epsilon^n}\phi(\frac{\lambda - x}{\epsilon})$ converges to δ_x.

Theorem 4 $\mathcal{D}(\Omega)$ *regarded as a class of regular distributions is dense in* $\mathcal{D}'(\Omega)$.

Proof Call the image of $\mathcal{D}(\Omega)$ in $\mathcal{D}'(\Omega)$ by M and assume that $\overline{M} \neq \mathcal{D}'(\Omega)$. Then there exists a continuous linear form F on $\mathcal{D}'(\Omega)$ such that $< F, T_\phi >= 0$, $\forall\phi \in \bar{M}$ and $F \neq 0$. But by Theorem 2, $\mathcal{D}(\Omega)$ is a Montel space and therefore it is reflexive . Hence there exists $\psi \in \mathcal{D}(\Omega)$ such that $< F, T >=< T, \psi >$ for every $T \in \mathcal{D}'(\Omega)$, $(F \in \mathcal{D}''(\Omega))$. In particular for $T = T_\phi$, $\int_\Omega \phi(x)\psi(x)dx = 0$ for all $\phi \in \mathcal{D}(\Omega)$. That means we have $< T_\psi, \phi >= 0$ for all $\phi \in \mathcal{D}(\Omega)$, i.e. $T_\psi = 0$. Since the map $\psi \to T_\psi$ is 1 to 1 we have $\psi = 0$, so that $< F, T >= 0$ $\forall T \in \mathcal{D}'(\Omega)$ implies $F \equiv 0$ which is a contradiction, which proves $\overline{M} = \mathcal{D}'(\Omega)$. $\qquad\square$

1.3 Differentiation of Distributions

If α is a multi-index, $\Omega \subset \mathbb{R}^n$ and $T \in \mathcal{D}'(\Omega)$, consider the formula

$$(D^\alpha T)(\phi) \doteq (-1)^{|\alpha|} < T, D^\alpha \phi >, \quad (\phi \in \mathcal{D}(\Omega)) \quad (*)$$

where D^α is a differential operator. Right-hand side makes sense since T is a distribution. In the case that T is a classical function sufficiently differentiable, the equality is verified by integration by parts in the sense of regular distributions so that the requirement Sect. 1.1 (b) of Introduction is met, (see also the next heading). By the fact that D^α also makes sense as a distribution, (*) defines a formula of differentiation for distributions, which is seen as follows:

By Sect. 1.2, Propostion 1 to every compact $K \subset\subset \Omega$ there correspond M and N such that for all $\phi \in \mathcal{D}_K$ we have $| < T, \phi > | \leq M \|\phi\|_N$. But then $|(D^\alpha T)(\phi)| \leq M \|D^\alpha \phi\|_N \leq M \|\phi\|_{N+|\alpha|}$ showing again by Sect. 1.2, Propostion 1 that $D^\alpha \in \mathcal{D}'(\Omega)$. Thus the derivative $D^\alpha T$ exists for every multi-index α and defines a distribution. Formula (*) in the proper distribution notation becomes

$$< D^\alpha T, \phi >= (-1)^{|\alpha|} < T, D^\alpha(\phi) >, \quad (\phi \in \mathcal{D}(\Omega)) \quad (\dagger)$$

For $n = 1$, the multi-index α becomes a positive integer m, thus ($\dagger$) takes the form $< D^m T, \phi >= (-1)^m < T, D^m \phi >$ or in alternative equivalent notation $< T^{(m)}, \phi >= (-1)^m < T, \phi^{(m)} >$.

The formula $D^\alpha D^\beta T = D^{\alpha+\beta} T = D^\beta D^\alpha T$ is valid for every distribution T and for all multi-indices α and β. This follows from the commutativity of D^α and D^β on $\mathcal{C}^\infty(\Omega)$:
$< D^\alpha D^\beta T, \phi >= (-1)^{|\alpha|} < D^\beta T, D^\alpha \phi >= (-1)^{|\alpha|+|\beta|} < T, D^\beta D^\alpha \phi >= (-1)^{|\alpha|+|\beta|} < T, D^{\alpha+\beta} \phi >=< D^{\alpha+\beta} T, \phi >$.

<u>Distribution derivative of functions</u>
If T_f is a regular distribution, then $D^\alpha f$ is the α-th distributional derivative of T_f. If $D^\alpha f$ also exists as classical functional derivative and is locally integrable, then it defines a regular distribution $T_{D^\alpha f}$. In this case the natural question will arise as to whether

$$D^\alpha T_f = T_{D^\alpha f} \quad (\sqrt{})$$

holds. Fortunately this consistency question has positive answer whenever the integrability requirement is satisfied. Because then

$$(-1)^{|\alpha|} \int_\Omega f(x) D^\alpha \phi(x) dx = \int_\Omega (D^\alpha f)(x) \phi(x) dx$$

is easily checked using the integration by parts and the compactness of the supports of test functions. However ($\sqrt{}$) is in general false, (c.f. [Rud], 6.14).

EXAMPLES

(i) Derivative of the Heaviside distribution; $(n = 1)$ $H(x) = \begin{cases} 1 & x > 0 \\ 0 & x < 0 \end{cases}$. We have
$< T_H, \phi >= \int_0^\infty \phi(x)dx$, then $< T_H', \phi >= - < T_H, \phi' >= - \int_0^\infty \phi'(x)dx = -(\phi(x)|_{x=0}^\infty = \phi(0) =< \delta_0, \phi >$ $(\delta_0 \equiv \delta)$.
Thus in the distribution sense $H' = \delta$.

(ii) $T = \delta$, $< \delta', \phi >= - < \delta, \phi'(x) >= -\phi'(0)$.

(iii) $f(x) = \ln |x|$ is locally integrable. Let us find the derivative of the regular distribution T_f:
$< T_f', \phi >= - < \ln |x|, \phi' >= - \int_{-\infty}^\infty \ln |x| \phi'(x)dx$. Decompose the domain of this integral and apply integration by parts

$$- \ln(-x)\phi'(x)dx = - \ln(\epsilon\phi(-\epsilon) + \int_{-\infty}^{-\epsilon} \frac{\phi(x)}{x}dx \quad (*)$$

$$- \int_\epsilon^\infty \ln(x)\phi'(x)dx = \ln(\epsilon)\phi(\epsilon) + \int_\epsilon^\infty \frac{\phi(x)}{x}dx \quad (**)$$

Adding (*) and (**) and passing to the limit as $\epsilon \to 0$ we find in view of Sect. 1.2, Example I

$$< T_f', \phi >= \lim_{\epsilon \to 0} \int_{|x|>\epsilon} \frac{\phi(x)}{x}dx =< v.p.\frac{1}{x}, \phi > .$$

Thus as distribution $(\ln |x|)' = v.p.\frac{1}{x}$.

(iv) Generalizing the previous example, for $n = 1$ let $f(x)$ be a function which is infinitely differentiable at every x except at $x = 0$ but the left and right limits exist of the derivative functions of any order. Let $\sigma_m = f^{(m)}(0+) - f^{(m)}(0-)$ (i.e. the jump (saltus) at the origin). Define $\tilde{f}^{(m)}$ as $\tilde{f}^{(m)}(x) = f^{(m)}(x)$ if $x \neq 0$ and $\tilde{f}^{(m)}(0)$ any real number. Then both f and $\tilde{f}^{(m)}$ are locally integrable, therefore they define regular distributions T_f and $T_{\tilde{f}^{(m)}}$. What is the relation between $(T_f)^{(m)}$ and $T_{\tilde{f}^{(m)}}$?

For $m = 1$; $< (T_f)', \phi >= - < T_f, \phi' >= - \int_{-\infty}^\infty f(x)\phi'(x)dx \quad (\checkmark)$

$$\int_{-\infty}^0 f(x)\phi'(x)dx = f(x)\phi(x)]|_{-\infty}^0 - \int_{-\infty}^0 f'(x)\phi(x)dx$$

$$= f(0-)\phi(0) - \int_{-\infty}^0 f'(x)\phi(x)dx \quad (\dagger)$$

and

$$\int_0^\infty f(x)\phi'(x)dx = -f(0+)\phi(0) - \int_0^\infty f'(x)\phi(x)dx \quad (\ddagger)$$

adding (†) and (‡):

$$< (T_f)', \phi >= [f(0+) - f(0-)]\phi(0) + \int_0^\infty f'(x)\phi(x)dx = \sigma_0\,\phi(0) + < T_{\tilde{f}'}, \phi >$$

Thus $(T_f)' = \sigma_0\delta + T_{\tilde{f}'}$ $(\sqrt{}\sqrt{})$.

This calculation is easily furthered:

$< (T_f)'', \phi > = (-1)^2 < T_f, \phi'' > = \int_{-\infty}^\infty f(x)\phi''(x)dx = f(x)\phi'(x)|_{-\infty}^0 + f(x)\phi'(x)|_0^\infty - \int_{-\infty}^\infty f'(x)\phi'(x)dx = -[f(0+) - f(0-)]\phi'(0) - \int_{-\infty}^\infty f'(x)\phi'(x)dx = < \sigma_0\,\delta' > - \int_{-\infty}^\infty f'(x)\phi'(x)$. Comparing   the   second term with $(\sqrt{})$ and $(\sqrt{}\sqrt{})$ we find $(T_f)'' = \sigma_1\delta + \sigma_0\delta' + T_{\tilde{f}''}$. More generally

$$(T_f)^{(m)} = \sigma_{m-1}\delta + \sigma_{m-2}\delta' + \cdots\cdots + \sigma_0\delta^{(m-1)} + T_{\tilde{f}^{(m)}}.$$

One of the objectives of the distribution theory is to enlarge the class of functions so that the objects in the enlarged class are infinitely differentiable in a certain sense. Conversely every distribution is locally $D^\alpha f$ of some continuous functions f for some multi-index α.

Proposition 1 *Suppose $T \in \mathcal{D}'(\Omega)$ and K is a compact subset of Ω. Then there exists a continuous function f on Ω and a multi-index α, such that $D^\alpha T_f = T|_K$.*

<u>Notes</u>: (i) Equivalently for every $\phi \in \mathcal{D}_K$, $< T, \phi >= (-1)^{|\alpha|} \int_\Omega f(x)(D^\alpha\phi)(x)dx$. (ii) An important interpretation of the Proposition is that $\mathcal{D}'$ is a minimal extension of C^0 in the sense that locally every distribution is weak partial derivative of a continuous function.

Proof Assume without loss of generality that $K \subset Q$ where Q is the unit cube in $\mathbb{R}^n$ i.e. $Q = \{x = (x_1, \ldots, x_n) : 0 \le x_i \le 1 \ (i = 1, \ldots, n)\}$. By the Mean Value Theorem for any $\psi \in \mathcal{D}_Q$

$$|\psi| \le \max_{x \in Q} |(D_i\psi)(x)|, \ \ i = 1, 2, \ldots, n \qquad (*)$$

Let $\Delta = D_1 D_2 \cdots D_n \equiv \dfrac{\partial^n}{\partial x_1 \cdots \partial x_n}$.

For $y \in Q$ let $Q(y) = \{x \in Q : 0 \le x_i \le y_i, \ 1 \le i \le n\}$, then $\psi(y) = \int_{Q(y)}(\Delta\psi)(x)dx$ $(**)$ $(dx = dx_1 \cdots dx_n$ and $\psi(0, \ldots, 0) = 0)$. If N is a non-negative integer, apply $(*)$ recursively

$$|D_j\psi| \le \max_{x \in Q} |(D_i D_j\psi)(x)| \le \cdots \le \max_{x \in Q} |\Delta^N(\psi)(x)|$$

$$D_k D_j\psi| \le \max_{x \in Q} |(D_i D_k D_j\psi)(x)| \le \cdots \le \max_{x \in Q} |\Delta^N(\psi)(x)|.$$

$$\|\psi\|_N \le \max_{x\in Q} |(\Delta^n \psi)(x)| \le \int_Q |\Delta^{N+1}\psi(x)|dx \quad (\forall \psi \in \mathcal{D}_Q) \qquad (\natural)$$

(in the last inequality (**) was used). As T is a distribution, there exists N and C by Sect. 1.2, Proposition 1 , such that $| < T, \phi > | \le C\|\phi\|_N, \quad (\phi \in \mathcal{D}_K)$. Hence by $(\natural)$

$$| < T, \phi > | \le C \int_K |(\Delta^{N+1}\phi)(x)|dx, \quad (\phi \in \mathcal{D}_K) \qquad (\natural\natural)$$

By (**) Δ is one-to-one on $\mathcal{D}_Q$, therefore also on $\mathcal{D}_K$, hence inductively Δ^{N+1} : $\mathcal{D}_K \to \mathcal{D}_K$ is one-to-one. Thus a functional $\tilde{T}$ can be defined on the range $\Re(\Delta^{N+1})$ by setting

$$< \tilde{T}, \phi > = < T, \phi >, \quad (\phi \in \mathcal{D}_K)$$

By $(\natural\natural)$: $| < \tilde{T}, \phi > | \le C \int_K |\psi(x)|dx, \quad \psi \in \Re(\Delta^{N+1})$,

Then by the Hahn-Banach theorem $\tilde{T}$ can be extended to a bounded linear functional on $L^1(K)$. That means according to the (L^1, L^∞) duality there is a bounded Borel measurable function g on K such that

$$< T, \phi > = < \tilde{T}, \Delta^{N+1}\phi > = \int_K g(x)(\Delta^{N+1}\phi)(x)dx \quad (\phi \in \mathcal{D}_K) \qquad (\sharp),$$

Defining $g(x) = 0$ outside K and setting

$$f(y) = \int_{-\infty}^{y_1} \cdots \int_{-\infty}^{y_n} g(x)dx_n \cdots dx_1 \quad (y \in \mathbb{R}^n)$$

we observe that f is continuous and n integrations by parts on $(\sharp)$ show that

$$< T, \phi > = (-1)^n \int_\Omega f(x)(\Delta^{N+2}\phi)(x)dx \quad (\phi \in \mathcal{D}_K)$$

(The integrated terms are 0 since $g = 0$ on K^c). This proves the assertion of the Proposition by the multi-index $\alpha = (N + 2, \ldots, N + 2)$. $\qquad\square$

1.4 Localization and Support of Distributions

First consider the global equality of distributions: As $E = \mathcal{D}(\Omega)$, $E' = \mathcal{D}'(\Omega)$ is a dual pair if T_1 and T_2 are two distributions satisfying $< T_1, \phi > = < T_2, \phi > \; \forall \phi \in E$ or $< T_1 - T_2, \phi > = 0 \; \forall \phi \in E$ will imply $T_1 = T_2$. On the other hand for regular distributions $< T_{f_1}, \phi > = < T_{f_2}, \phi > \; \forall \phi \in E$ would mean $f_1 = f_2$ $a.e.$, (Almost everywhere equality), (c.f. Sect. 1.2, Example 2).

We can also consider the local equality of distributions, Let G be an open subset of Ω. For $T_i \in \mathcal{D}'(\Omega)$ $(i = 1, 2)$, the equality $T_1 = T_2$ in G means, by definition that $T_1\phi = T_2\phi$ for every $\phi \in \mathcal{D}(G)$.

If f is a locally integrable function, then $T_f = 0$ in $G \Leftrightarrow f(x) = 0$ for almost every $x \in G$. Similarly $T_\mu = 0 \Leftrightarrow \mu(A) = 0$ for every Borel set $A \subset G$, (c.f. Sect. 1.2, Example 3).

If the local behavior of a distribution is known, it can also be described globally. This requires a special device called the partition of unity:

Let $\mathcal{G}$ be a collection of open sets in $\mathbb{R}^n$ the union of which is Ω, then there exists a sequence $\{\psi_j\} \subset \mathcal{D}(\Omega)$ satisfying

(a) $\psi_j \geq 0$ and each ψ_j has its support A_j in some member of $\mathcal{G}$,
(b) $\sum_{j=0}^{\infty} \psi_j(x) = 1$, $\forall x \in \Omega$,
(c) For every $K \subset\subset \Omega$, there exists an integer m and an open set $G \supset K$ such that
$\psi(x) + \cdots + \psi_m(x) = 1$ for all $x \in G$.

<u>Note</u>. Such a collection $\{\psi_j\}$ is called a <u>locally finite partition of unity</u> in Ω subordinate to the open covering $\mathcal{G}$ of Ω. This terminology results from the consistency of (b) and (c) which imply that every point $x \in \Omega$ has a neighborhood intersecting only a finite number of the support of $\{\psi_j\}$.

Partition of unity is a tool utilized in the proofs of many theoretical results. It helps us paste together the localizations of a distributions to represent it globally.

Proof Let S be a countable dense subset of $\Omega \subset \mathbb{R}^n$: Consider the countable collection $B_1, B_2, B_3 \cdots$ of closed balls such that the center c_j of B_j is in S, its radius r_j is a rational and B_j lies in some member of $\mathcal{G}$. Let U_j be the open ball with center at c_j and radius $r_j/2$. Using the denseness of S we easily verify that if $x \in \Omega$, then it will be included in some U_j, thus $\bigcup_j U_j = \Omega$.

There are smooth functions $\phi_j \in \mathcal{D}(\Omega)$ such that $\phi_j \geq 0$, $\phi_j = 1$ on U_j and $\phi_j = 0$ off B_j, ([Hor], Proposition 2.12.5). Let $\psi_1 = \phi_1$ and define inductively

$$\psi_2 = (1 - \phi_1)\phi_2, \cdots\cdots , \psi_{j+1} = (1 - \phi_1)(1 - \phi_2)\cdots(1 - \phi_j)\phi_{j+1}, \quad (j \geq 2)$$

Obviously $\psi_j \geq 0$ and $\psi_j = 0$ outside B_j. This proves (a).

We easily establish the relation by induction

$$\psi_1 + \psi_2 + \cdots + \psi_j = 1 - (1 - \phi_1)(1 - \phi_2)\cdots(1 - \phi_j) \qquad (*)$$

for it is trivial for $j = 1$. If it holds for j, then

$$\psi_1 + \cdots + \psi_j + \psi_{j+1} = 1 - (1 - \phi_1)(1 - \phi_2)\cdots(1 - \phi_j) + (1 - \phi_1)(1 - \phi_2)\cdots(1 - \phi_j)\phi_{j+1}$$

$= 1 - (1 - \phi_1)\cdots(1 - \phi_j)[1 - \phi_{j+1}]$.

If $x \in U_j$, then $\phi_j(x) = 1$ and by $(*)$ we have $\psi_1(x) + \cdots + \psi_j(x) = 1$. If $x \in U_1 \cup U_2 \cup \cdots \cup U_j$ for any j, then by $(*)$ we still have $\psi_1(x) + \cdots + \psi_j(x) = 1$ $(\dagger)$.

This yields (b). On the other hand by the definition of compactness given any $K \subset\subset \Omega$, there exists m such that $K \subset U_1 \cup U_2 \cup \cdots \cup U_m = G$ proves (c). $\qquad\square$

Following theorem illustrates the utilization of the partition of unity for the passage from the local to the global behaviour of a distribution.

Theorem 1 *Let $\mathcal{G}$ be an open cover of $\Omega \subset \mathbb{R}^n$ and suppose that for each $G \in \mathcal{G}$ there is a distribution $T_G \in \mathcal{D}'(G)$ satisfying*

$$T_G = T_H \quad on \quad G \cap H$$

whenever $G \cap H \neq \emptyset$, $H \in \mathcal{G}$ and $T_H \in \mathcal{D}'(H)$. Then there exists a unique $T \in \mathcal{D}'(\Omega)$ such that $T = T_G$ on G

Proof Let $\{\psi_j\}$ be a locally finite partition of unity subordinate to $\mathcal{G}$ and associate to each j, a set $G_j \in \mathcal{G}$ such that G_j contains the support of ψ_j.

If $\phi \in \mathcal{D}(\Omega)$, then by the property (b) of a partition of unity $\phi = \sum_j \psi_j \phi$. But as the support of ϕ is compact and the partition of unity is locally finite there are finitely many non-zero terms in this summation. Define

$$T(\phi) = \sum_{j=1}^{\infty} <T_{G_j}, \psi_j\phi> \qquad (\bullet)$$

By property (c) of a partition of unity , the compact support of ϕ intersects the supports of only a finite number of ψ_j. Therefore the summation is well-defined. Furthermore $\phi \longrightarrow T(\phi)$ is clearly linear.

Let us show that T is continuous. Suppose $\phi_j \to 0$ as $n \to \infty$ in $\mathcal{D}(\Omega)$. According to Sect. 1.2, Theorem 1 $\exists K \subset\subset \Omega$ containing the supports of all ϕ_j. Fixing m as in (c) , then by putting $\phi = \phi_r$ in $(\bullet)$

$$T(\phi_r) = \sum_{j=1}^{m} <T_{G_j}, \psi_j\phi_r> \quad (r = 1, 2, \ldots), \qquad (\bullet\bullet)$$

then $\psi_j\phi_r \to 0$ in $\mathcal{D}(G_j)$ as $r \to \infty$, so that by taking limit of the finite sum on the right-hand side of $(\bullet\bullet)$ we find $T(\phi_r) \to 0$ as $r \to \infty$. Hence according to Sect. 1.2, Proposition 2, T is continuous and a distribution $T \in \mathcal{D}'(\Omega)$, $T(\phi) \equiv <T, \phi>$.

To show $T = T_G$ in G : Let $\phi \in \mathcal{D}(G)$. $\phi_j\phi \in \mathcal{D}(G_j \cap G)$, $(j = 1, 2, \ldots)$. Then by the assumption of the theorem $<T_{G_j}, \phi_j\phi> = <T_G, \phi_j\phi>$, hence $<T, \phi> = \sum_j <T_G, \psi_j\phi> = <T_G, \sum_j \psi\phi> = <T_G, \phi>$ which proves the existence. For the uniqueness if in $T = T_G$ we replace $G_j \leftrightarrow G$ then T should satisfy $(\bullet)$. $\square$

Note. The most important special case is $\Omega = \mathbb{R}^n$.

Support of a Distribution.

Suppose $T \in \mathcal{D}'(\Omega)$ and G be an open subset of Ω. If $<T, \phi> = 0$ for all $\phi \in \mathcal{D}(G)$, then T is said to vanish in G. let W be the union of all open sets $G \subset \Omega$ in which T vanishes. The complement of W relative to Ω is the support of T, denoted $supp T$.

Lemma *If W is as described above, then T vanishes in W.*

Proof Let $\mathcal{G}$ be the collection of all open sets in which T vanishes and let $\{\psi_j\}$ be a locally finite partition of unity subordinate to $\mathcal{G}$. If $\phi \in \mathcal{D}(W)$, then $\phi = \sum \psi_j \phi$. As only finitely many terms in this sum is non-zero $< T, \phi > = \sum < T, \psi_j \phi >$ since each ψ_j has its support contained in some $G \in \mathcal{G}$.

EXAMPLES

(1) Let $f \in \mathcal{C}(\Omega)$ and T_f be the corresponding regular distribution. Then $supp T_f = suppf$.
(2) $supp \delta_x = \{x\}$ since if $a \notin G$ and $\phi \in \mathcal{D}(G)$, then obviously $< \delta_x, \phi > = 0$. In particular $supp\delta = \{0\}$.

In the case of distributions with compact support the local result proved in Proposition 1, Sect. 1.3 can be turned into a global one. But firstly

Theorem 2 *Suppose K is compact, V and Ω are open on $\mathbb{R}^n$ and $K \subset \Omega$. Suppose also that $T \in \mathcal{D}'(\Omega)$, that K is the support of T and that T has order N. Then there exist finitely many continuous functions f_β in Ω), one for each multi-index β with $\beta_i \leq N + 2$ for $i = 1, \cdots, n$ with support in V such that $T = \sum_\beta D^\alpha f_\beta$. (The derivatives are to be understood in the distribution sense).*

Proof Choose an open set W with compact closure $\bar{W}$ such that $K \subset W$ and $\bar{W} \subset V.$, Then apply Proposition 1, Sect. 1.3 with $\bar{W} \subset V$. Put $\alpha = (N + 2, \ldots, N + 2)$. The proof of Proposition 1 shows that there is a continuous function f in Ω such that

$$T\phi = (-1)^{|\alpha|} \int_\Omega f(x)(D^\alpha \phi)(x) dx, \quad \phi \in \mathcal{D}(W) \quad (\dagger).$$

We may multiply f by a continuous function which is 1 on $\bar{W}$ and whose support lies in V without disturbing $(\dagger)$. Fix $\psi \in \mathcal{D}(\Omega)$ with support in W, such that $\psi = 1$ on some neighborhood containing K. Then $(\dagger)$ implies , for every $\phi \in \mathcal{D}(\Omega)$ that

$$T\phi = T(\psi\phi) = (-1)^{|\alpha|} \int_\Omega f.D^\alpha(\psi\phi)$$

$$= (-1)^{|\alpha|} \int_\Omega f \sum_{\beta \leq \alpha} c_{\alpha\beta} D^{\alpha-\beta}\psi D^\beta \phi.$$

This is the assertion of the theorem when the differentiation is in the distribution sense with $f_\beta = (-1)^{|\alpha-\beta|} c_{\alpha\beta} f.D^{\alpha-\beta}\psi$ $(\beta \leq \alpha)$.

Theorem 3 (Global structure of distributions) *Suppose $T \in \mathcal{D}'(\Omega)$. There exist continuous functions g_α, one for each multi-index α, such that*

(a) Each compact $K \subset \Omega$ intersects the supports of only a finitely many g_α, and
(b) $T = \sum_\alpha D^\alpha g_\alpha$.

If T has finite order, then the functions g_α can be chosen so that only finitely many are different from 0 ([Rud], Theorem 6.28).

Proof There are compact cubes Q_i and open sets V_i $(i = 1, 2, 3, \ldots)$ such that $Q_i \subset V_i \subset \Omega$, Ω is the union of the Q_i and no compact subset of Ω intersects infinitely many V_i. there exist $\phi_i \in \mathcal{D}^{(}V_i)$ such that $\phi_i = 1$ on Q_i. Use this sequence $\{\phi_i\}$ to construct a partition of unity $\{\psi_i\}$, each ψ_i has its support in V_i. Theorem 2 applies to each $\psi_i T$. It shows that there are finitely many continuous functions $f_{i,\alpha}$ in V_i such that

$$\psi T = \sum_\alpha D^\alpha f_{i,\alpha} \quad (1)$$

Define

$$g_\alpha = \sum_{i=1}^{\infty} f_{i,\alpha}. \quad (2)$$

These sums are locally finite, in the sense that each compact $K \subset \Omega$ intersects the supports of only finitely many $f_{i,\alpha}$. It follows that each g_α is continuous in Ω and that (a) holds.

Since $\phi = \sum \psi_i \phi$, for every $\phi \in \mathcal{D}(\Omega)$, we have $T = \sum \psi_i T$ and therefore (1) and (2) give (b).

The final assertion follows from Theorem 2.

1.5   Tempered Distributions and Distributions of Compact Support

Reconsider the spaces of test functions $\mathcal{D}(\mathbb{R}^n)$, $\mathcal{S}(\mathbb{R}^n)$ and $\mathcal{C}^\infty(\mathbb{R}^n)$; $\mathbb{R}^n$ may be replaced by any open subset of Ω.

We have the following obvious inclusions

$$\mathcal{D} \subset \mathcal{S} \subset \mathcal{C}^\infty,$$

this naively suggests that

$$(\mathcal{C}^\infty)' \subset \mathcal{S}' \subset \mathcal{D}'.$$

However $\mathcal{D}$, $\mathcal{S}$ and $\mathcal{C}^\infty$ have different topologies, therefore the relation between the duals which consist of linear functionals continuous in their respective topologies should be carefully studied.

Tempered Distributions

$\mathcal{S}_n^* \equiv (\mathcal{S}(\mathbb{R}^n))'$ plays a very important role in the Fourier analysis of distributions and takes the name of 'tempered distributions' once its relation to $\mathcal{D}'$ (Schwartz distributions) is established.

Theorem 1 *(a) $\mathcal{D}(\mathbb{R}^n)$ is dense in $\mathcal{S}(\mathbb{R}^n)$.*
(b) The identity mapping $i : \mathcal{D}(\mathbb{R}^n) \to \mathcal{S}(\mathbb{R}^n)$ is continuous.

Proof (a) The inclusion $\mathcal{D} \subset \mathcal{S}$ is obvious. Choose $f \in \mathcal{S}(\mathbb{R}^n)$, $\phi \in \mathcal{D}(\mathbb{R}^n)$ so that $\phi(x) = 1$ on the unit ball of $\mathbb{R}^n$. Put $f_t(x) = f(x)\phi(tx)$, $(x \in \mathbb{R}^n, t > 0)$. Then $f_t \in \mathcal{D}(\mathbb{R}^n)$. If P is a polynomial and α is a multi-index, then

$$P(x)D^\alpha(f - f_t)(x) = P(x)D^\alpha[f(x)(1 - \phi(tx)];$$

by the Leibniz formula for the differentiation of products

$$= P(x) \sum_{\beta \leq \alpha} c_{\alpha\beta} t^{|\beta|} D^\beta [1 - \phi(tx)].$$

But $\phi(tx) = 1$ when $t|x| \leq 1$, thus $D^\beta[1 - \phi(tx)] = 0$ for all multi-indices β when $|x| \leq 1/t$. Since $f \in \mathcal{S}_n$, $P.D^{\alpha-\beta}f$ is bounded for all $\beta \leq \alpha$. It follows that the above sum tends to 0 uniformly on $\mathbb{R}^n$ when $t \to 0$, in another words the test functions $f_t \in \mathcal{D}(\mathbb{R}^n)$ converges in the topology of $\mathcal{S}_n$ to $f \in \mathcal{S}_n$ as $t \to 0$.

(b) If K is a compact set in $\mathbb{R}^n$, the topology induced on $\mathcal{D}_K$ by $\mathcal{S}_n$ is the same as its usual Fréchet topology . This follows from the boundedness of $(1 + \|x\|^2)^N$ on K. The identity function $i : \mathcal{D}_K \to \mathcal{S}_n$ is therefore continuous , even a homeomorphism. As $\mathcal{D}$ is the inductive limit of $\mathcal{D}_K$'s , the continuity of $i : \mathcal{D}_K \to \mathcal{S}_n$ is sufficient for the continuity of $i : \mathcal{D} \to \mathcal{S}$.

CONCLUSIONS

(1) By the continuity of $i : \mathcal{D}(\mathbb{R}^n) \to \mathcal{S}_n$ the inverse image of an open set in $\mathcal{S}_n$ is also open in $\mathcal{D}$. Therefore the topology of $\mathcal{D}(\mathbb{R}^n)$ is finer than the topology of $\mathcal{S}(\mathbb{R}^n)$.

(2) Because of (1), if $h : \mathcal{S}_n \to E$ is a continuous map into a locally convex space E, then its restriction $h|_\mathcal{D}$ to $\mathcal{D}(\mathbb{R}^n)$ is also continuous. In particular if $U \in \mathcal{S}_n'$, then $U|_\mathcal{D} = T$ is continuous and linear, therefore $< T, \phi > = < U, i\phi >$ defines a distribution , i.e. an element of $\mathcal{D}'(\mathbb{R}^n)$. Two different members of $\mathcal{S}_n'$ can not give rise to the same distribution in $\mathcal{D}'$: If $U_1 \neq U_2$, $\exists \psi \in \mathcal{S}_n$ such that $< U_1, \psi > \neq < U_2, \psi >$. If $\psi \in \mathcal{D}$, then $T_1 = U_1|_\mathcal{D} \neq U_2|_\mathcal{D} = T_2$. Otherwise by Theorem 1a there exists a sequence $\{\phi_j\} \subset \mathcal{D}(\mathbb{R}^n)$, $\phi_j \xrightarrow{\mathcal{S}_n} \psi$. $< U_1, i\phi_j > = < U_2, i\phi_j >$ can not hold for all j, otherwise it implies $< U_1, \psi > = < U_2, \psi >$ contradicting the assumption. Then for some j_0, $< U_1, i\phi_{j_0} > = < T_1, \phi_{j_0} > \neq < U_2, i\phi_{j_0} > = < T_2, \phi_{j_0} >$ so that $T_1 \neq T_2$ holds. In this way we have a vector space isomorphism, say γ, such that $\gamma(\mathcal{S}_n^*) \subset \mathcal{D}'(\mathbb{R}^n)$. If $T \in \gamma(\mathcal{S}_n^*)$, then $U = \gamma^{-1}(T)$ is the continuous extension of T to $\mathcal{S}_n$. Identifying $\gamma(\mathcal{S}_n^*)$ with $\mathcal{S}_n^*$ we arrive at the following definition :

Definition 1 The elements of the dual space $\mathcal{S}_n^* \equiv \mathcal{S}'(\mathbb{R}^n)$ are called <u>tempered distributions</u>.

The tempered distributions are precisely those $T \in \mathcal{D}'(\mathbb{R}^n)$ that have continuous extensions to $\mathcal{S}_n$.

(During the text the symbols $\mathcal{S}_n^*, \mathcal{S}_n'(\mathbb{R}^n), \mathcal{S}'(\mathbb{R}^n)$ are used interchangeably for tempered distributions depending on the context and other symbols used).

<u>Note</u>. What does the qualifier "tempered" (referred to as 'temperated' in some sources) signify? In fact it indicates a growth restriction at infinity. For this reason an alternative terminology is <u>slowly increasing distributions</u> (or distributions of slow growth).

Definition 2 A function $f \in \mathcal{C}^\infty(\mathbb{R}^n)$ is called <u>slowly increasing</u> if for any multi-index α, $\exists$ a non-negative integer N such that $\lim_{x \to \infty} \|x\|^{-N} |D^\alpha f(x)| = 0$.

The totality of slowly increasing functions is denoted by $\mathcal{O}(\mathbb{R}^n)$. It becomes a locally convex space by the usual algebraic operations and the seminorms $p_{h,\alpha}(f) = \sup_{x \in \mathbb{R}^n} |h(x) D^\alpha f(x)|$, $f \in \mathcal{O}(\mathbb{R}^n)$, where $h(x)$ is an arbitrary function in $\mathcal{S}_n$ and α is a multi-index. We have $p_{h,\alpha} < \infty$, e.g. this is easily seen for $n = 1$ by $p_{h,j}(f) = \sup_{x \in \mathbb{R}} |(x^N h(x))(\dfrac{D^j f(x)}{x^N})|$ and use the definitions of functions in $\mathcal{S}(\mathbb{R})$ and $\mathcal{O}(\mathbb{R})$, for any n use Leibniz's formula of differentiation to show $h(x) D^\alpha f(x) \in \mathcal{S}_n$. Besides by a proof similar to that of Theorem 1a it is shown that $\mathcal{D}(\mathbb{R}^n)$ is dense in $\mathcal{O}(\mathbb{R}^n)$.

Using any function $f \in \mathcal{O}(\mathbb{R}^n)$ we obtain a regular distribution T_f which is actually tempered. In fact $T_f(\phi) = <f, g> = \int_{\mathbb{R}^n} f(x)\phi(x)dx$ $(\phi \in \mathcal{S}_n)$. The integral converges absolutely by the assumptions on f and ϕ. As linearity is clear, it is sufficient to show continuity: Let $\phi_j \to 0$ in the topology of $\mathcal{S}_n$, then considering

$$|\int_{\mathbb{R}^n} f(x)\,\phi_j(x)\,dx| = |\int_{\mathbb{R}^n} \frac{f(x)}{(1 + \|x\|^2)^N}(1 + \|x\|^2)^N \phi_j(x)\,dx|$$

for N sufficiently large $\dfrac{f(x)}{(1 + \|x\|^2)^N}$ is absolutely integrable since f is slowly increasing. For such large N, the integral on the right is dominated by

$$\sup_{x \in \mathbb{R}^n}[(1 + \|x\|^2)^N \phi_j(x)] \int_{\mathbb{R}^n} \frac{|f(x)|}{(1 + \|x\|^2)^N} dx \,,$$

as $\phi_j \to 0$ in $\mathcal{S}_n$, the supremum also tends to 0, thus continuity.

As in the case of a distribution in $\mathcal{D}'(\mathbb{R}^n)$, we can define the generalized derivative of a tempered distribution T by

$$D^\alpha T(\phi) = (-1)^{|\alpha|} < T, D^\alpha \phi >, \quad \phi \in \mathcal{S}_n)$$

Lemma 1 *If P is a polynomial and $h \in S_n$, then for any tempered distribution T, $D^\alpha T$, $P.T$ and $h.T$ are also tempered distributions.*

Proof $\phi \to D^\alpha \phi$, $\phi \to P.\phi$ and $h \to h.\phi$ are continuous, linear mappings of S_n into S_n in the topology of S_n. Then the conclusion follows from the above definition of the derivative of a tempered distribution and the operations on distributions (c.f. Sect. 1.6, V) and the Note given there). According to these definitions of operations: $(P.T)(\phi) = < T, P.\phi >$ and $(h.T)(\phi) = < T, h, \phi >$, $(\phi \in S_n)$. $\qquad\square$

EXAMPLES

(a) Let μ be positive Borel measure on $\mathbb{R}^n$ such that $I_{n_0} \doteq \int_{\mathbb{R}^n} (1 + \|x\|^2)^{-n_0} d\mu(x) < \infty$ for some positive integer n_0. Then T_μ defined by $< T_\mu, \phi > = \int_{\mathbb{R}^n} \phi \, d\mu$ is a tempered distribution, $(\phi \in S_n)$. To see this suppose $\phi_j \to 0$ in S_n.

Then $| < T_\mu, \phi_j > | = | \int_{\mathbb{R}^n} (1 + \|x\|^2)^{n_0} \dfrac{\phi_j(x)}{(1 + \|x\|^2)^{n_0}} d\mu |$

$\leq \sup_{x \in \mathbb{R}^n} (1 + \|x\|^2)^{n_0} |\phi_j(x)| \int_{\mathbb{R}^n} (1 + \|x\|^2)^{-n_0} d\mu(x) = \sup_{x \in \mathbb{R}^n} (1 + \|x\|^2)^{n_0} |\phi_j(x)|.I_{n_0} \longrightarrow 0$ since $\phi_j \in S_n$. Thus $< T_\mu, \phi_j > \to 0$ as $j \to \infty$ and T_μ is continuous on S_n defining a tempered distribution.

(b) Every $f \in L^p(\mathbb{R}^n)$ $(1 \leq p \leq \infty)$ defines a tempered distribution. In fact this is a special case of (a) when we replace $d\mu$ by $|f| dx$. To show that $|f| dx$ satisfies the condition in (a) consider the inequality

$$\int_{\mathbb{R}^n} (1 + \|x\|^2)^{-k} |f(x)| dx \leq \left(\int_{\mathbb{R}^n} |f(x)|^p dx \right)^{1/p} \left(\int_{\mathbb{R}^n} (1 + \|x\|^2)^{-kq} dx \right)^{1/q}$$

resulting from the Hölder's inequality where q is the exponent conjugate to p, $(\frac{1}{p} + \frac{1}{q} = 1)$. For large k say $k = n_0$ the right side is finite , thus the condition of (a) is satisfied. Hence $< T, \phi > = \int_{\mathbb{R}^n} f(x) \phi(x) dx$, $(\phi \in S_n)$ defines a tempered distribution.

(c) As a consequence of (b) every polynomial and every measurable function whose absolute value is majorized by some polynomial defines a tempered distribution.

(d) Similar to $\mathcal{D} \subset \mathcal{D}'$ we have $S_n \subset S_n'$ and the imbedding is injective and dense. This is shown with a proof parallel to that of Sect. 1.2, Theorem 4.

Theorem 2 (Structure theorem for tempered distributions) *Every distribution $T \in S'(\mathbb{R}^n)$ is a finite mixed partial derivative of a slowly increasing continuous function. ([Rich-Youn], Appendix 2).*

Distributions of Compact Support

The support of a distribution was defined in Sect. 1.4. The prototype of distributions having a compact support is a regular distribution generated by $f \in \mathcal{D}(\mathbb{R}^n)$. By the Example (1) of the support of a distribution, $supp T_f = supp f$ hence this T_f has a compact support. However the domain of this T_f is larger than $\mathcal{D}$. In fact if $\phi \in C^\infty(\mathbb{R}^n)$, $< T_f, \phi > = \int_{\mathbb{R}^n} f(x) \phi(x) dx$. The integral is absolutely convergent since f has compact support. For the continuity of T_f: Let $supp f = K_0$ and let $\{\phi_j\}$

be a sequence converging to 0 in the topology of $C^\infty(\mathbb{R}^n)$, that means $D^\alpha\phi_j \to 0$ uniformly on every compact set , in particular $|\phi_j| \to 0$ on compacts. We have

$$| < T_f, \phi_j > | \le \sup_{x\in K_0} |f(x)|| \int_{K_0} \phi_j(x)dx| \le \sup_{x\in K_0} |f(x)| \sup |\phi_j(x)| m(K_0)$$

(m: Lebesgue measure in $\mathbb{R}^n$), the right-hand side tends to zero, thus T_f is a continuous, linear functional on $C^\infty(\mathbb{R}^n)$.

<u>Note on the Notation.</u> In all of the discussions of this Subsection on the tempered distributions and the distributions of compact support $\mathbb{R}^n$ can be replaced by an open subset $\Omega \subseteq \mathbb{R}^n$. It is customary to denote $C^\infty(\Omega)$ by $\mathcal{E}(\Omega)$, thus the regular distribution T_f just discussed will be in $\mathcal{E}'$.

As the boundedness of $P(x)D^\alpha f(x)$ on Ω (in particular on compact subsets K) implies the boundedness of $D^\alpha f(x)$, the topology of $\mathcal{S}_n(\Omega)$ is finer than the topology of $\mathcal{E}(\Omega)$. Hence a linear functional continuous on $\mathcal{E}(\Omega)$ is also continuous when restricted to $\mathcal{S}_n(\Omega)$. Thus we have $\mathcal{E}'(\Omega) \subset \mathcal{S}'(\Omega)$.

As for the relation of $\mathcal{E}'(\Omega)$ to the distributions with compact support:

Theorem 3 *A distribution $T \in \mathcal{D}'(\Omega)$ with compact support can be extended uniquely to a continuous, linear functional $\tilde{T}$ on $\mathcal{E}(\Omega)$ such that $\tilde{T}(f) = 0$ if $f \in \mathcal{E}(\Omega)$ vanishes in a neighborhood containing $suppT$.*

<u>Note.</u>The condition implies that if $f_i \in \mathcal{E}(\Omega)$ ($i = 1, 2$) satisfies $f_1 = f_2$, then $\tilde{T}(f_1) = \tilde{T}(f_2)$.

Proof Let $K = suppT$. For $\epsilon > 0$ sufficiently small, let us cover the compact set K by a finite union of open spheres $S(x; \epsilon)$ with center at $x \in K$ and having radius ϵ. Let $\{\psi_j(x)\}$, ($j \in J$) be a partition of unity subordinate to this finite system of spheres. Then define the function $\phi(x) = \sum \psi_j$, the sum being over ψ_j with $supp\psi_j \cap K' \neq \emptyset$, where K' is a compact neighborhood of K contained in the interior of the above system of spheres. Then $\phi(x) \in \mathcal{D}(\Omega)$ and $\phi(x) = 1$ in a neighborhood of K. Define $\tilde{T}(f)$ for $f \in \mathcal{E}(\Omega)$ by $\tilde{T}(f) = T(\phi f)$. This definition is independent of the choice of ϕ, for if $\phi_1 \in \mathcal{D}(\Omega)$ equals 1 in a neighborhood of K, then for any $f \in \mathcal{E}(\Omega)$ the function $(\phi - \phi_1)f \in \mathcal{D}(\Omega)$ vanishes in a neighborhood of K, so that $T(\phi(x)f) - T(\phi_1 f) = T[(\phi - \phi_1)f] = 0$.
$\tilde{T}$ is an extension of T since on K, $\phi = 1$ thus $\tilde{T} = T$ on $suppT$).

By applying the Leibniz's formula of differentiation to $D^\alpha(\phi f)$ we see that $\{\phi f\}$ ranges over a bounded set of $\mathcal{D}(\Omega)$ when f ranges over a bounded set of $\mathcal{E}(\Omega)$, (c.f. Sect. 1.2, Property II for bounded sets in $\mathcal{D}(\Omega)$). As a continuous, linear functional is bounded on bounded sets $\tilde{T}(f) = T(\phi f)$ is bounded on bounded sets of $\mathcal{E}(\Omega)$. Since $\mathcal{E}(\Omega)$ is metrizable ,it is bornological, thus $\tilde{T}$ being bounded on bounded sets of $\mathcal{E}$ is continuous.

Letting $f \in \mathcal{E}(\Omega)$ vanish in a neighborhood , say $V(K)$ of K, then by choosing a $\phi \in \mathcal{D}(\Omega)$ that vanish in $\Omega \setminus V(K)$, we have $\tilde{T}(f) = T(\phi f) = 0$.   □

Remark Following the construction in the proof we can also assert the following :
For some constants C and N

$$| < \tilde{T}, f > | \leq C. \sup\{|D^\alpha f(x) : |\alpha| \leq N, \, x \in K'\} \text{ for all } f \in \mathcal{E}(\Omega)$$

This is because as $T \in \mathcal{D}'(\Omega)$ and K' is compact , there exists, according to Sect. 1.2, Proposition 1, constants M and N such that $| < T, \psi > || \leq M \sup\{|D^\alpha \psi(x)| : |\alpha|| \leq N, x \in K'\}$. For any $f \in \mathcal{E}(\Omega)$ let $\psi = \phi f \in \mathcal{D}_{K'}$ (ϕ as in the proof). By again the Leibniz's formula and recalling that $\phi = 1$ in K' we have

$$\sup\{|D^\alpha(\phi f)| : |\alpha| \leq N, x \in K'\} \leq C \sup\{|D^\alpha f(x)| : |\alpha| \leq N, x \in K'\}$$

or $| < T, \phi f > | = | < \tilde{T}, f > || \leq C \sup\{|D^\alpha f(x)| : |\alpha| \leq N, x \in K'\}$ $f \in \mathcal{E}(\Omega)$.

Theorem 4 *Let S be a linear functional on $\mathcal{E}(\Omega)$ such that for some constant C, a non-negative integer N and a compact subset K of Ω*

$$| < S, f > | \leq C \sup\{|D^\alpha f(x)| : |\alpha| \leq N, x \in K\} \; \forall f \in \mathcal{E}(\Omega)$$

holds. Then the restriction of S to $\mathcal{D}(\Omega)$ is a distribution T with compact support contained in K.

Proof We first observe that the inequality implies $< S, f >= 0$ if f vanishes identically in a neighborhood of K. Thus if $\phi \in \mathcal{D}(\Omega)$ equals 1 in a neighborhood of K, then as $f - \phi f$ vanishes in a neighborhood of K we have $< S, f >=< S, \phi f$ for all $f \in \mathcal{E}(\Omega)$. Then by virtue of the Leibniz's formula we can conclude that $< S, \phi f >= T(f)$ is bounded on bounded sets of $\mathcal{D}(\Omega)$, so that $\mathcal{D}(\Omega)$ being bornological is a continuous linear functional on $\mathcal{D}(\Omega)$ with support contained in K.

Thus we have established:

Proposition 1 *The set of all distributions in Ω with compact support is identical with the space $\mathcal{E}'(\Omega)$ of all continuous, linear functionals on $C^\infty(\Omega)$. A linear functional T on $C^\infty(\Omega)$ belongs to $\mathcal{E}'(\Omega)$ iff for some constant C, a non-negative integer N and a compact subset $K \subset\subset \Omega$*

$$| < T, f > | \leq C \sup\{|D^\alpha f(x)| : |\alpha| \leq N, x \in K\} \text{ for all } f \in C^\infty(\Omega).$$

<u>Notes</u>:
(1) The following commutative diagram holds:
   All arrows represent algebraic inclusions which are also continuous and dense (see also Sect. 1.2, Theorem 4).
(2) All compactly supported distributions are tempered.
(3) Spaces on the left are (topologically) compactly supported versions of the spaces on the right.

Fig. 1.2 Commutativity of
the imbedding of test
functions and smooth
functions

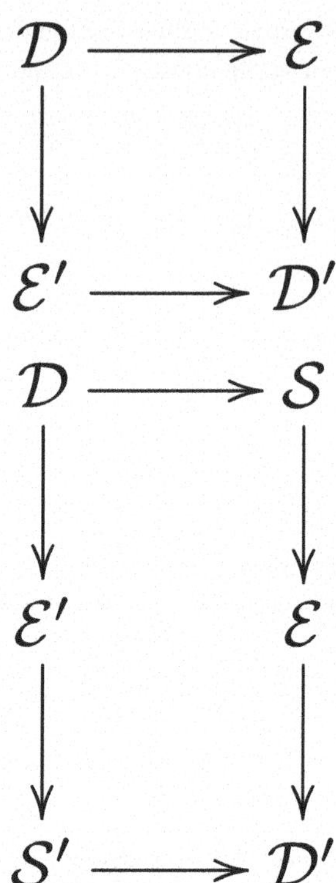

Fig. 1.3 Commutativity of
the imbedding of test
functions and rapidly
decreasing functions into
Schwartz distributions via
compactly supported or
tempered distributions

(4) Generalizing the diagram in the Fig. 1.2 so as to include rapidly decreasing functions and their dual space tempered distributions we have the following extension (5) The proposition also implies that a distribution with compact support is of finite order (Fig. 1.3).

1.6   Operations on Distributions

1.6.1   Monic Operations

In this section we review new distributions obtained from a single distribution by utilizing some operations or algebraic rules. These operations and rules are applied to the distributions in $\mathcal{D}'(\Omega)$ as well as to the tempered distributions and distributions with compact support either with no alteration or some slight modification.

   (I) Multiplication by a scalar: λT,
  (II) Translation: $T(x - a)$,
 (III) Transposition: $T(-x)$,
  (IV) Change of the unit: $T(ax)$,

(V) Multiplication by an infinitely differentiable function: ψT
(VI) Derivative: T', $T^{(m)}$, D^α,
(VII) Fourier transform: $\mathcal{F}[T]$ (or $\hat{T}$),

Among these (VI) has already been studied in Sect. 1.3 (T' and $T^{(m)}$ refer to the differentiation of distributions in $\mathbb{R}$). Fourier transform deserves a separate section (c.f. Sect. 1.8).

The definitions of all of these operations (including the Fourier transform) obey the following scheme:

$$< \text{Operation } (T), \phi >=< T, \overbrace{\text{Operation } (\phi)} > .$$

Of course the $\overbrace{\text{Operation}}$ should correspond to a meaningful operation on test functions.

In (VI) $< D^\alpha T, \phi >=< T, (-1)^{|\alpha|} D^\alpha \phi >$. Thus Operation $\equiv D^\alpha$ (the distributional derivative) ; $\overbrace{\text{Operation}} \equiv (-1)^{|\alpha|} D^\alpha$ (the functional derivative of test functions). As for the other operations

(I) $< \lambda T, \phi >=< T, \lambda\phi >= \lambda < T, \phi >$ $(\lambda \in \mathbb{R})$,
(II) $< T(x - a), \phi(x) >=< T(x), \phi(x + a) >$, (see also Sect. 1.7, Definition 4),
(III) $< T(-x), \phi(x) >=< T(x), \phi(-x) >$,
(IV) $< T(ax), \phi(x) >=< T(x), \frac{1}{|a|^n} \phi(\frac{x}{a}) >$,
(V) $< \psi T, \phi >=< T, \psi\phi >$,   ($\psi$ will be in $\mathcal{D}$, $\mathcal{S}_n$ or $\mathcal{C}^\infty$ depending whether T is a Schwartz distribution , a tempered distribution or a distribution with compact support respectively),
(VII) $< \mathcal{F}[T], \phi >=< T, \mathcal{F}[\phi] >$ (in alternative notation $< \hat{T}, \phi >=< T, \hat{\phi} >$).

<u>Notes:</u>
(1) In the case of a regular distribution T_f where f is locally integrable and $< T_f, \phi >= \int f \phi \, dx$, the rules (I)–(V) are easily verified by the rules of integration via the change of variable.
(2) Distributions are functions of the form Test Functions $\to \mathbb{R}$, they do not have point values. Therefore symbols like $T(x), T(x - a)$ do not denote numbers. Therefore they indicate either (i) one of the above operations to be performed on distributions or (ii) different spaces that the distributions belong to. With this symbolization we may also introduce the concept of odd and even distributions. Namely a distribution $T(x)$:
(a) is <u>odd</u> if $T(-x) = -T(x)$
(b) is <u>even</u> if $T(x) = T(-x)$
Besides $T(x)$ is <u>positive</u> if $< T(x), \phi >\geq 0$ for all test functions ϕ.
(3) The fact that the operations (I)–(V) yield new distributions can be shown easily. The linearity part is obvious.
Sample proof of the continuity: Consider Operation II with the test function space $\mathcal{D}(\Omega)$. Let $\{\phi_j\}$ be a sequence of $\phi_j \to 0$ in the topology of $\mathcal{D}(\Omega)$. This means by Sect. 1.2, Theorem 1 that the supports of ϕ_j are contained in a compact set $K \subset \Omega$

and $\phi_j \to 0$ uniformly along with all of its partial derivatives. But this implies that $\phi_j(x + a)$ have supports contained in $K - a$ and $\phi_j(x + a) \to 0$ also in the topology of $\mathcal{D}(\Omega)$. Hence as T is a continuous functional on $\mathcal{D}(\Omega)$ we have $< T(x), \phi_j(x + a) > \to 0$.

(4) If T is a regular distribution T_f, then the consistency of the operations , i.e. Operation $(T_f) = T_{Operation(f)}$ clearly holds noticing that Operation(f) also produces locally integrable functions in (I)–(V). The consistency of the operation of differentiation (VI) was shown in Sect. 1.3.

1.6.2 *Operations with More Than One Distribution*

In this category we list the following operations:

  (I) Sum of a finite number of distributions
 (II) Series of distributions
(III) The tensor (direct) product of distributions
(IV) The convolution product
       (More involved operation of the convolution product is discussed separately in Sect. 1.7.

(I) If S and T are distributions in $\mathcal{D}'(\Omega)$, $(\Omega \subset \mathbb{R}^n)$, define $S + T$ by $< S + T, \phi >=< S, \phi > + < T, \phi >$, $\phi \in \mathcal{D}(\Omega)$. The right-hand side is a distribution since $\mathcal{D}'(\Omega)$ is a vector space. The consistency for the sum of regular distributions holds as it reduces to the additivity of integrals. The rule can be immediately generalized to the sum of a finite number of distributions.

(II) Let $\{S_m\}$, $(m = 1, 2, \ldots)$ be a sequence of distributions. Form the sequence $T_j = \sum_{m=1}^{j} S_m$ of partial sums of the distributions. If $\{T_j\}$ is a convergent sequence of distributions then the series $\sum_{m=1}^{\infty} S_m$ is defined as the limit $\lim_{j\to\infty} T_j$. According to Sect. 1.2, Proposition 2 to check the convergence of the sequence of numbers $< T_j, \phi >$ for every ϕ will be sufficient , i.e. the convergence of $\{T_j\}$ in the strong topology and in the weak star topology $\sigma(E', E)$ $(E = \mathcal{D}(\Omega))$ are equivalent.

(III) Tensor (direct) product of distributions.

The lack of a suitable distribution extension of the product $f(x)g(x)$ in classical analysis is one of the deficiencies of the linear theory. In fact a first hand generalization of the classical product $f(x)g(x)$ as a regular distribution fails: the product of two locally integrable functions need not be locally integrable , e.g. $f(x) = g(x) = \frac{1}{\sqrt{x}}$ $(x \neq 0)$, then $f(x)g(x) = \frac{1}{x}$ is not locally integrable. However even $f(x)g(x)$ is locally integrable , if we want to define the product of regular distributions in a natural way, i.e. pointwise as $(T_f . T_g)(\phi) = T_f(\phi)T_g(\phi)$ the consistency will be violated since in general $\int f(x)g(x)\phi(x)dx \neq \int f(x)\phi(x)dx \int g(x)\phi(x)dx$, i.e. $T_{fg} \neq T_f T_g$.

Also $\delta^2(x)$ has no distributional meaning. These observations will be the starting point of the arguments in Chap. 3.

Two particular cases where the products of distributions can be defined within the established theory are:

(a) multiplication of a distribution with a smooth function (the latter regarded as a regular distribution), this is the operation (V) in Sect. 1.6.1.

(b) the direct product (tensor product); the above inconsistency within the framework of regular distributions disappears if we consider the equality

$$\int f(x)\, g(x)\, \phi(x)\, \psi(y)\, dx\, dy = \int f(x)\, \phi(x)\, dx \int g(y)\, \psi(y)\, dy, \qquad (\sqrt{})$$

which makes sense also distributionally . Let $\kappa(x, y) = f(x)g(y)$, $\Phi(x, y) = \phi(x)\psi(y)$. Then we can regard the equality $(\sqrt{})$ as
$< \kappa(x, y), \Phi(x, y) >=< T_g(y), < T_f(x), \Phi_y(x) >>$, $(\Phi_y(x)$ denotes the section of $\Phi(x, y)$ at fixed y). We can consider $\kappa(x, y)$ as the direct product $T_f(x)T_g(y)$ of regular distributions, (note that the right-hand side also makes sense). But notice that now the distribution κ and the test function Φ have been defined on the product space which has higher dimension.

To generalize $(\sqrt{})$ and also treat it in a rigorous manner in such a way to dispose its relation to the tensor products of the locally convex spaces, we proceed as follows: Let G and H be open subsets of $\mathbb{R}^k$ and $\mathbb{R}^l$ respectively. Consider the bilinear map from the space $\mathcal{D}(G) \times \mathcal{D}(H)$ into $\mathcal{D}(G \times H)$ by associating to every pair $\phi(x) \in \mathcal{D}(G)$, $\psi(y) \in \mathcal{D}(H)$, the function $(x, y) \longrightarrow \phi(x)\psi(y)$. To this bilinear map there correspond a unique linear map $J : \mathcal{D}(G) \bigotimes \mathcal{D}(H) \longrightarrow \mathcal{D}(G \times H)$. It is easily shown that J is injective and we usually identify $\mathcal{D}(G) \bigotimes \mathcal{D}(H)$ with its image in $\mathcal{D}(G \times H)$. Also by utilizing a result of Henri Cartan , it is shown that this image is dense in $\mathcal{D}(G \times H)$ (c.f. [Hor], 4.8, Lemma 1).

Definition 1 Let $S \in \mathcal{D}'(G)$, $T \in \mathcal{D}'(H)$, then $S \times T$ denotes the linear form on $\mathcal{D}(G) \bigotimes \mathcal{D}(H)$ which to $\phi \in \mathcal{D}(G)$ and $\psi \in \mathcal{D}(H)$ associates $< S \times T, \phi \otimes \psi >=< S, \phi >< T, \psi >$.

<u>Note.</u> Before showing the continuity of this linear form, assume that $\zeta(x, y) \in \mathcal{D}(G \times H)$ and $supp\zeta \subset K \times L$ where K and L are compact subsets of G and H respectively. Then we note that $supp\zeta_y(x) \subset K$ and $supp\zeta_x(y) \subset L$; also $\zeta_y(x) \in \mathcal{D}(G)$, $\zeta_x(y) \in \mathcal{D}(H)$.

Lemma 1 If $\zeta \in \mathcal{D}(K \times L)$ and $S \in \mathcal{D}'(G)$, then $\Theta(y) =< S(x), \zeta_y(x) > belongs$ to $\mathcal{D}_L$.

Proof If $y \notin L$, then $\zeta(x, y) = 0$ for every x and $< S(x), \zeta_y(x) >= 0$, hence $suppTheta(y) \subset L$.
For the continuity of $\Theta(y)$: Firstly the map $y \to \zeta_y(x)$ from H into $\mathcal{D}_K$ is continuous. For considering the topology in $\mathcal{D}_K$ induced by the norms $\| \cdot \|_N$ (c.f. Sect. 1.2) functions $y \to D_x^\alpha \zeta(x, y)$ are uniformly equi-continuous as a consequence of the uniform continuity of functions $(x, y) \to D_x^\alpha \zeta(x, y)$, (the lower-case x denotes derivative with respect to x). Since S is continuous on $\mathcal{D}(G)$ we obtain the continuity of $\Theta(y)$. Next if we show

$$D_y^\beta < S(x), \zeta_y(x) =< S(x), D_y^\beta \zeta(x, y) > \qquad (\dagger)$$

it will imply that $\Theta(y)$ has continuous partial derivatives using the continuity of $\Theta(y)$ which was already proved.

To prove $(\dagger)$ we differentiate recursively , ∂_j denoting the differentiation of order 1 with respect to one variable y_j. Let $\bar{h}$ be a vector in $\mathbb{R}^l$ parallel to the y_j-axis with $\|h\| = h_j$ and $y_0 \in H$. The function

$$x \longrightarrow h_j^{-1}(\zeta(x, y_0 + \bar{h}) - \zeta(x, y_0)) \qquad (\ddagger)$$

tends to the function $x \longrightarrow (\partial_j \zeta)(x, y_0)$ in $\mathcal{D}(G)$ as $\|h\| \to 0$. To see this we should again consider the topology in $\mathcal{D}(G)$ determined by the norms given by 1.2, (*) . Differentiating on the right side of $(\ddagger)$ and applying the Mean Value Theorem we find

$$h_j^{-1} \left\{ D_x^\beta \zeta(x, y_0 + \bar{h}) - D_x^\beta (\partial_j \zeta)(x, y_0) \right\}$$

$$= D_x^\beta \partial_j \zeta(x, y_0 + t\bar{h}) - D_x^\beta (\partial_j \zeta)(x, y_0) \quad (0 < t < 1).$$ The difference can be made arbitrarily small in x by taking $\|\bar{h}\|$ sufficiently small since the function $(x, y) \longrightarrow D_x^\beta \partial_j \zeta(x, y)$ is uniformly continuous.

We have therefore $\partial_j < S(x), \zeta(x, y_0) >= \lim_{\|\bar{h}\| \to 0} < S(x), \dfrac{\zeta(x + y_0 + \bar{h} - \zeta(x, y_0)}{h_j} >=<$ $S(x), \lim_{\|\bar{h}\| \to 0} \dfrac{\zeta(x, y_0 + \bar{h} - \zeta(x, y_0)}{h_j} >=< S(x), \partial_j \zeta(x, y_0) >$ showing $(\dagger)$ for the derivative ∂_j and will be sufficient for showing $(\dagger)$ by recursive differentiation. $\square$

Theorem 1 *If $S \in \mathcal{D}'(G)$ and $T \in \mathcal{D}'(H)$, then the linear form $\zeta \longrightarrow < T(y), < S(x), \zeta_y(x) >>$ is continuous on $\mathcal{D}(G \times H)$.*

Proof By Sect. 1.2, Proposition 1, for each compact $K \subset G$ there exists a constant $M_K > 0$ and a non-negative integer N such that

$$| < S, \phi > | \le M_K \sup\{|D^\alpha \phi(x)| : |\alpha| \le N, x \in K\} \text{ for all } \phi \in \mathcal{D}(G).$$

Similarly for each compact subset $L \subset H$, there is $M_L > 0$ and a non-negative integer ν satisfying

$$| < T, \psi > | \le M_L \sup\{|D^\beta \psi(y)| : |\beta| \le \nu, y \in L\} \text{ for all } \psi \in \mathcal{D}(H).$$

If $\zeta \in \mathcal{D}(K \times L)$ then we have

$$| < S(x), D_y^\beta \zeta(x, y) > | \le M_K \sup\{|D_x^\alpha D_y^\beta \zeta(x, y)| : |\alpha| \le N, y \in L\} \quad \forall y \in L \text{ and } |\beta| \le \nu.$$

Using the Lemma 1 we can write

$$| < T(y), < S(x), \zeta(x, y)| \le M_L \sup\{|D_y^\beta < S(x), \zeta(x, y)| : |\beta| \le \nu, y \in L\}$$

$$\stackrel{(\dagger)}{=} M_L \sup\{| < S(x), D_y^\beta \zeta(x, y)| : |\beta| \le \nu, y \in L\}$$

$$\le M_K M_L \sup\{|D_x^\alpha D_y^\beta \zeta(x, y)| : |\alpha| \le N, |\beta| \le \nu, (x, y) \in K \times L\}$$

$$\le M_K M_L \sup\{|D_x^\alpha D_y^\beta \zeta(x, y)| : |\alpha + \beta| \le N + \nu, (x, y) \in K \times L\}.$$

Again resorting to Sect. 1.2, Proposition 1, the theorem is proved. $\qquad\square$

Note. To relate to $S \times T$ of Definition 1: If $\zeta(x, y) = \phi(x)\psi(y)$ with $\phi \in \mathcal{D}(G)$ and $\psi(y) \in \mathcal{D}(H)$, that means $\zeta = J(\phi \otimes \psi)$, then $< T(y), < S(x), \zeta(x, y) >>=< T(y), \psi(y) < S(x), \phi(x) >>=< S(x), \phi(x) >< T(y), \psi(y) >=< S \times T, \phi \otimes \psi >$.

As $J(\mathcal{D}(G)) \otimes \mathcal{D}(H))$ is dense in $\mathcal{D}(G \times H)$, $S \times T$ will have a unique extension (by the well-known results of analysis) to $\mathcal{D}(G \times H)$ which will be called the tensor product of the distributions S and T, (J is given under III).

Definition 2 Given two distributions $S \in \mathcal{D}'(G)$ and $T \in \mathcal{D}'(H)$ their tensor product $S \otimes T$ is the unique distribution on $\mathcal{D}(G \times H)$ satisfying

$$< S \otimes T, \phi \otimes \psi >=< S, \phi >< T, \psi > \text{ for all } \phi \in \mathcal{D}(G) \text{ and } \psi \in \mathcal{D}(H).$$

Then by the Theorem 1 and above note we have

Proposition 1 *If* $S \in \mathcal{D}'(G)$ *and* $T \in \mathcal{D}'(H)$, *then for every* $\zeta \in \mathcal{D}(G \times H)$ *we have*

$$< S \otimes T, \zeta >=< S(x), < T(y), \zeta(x, y) >>=< T(y), < S(x), \zeta(x, y) >>=< T \otimes S, \zeta > .$$

Proof By the Note $< S(x), < T(y), \zeta(x, y) >>$ is the extension of $S \times Y$ from $J(\mathcal{D}(G)) \otimes \mathcal{D}(H))$ to $\mathcal{D}(G \times H)$, therefore it should be equal to $< S \otimes T, \zeta(x, y) >$.

For the commutativity part we replace x by y and S by T in the proof of Theorem 1. $\qquad\square$

Theorem 2 *If* $S \in \mathcal{D}'(G)$ *and* $T \in \mathcal{D}'(H)$, *then* $supp(S \otimes T) = supp S \times supp T$.

Proof If $(x, y) \in supp S \times supp T$, then for every neighborhood U of x, there exists a function $\phi \in \mathcal{D}(G)$ with support contained in U such that $< S, \phi > \ne 0$, (otherwise if $\forall \phi \in \mathcal{D}(G)$, $< S, \phi >= 0$, then S would vanish in U so x can not be in the support of S). Similarly for every neighborhood V of y, $\exists \psi \in \mathcal{D}(H)$, such that $< T\psi > \ne 0$. Then $\phi \otimes \psi$ would have support in the neighborhood $U \times V$ of (x, y) and $< S \otimes T, \phi \otimes \psi >=< S, \phi >< T, \psi > \ne 0$, thus $(x, y) \in supp S \otimes T$.

Conversely suppose $(x, y) \notin supp S \times supp T$. Then either $x \notin supp S$ or $y \notin supp T$. Suppose the first is true. Then there exists an open neighborhood U of x which does not meet supp S. If $\zeta \in \mathcal{D}(G \times H)$ has its support contained in $U \times H$, then for each $y \in H$, the function $\zeta(., y)$ has its support in U, therefore $< S \otimes T, \zeta >=< T(y), < S(x), \zeta(x, y) >>= 0$ Since $U \times H$ is a neighborhood of (x, y) we have $(x, y) \notin supp S \otimes T$. $\qquad\square$

EXAMPLES

(1) Let $f \in \mathcal{C}(G)$, $g \in \mathcal{C}(H)$ and T_f, T_g be corresponding regular distributions. Then $< T_f \otimes T_g, \phi \otimes \psi >=< T_f, \phi >< T_g, \psi >= \int f(x)\phi(x)\,dx \int g(y)\psi(y)\,dy = \int \int f(x)g(y)\phi(x)\psi(y)\,dx\,dy$. If we identify $f \otimes g$ by $J(f \otimes g) = f(x)g(x)$ which is defined on $G \times H$ we have the consistency $T_f \otimes T_g = T_{f \otimes g}$.

(2) Let $\delta(x)$ be the Dirac measure (Sect. 1.2, Example 1a) on $\mathbb{R}^k$ and $\delta(y)$ be the Dirac measure on $\mathbb{R}^l$. Then $< \delta(x) \otimes \delta(y), \phi \otimes \psi >=< \delta(x), \phi(x) >< \delta(y), \psi(y) >= \phi(0)\psi(0)$, that is $< \delta(x) \otimes \delta(y) >$ extends to be the delta measure on $\mathbb{R}^{k+l}$.

(3) Consider $< \delta(x) \otimes 1 >$ where 1 is regarded as a regular distribution :
$< \delta(x) \otimes 1, \zeta(x, y) >=< 1, < \delta(x), \zeta(x, y,) >>=< 1, \zeta(0, y) >= \int \zeta(0, y)\,dy$.

Corollary *(a) The tensor product of two tempered distributions is tempered.*

(b) The tensor product of two distributions with compact supports has compact support.

Proof Let S and T be two tempered distributions in G and H. Let $\zeta \in \mathcal{S}(\mathbb{R}^{k+l})$. Then $\eta(y) =< S(x), \zeta(x, y) >$ belongs to $\mathcal{S}(\mathbb{R}^l)$. To see this first note that

$$y \longrightarrow D_y^\beta \zeta(x, y) \text{ lower case } y \text{ indicates differentiation on } y \quad (\sqrt{})$$

is uniformly continuous. Using the equality (†) in the proof of the Lemma 1 we have:
$y^\alpha D_y^\beta \eta(y) = y^\alpha D_y^\beta < S(x), \zeta(x, y) >= y^\alpha < S(x), D_y^\beta \zeta(x, y) >$.   For   fixed y, $\zeta_y(x) \equiv \zeta(x, y)$ is also a rapidly decreasing function and obviously the class of rapidly decreasing functions is closed under partial differentiation, thus $< S(x), D_y^\beta \zeta(x, y) >$ makes sense.

Now $\lim_{y \to \infty} D_y^\beta \eta(y) = \lim_{y \to \infty} < S(x), y^\alpha D_y^\beta \zeta(x, y) >=< S(x), \lim_{y \to \infty} y^\alpha D_y^\beta \zeta(x, y) >$ (by continuity given in $(\sqrt{})$) is bounded as $\zeta(x, y)$ is rapidly decreasing. Thus $\eta(y) \in \mathcal{S}(\mathbb{R}^l)$ and since $T \in \mathcal{S}'(\mathbb{R}^l)$, $< T(y), \eta(y) >=< T(y), < S(x), \zeta(x, y) >>< \infty$ yielding $S \otimes T$ as a tempered distribution.

(b) This is also clear since when $S \in (\mathcal{C}^\infty)'$, $T \in (\mathcal{C}^\infty)'$ and $\zeta(x, y) \in \mathcal{C}^\infty(\mathbb{R}^{k+l})$, then $< S(x), \zeta(x, y) >\in \mathcal{C}^\infty(\mathbb{R}^l)$.   □

Remark 1 For $G = \mathbb{R}^k$ and $H = \mathbb{R}^l$ we have by this Subsection

$$\mathcal{S}'(\mathbb{R}^{k+l}) \cong \mathcal{S}'(\mathbb{R}^k) \widetilde{\otimes} \mathcal{S}'(\mathbb{R}^l) \cong \left(\mathcal{S}(\mathbb{R}^k) \widetilde{\otimes} \mathcal{S}(\mathbb{R}) \right)'$$

where $\widetilde{\otimes}$ denotes the completion of the tensor product spaces spanned by the linear combinations, e.g. in the first congruence the completion of the space spanned by the linear combinations of $S \otimes T$, $S \in \mathcal{S}'(\mathbb{R}^k)$, $T \in \mathcal{S}'(\mathbb{R}^l)$.

Similarly $\mathcal{D}'(\mathbb{R}^{k+l}) \cong \mathcal{D}'(\mathbb{R}^k) \widetilde{\otimes} \mathcal{D}'(\mathbb{R}^l) \cong \left(\mathcal{D}(\mathbb{R}^k) \widetilde{\otimes} \mathcal{D}(\mathbb{R}^l) \right)'$.

Remark 2 The completion of he tensor product spaces of distributions, e.g. that of $\mathcal{D}'(\mathbb{R}^k) \otimes \mathcal{D}'(\mathbb{R}^l)$ is based on the following topology (c.f. [Rob-R] , Chap. VII.2, Proposition 7) : There is a finest locally convex topology on $\mathcal{D}'(\mathbb{R}^k) \otimes \mathcal{D}'(\mathbb{R}^l)$ under which the canonical mapping of $\mathcal{D}'(\mathbb{R}^k) \times \mathcal{D}(\mathbb{R}^l)$ into $\mathcal{D}'(\mathbb{R}^k) \otimes \mathcal{D}'(\mathbb{R}^l)$ is continuous.

1.6.3 Schwartz Kernel Theorem

Every function $\kappa \in \mathcal{C}(\mathbb{R}^2)$ defines an integral operator $\mathcal{K}$ from $\mathcal{D}(\mathbb{R})$ into $\mathcal{C}(\mathbb{R})$ by the formula

$$(\mathcal{K}\phi)(x) = \int \kappa(x, y)\, \phi(y)\, dy, \quad \phi \in \mathcal{D}(\mathbb{R}), \quad x \in \mathbb{R} \quad (*)$$

More generally any square integrable kernel $\kappa \in L^2(\mathbb{R}^2)$ defines a bounded linear operator on L^2 by the integral in (*). However, not every bounded linear operator has a square integrable kernel representation. For example the identity operator I has no expression like (*) unless $\kappa(x, y) = \delta(x - y)$, i.e. a distribution in $\mathbb{R}^2$, not a member of $L^2(\mathbb{R}^2)$.

For continuous kernels it is not easy to characterize the resulting operators according to (*). However the distributional analogue has more affirmative assertions as stated in the so-called Schwartz Kernel Theorems. In different versions of this theorem bounded linear operators $\mathcal{C}^\infty \to \mathcal{E}'$, $\mathcal{S} \to \mathcal{S}'$ or $\mathcal{D} \to \mathcal{D}'$ are characterized as distributional kernels in product spaces.

We specify the following version of the theorem related to Schwartz distributions :

Theorem 3 (The Schwartz Kernel Theorem)

(i) Every $\kappa \in \mathcal{D}'(\mathbb{R}^{k+l})$ defines a linear, bounded map $\mathcal{K} : \mathcal{D}(\mathbb{R}^k) \longrightarrow \mathcal{D}'(\mathbb{R}^l)$ according to $< \mathcal{K}\phi, \psi > = < \kappa, \psi \otimes \phi >$; $\phi \in \mathcal{D}(\mathbb{R}^k)$, $\psi \in \mathcal{D}(\mathbb{R}^l)$ where formally $(\mathcal{K}\phi)(x) = \int_{\mathbb{R}^k} \kappa(x, y)\, \phi(y)\, dy$ (†).

(ii) Conversely, to every such linear, bounded map $\mathcal{K}$ there is a unique distribution κ such that (†) is valid.

(i) and (ii) can be restated as $\mathcal{L}(\mathcal{D}(\mathbb{R}^k), \mathcal{D}'(\mathbb{R}^l)) \cong \mathcal{D}'(\mathbb{R}^{k+l}) \cong \mathcal{D}'(\mathbb{R}^k) \widetilde{\otimes} \mathcal{D}'(\mathbb{R}^l)$.

<u>Outline of the Proof.</u>
(i) If $\kappa \in \mathcal{D}'(\mathbb{R}^{k+l})$, then (†) defines a distribution $\mathcal{K}\phi$ since $\psi \to < \kappa, \psi \otimes \phi >$ is continuous. Also $\mathcal{K}$ is continuous since $\phi \to < \kappa, \psi \otimes \phi >$ is continuous.
(ii) For the converse, the uniqueness is the result of the uniqueness in Definition 2. For the existence consider open sets $O_1 \subset \mathbb{R}^k$, $O_2 \subset \mathbb{R}^l$ and choose compact sets K_1, K_2 as the neighborhoods of $\bar{O}_1, \bar{O}_2$. For $(x_1, x_2) \in O_1 \times O_2$ and small $\epsilon > 0$ set

$$\kappa_\epsilon(x_1, x_2) = \epsilon^{-k-l} < \mathcal{K}\psi_2((x_2 - .)/\epsilon), \psi_1((x_1 - .)/\epsilon) >, \quad \psi \in \mathcal{D}(K_1), \ \psi_2 \in \mathcal{D}(K_2).$$

We observe that if there is already a distribution κ satisfying ($\dagger$), then κ_ϵ would be the convolution $\kappa * \vec{\psi}_\epsilon$ ($\vec{\psi} = (\psi_1, \psi_2)$) and would converge to κ as $\epsilon \to 0$ according Sect. 1.7, Theorem 7 of the next section. Now the rest of the proof would consist of showing that κ_ϵ does have a limit in $\mathcal{D}'(O_1 \times O_2)$ when $\epsilon \to 0$ and O_1, O_2 being arbitrarily selected, ($\dagger$) is fulfilled for the limit. For the lengthy technical details, see [Schw-3, Ehr].

1.7 Convolution of Distributions

The convolution product of two locally integrable functions $f(x)$ and $g(y)$ is a function $h(x)$ defined by $h(x) = \int f(t)g(x - t)\, dt$ whenever the integral exists. We denote h by the symbol (*) : $h(x) = f(x) * g(x)$.

The word 'product' is justified by the observations:

(1) Calling $\theta = x - t$, $h(x) = \int f(x - \theta)g(\theta)\, d\theta = g(x) * f(x)$, thus commutativity holds,

(2) The distributivity with respect to addition $f(x) * [g_1(x) + g_2(x)] = f(x) * g_1(x) + f(x) * g_2(x)$.

Physical motivation. If $f(x)$ is the measure of a certain physical quantity and $g(x)$ indicates the performance of the measuring instrument , g should be regarded either rapidly decreasing or having compact support. In fact usually the instruments are not capable of perceiving rapid variations of the entity to be measured , as a result what is measured at x is rather an average of the form $\int f(t)g(x - t)\, dt$. As a simple illustration of this point consider the function given in Sect. 1.2, Example II) (applications of Proposition 2)

$$\Pi(x) = \begin{cases} 0 & \text{for } |x| > 1/2 \\ 1 & \text{for } |x| < 1/2 \\ 1/2 & \text{for } |x| = 1/2 \end{cases}$$

take $\quad g(x) = \frac{1}{a}\Pi(\frac{x}{a})$ $(a > 0)$. $\quad$ In $\quad$ this $\quad$ case $\quad$ $h_a(x) = f(x) * \frac{1}{a}\Pi(\frac{x}{a}) = \frac{1}{a}\int_{\mathbb{R}} f(t)\Pi(\frac{x-t}{a})\, dt = \frac{1}{a}\int_{x-\frac{a}{2}}^{x+\frac{a}{2}} f(t)\, dt$.

Thus $h_a(x)$ simply represents the average value of $f(x)$ between $x - \frac{a}{2}$ and $x + \frac{a}{2}$.

We also note that by Sect. 1.2, Example II, $\frac{1}{a}\Pi(\frac{x}{a})$ is a delta sequence as $a \to 0$. Then using some properties of convolutions (c.f. Example 1) before the Definition 3 and Theorem 7 ahead) and regarding $f(x)$ as a regular distribution we find $\lim_{a \to 0} h_a(x) = f(x) * \lim_{a \to 0} \frac{1}{a}\Pi(\frac{x}{a}) = f(x) * \delta(x) = f(x)$.

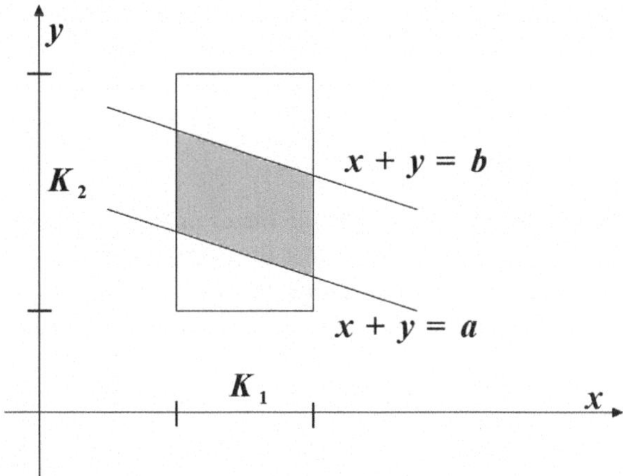

Fig. 1.4 The verification of the condition of compactness for the convolution product of two ditributions

Convolution product of distributions

To motivate the definition of such a product consider locally integrable functions f and g. Then

$$< f * g, \phi >= \int (f * g)(t)\, \phi(t)\, dt = \int \phi(t)\Big(\int f(\tau) g(t - \tau)\, d\tau \Big) dt$$

by change of variables $x = \tau$ and $y = t - \tau$:
$< f * g, \phi >= \int \int f(x)\, g(y)\, \phi(x + y)\, dx\, dy.$

Thus viewing $f.g$ as the direct product of distributions we arrive at a possible definition of the convolution of two distributions.

Definition 1 Let $S \in \mathcal{D}'(\mathbb{R}^n)$, $T \in \mathcal{D}'(\mathbb{R}^n)$ be two distributions , then the convolution $S * T$ is defined as

$$< S * T, \phi >=< S \bigotimes T, \phi^\Delta > \quad \phi \in \mathcal{D}(\mathbb{R}^n)$$

subject to the existence where ϕ^Δ denotes the function belonging to $\mathcal{E}(\mathbb{R}^{2n})$ given by $(x, y) \rightarrow \phi(x + y)$.

To explain the difficulties for the existence of a convolution product in the definition , consider the case $n = 2$. If $\phi \in \mathcal{D}(\mathbb{R})$, let $supp\,\phi = [a, b]$, then the function of two variables ϕ^Δ will not have compact support in $\mathbb{R}^2$ since $(x, y) : a \le x + y \le b$ is an unbounded strip in two dimensions, so that it is not a test function in $\mathbb{R}^2$. However let us suppose that $supp\,S = K_1$, $supp\,T = K_2$ where K_1 and K_2 are two compact sets in $\mathbb{R}^2$, then $\{(x, y) : a \le x + y \le b\} \cap (K_1 \times K_2)$ is also compact as in the Fig. 1.4 so that $< S \bigotimes T, \phi(x + y) >$ makes sense.

Theorem 1 *Let $S \in \mathcal{E}'(\mathbb{R}^n)$ and $T \in \mathcal{E}'(\mathbb{R}^n)$. If $\phi \in \mathcal{D}(\mathbb{R}^n)$, then*

$$< S * T, \phi >=< S \bigotimes T, \phi^\Delta >$$

exists and defines a distribution.

Proof First we prove that the linear map $\phi \longrightarrow \phi^\Delta$ is continuous from $\mathcal{D}(\mathbb{R}^n)$ into $\mathcal{E}(\mathbb{R}^{2n})$. We should show that for every compact $K \subset \mathbb{R}^n$ the map $\phi \longrightarrow \phi^\Delta$ from $\mathcal{D}_K$ into $\mathcal{E}(\mathbb{R}^{2n})$ is continuous. In view of the topologies in $\mathcal{C}^\infty$ discussed in Sect. 1.2, a neighborhood V of 0 in $\mathcal{E}(\mathbb{R}^{2n})$ contains a set of the form

$$\{\psi : |D^\alpha \psi(\zeta)| \leq \epsilon, \ \alpha \in \mathbb{N}^{2n}, \ |\alpha| \leq m, \ \zeta \in M\}$$

for some $m \in \mathbb{N}$ and a compact subset M of $\mathbb{R}^{2n}$. If the set $U = \{\phi : |D^\beta \phi(x), |\beta| \leq 2mx \in K\}$ is a neighborhood of 0 in $\mathcal{D}_K$, then $\phi \in U$ will imply $\phi^\Delta \in V$. To see this let α and m be as in the above set. Write α explicitly as $\alpha = \gamma + \delta$, where $\gamma(\gamma_1, \ldots, \gamma_n, 0, \ldots, 0)$ and $\delta = (0, \ldots, 0, \delta_1, \ldots, \delta_n)$. Then $D^\alpha \phi^\Delta(x, y) = D_y^\delta D_x^\gamma \phi^\Delta \phi^\Delta(x, y) = D^{\gamma+\delta} \phi(x + y)$. Since $x, y \in \mathbb{R}^n$, $(x, y) \in \mathbb{R}^{2n}$, $x + y \in \mathbb{R}^n$ and as $\gamma + \delta| \leq m < 2m$ we have $D^{\gamma+\delta} \phi(x + y) \leq \epsilon$. Thus $\phi^\Delta \in V$ and $\phi \to \phi^\Delta$ is continuous taking value in $\mathcal{E}^{2n}$. But $S \bigotimes T$ is a distribution with compact support as the tensor product of two distributions with compact support, (c.f. Sect. 1.6.2, Corollaray (b) to Proposition 1). Hence $\phi \longrightarrow < S \bigotimes T, \phi^\Delta >$ is continuous and linear on $\mathcal{D}(\mathbb{R}^n)$, thus defines a distribution in $\mathcal{D}'(\mathbb{R}^n)$. $\square$

With the notation of Sect. 1.6.2 we can write

$$< S * T, \phi >=< S(x), < T(y), \phi(x + y) >>=< T(y), < S(x), \phi(x + y) >> .$$

In fact the requirement that both S and T are distributions with compact support like in this theorem is a bit too strong and can be replaced by the following.

For any compact subset $K \subset \mathbb{R}^n$ denote by K^Δ the subset of $\mathbb{R}^n \times \mathbb{R}^n$ formed by all points (x, y) such that $x + y \in K$.

Condition Γ : Let F_1 and F_2 be two closed subsets of $\mathbb{R}^n$. The condition is satisfied if for every compact subset K of $\mathbb{R}^n$, the set $(F_1 \times F_2) \cap K^\Delta$ is compact in $\mathbb{R}^n \times \mathbb{R}^n$. (Note. A symbolic representation will be similar to the Fig. 1.4 except that the sets K_1 and K_2 in Fig. 1.4 are compact while in here F_1, F_2 are only closed. The shaded region may represent K^Δ).

Definition 2 Let S and T be two distributions on $\mathbb{R}^n$ such that $F_1 = supp S$ and $F_2 = supp T$ satisfy the condition Γ. Then the convolution $S * T$ is a distribution defined by

$$< S * T, \phi >=< S \bigotimes T, \phi^\Delta > \quad \text{for all} \ \phi \in \mathcal{D}(\mathbb{R}^n).$$

It should be proved that $S * T$ is indeed a continuous linear form on $\mathcal{D}(\mathbb{R}^n)$, i.e. a distribution.
We first state a lemma.

Lemma *Let $T \in \mathcal{D}'(\mathbb{R}^n)$ and $\phi \in \mathcal{E}(\mathbb{R}^n)$ and assume that $K = supp T \cap supp \phi$ is compact. In this case $< T, \phi >$ can be given a meaning as follows : Consider $\xi \in \mathcal{D}(\mathbb{R}^n)$ such that $\xi(x) = 1$ in a neighborhood of K. Define $< T, \phi >=< T, \xi\phi >$. Then $< T, \phi >$ is well-defined, i.e. it is independent of the choice of ξ.*

Note. Such a function is called a <u>mesa function</u>.

Proof $\xi\phi \in \mathcal{D}(\mathbb{R}^n)$, thus $< T, \xi\phi >$ makes sense. Let $\psi \in \mathcal{D}(\mathbb{R}^n)$ be another function which is equal to 1 in a neighborhood of K. Then $supp(\xi - \psi)\phi = supp(\xi - \psi) \cap supp\phi \subset K^c \cap supp\phi = [(supp T)^c \cap supp\phi] \cup [supp\phi \cap (supp\phi)^c] \subset (supp T)^c$. Therefore $< T, (\xi - \psi)\phi >= 0$, i.e. $< T, \xi\phi >=< T, \psi\phi >$.

Notes. (1) The definition of the action of $T \in \mathcal{D}'$ on a function in $\mathcal{E}(\mathbb{R}^n)$ given in the Lemma is consistent with the usual $< T, \phi >$ whenever $\phi \in \mathcal{D}(\mathbb{R}^n)$, since in this case we can freely choose ξ by taking it equal to 1 on the $supp\phi$.
(2) Under the conditions of the Lemma the derivative formula of the distributions still holds. In fact letting $\xi \in \mathcal{D}(\mathbb{R}^n)$ be equal to 1 in an open set U containing K and noticing that $supp D^\alpha T \subset supp T$ and $supp D^\alpha \phi \subset supp\phi$ we have $< D^\alpha T, \phi >=< D^\alpha T, \xi\phi >= (-1)^{|\alpha|} < T, D^\alpha(\xi\phi) >= (-1)^{|\alpha|} < T, D^\alpha\phi >$, (the last equality is by the fact that ξ is constant on U, so that the only non-zero term in the Leibniz formula $D^\alpha(\xi\phi) = \sum_{\beta \leq \alpha} c_{\alpha\beta}(D^{\alpha-\beta}\xi)(D^\alpha\phi)$ is $D^\alpha\phi$, $c_{\alpha\alpha} = \binom{\alpha}{\alpha} = 1$).
(3) We can similarly prove $< \psi T, \phi >=< T, \psi\phi >$, $\psi, \phi \in \mathcal{E}(\mathbb{R}^n)$.

Returning to the continuity of the linear form given in Definition 2:

It is sufficient to show the continuity on $\mathcal{D}_K$ for every $K \subset\subset \mathbb{R}^n$. As $supp \otimes T = F_1 \times F_2$, where $F_1 = supp$, $F_2 = supp T$ (Sect. 1.6.2, Theorem 2) and $supp\phi^\Delta \subset K^\Delta$, we have by the assumed condition Γ and closedness of any support that $K_0 = (F_1 \times F_2) \cap supp\phi^\Delta$ is compact. Thus the requirement of Lemma 1 holds for $T \longleftrightarrow S \otimes T$ and $\phi \longleftrightarrow \phi^\Delta$, so that $< S \otimes T, \phi^\Delta >$ can be evaluated as $< S \otimes T, \gamma\phi^\Delta >$ for some $\gamma(x, y) \in \mathcal{D}(\mathbb{R}^{2n})$ taking value 1 on K_0. Then we have to show that $\phi \longrightarrow < S \otimes T, \gamma\phi^\Delta >$ is a continuous linear form on $\mathcal{D}_K$. But following the proof of Theorem 1 the map $\phi \longrightarrow \phi^\Delta$ is a continuous map from $\mathcal{D}_K$ into $\mathcal{E}(\mathbb{R}^{2n})$. As the map $\phi^\Delta \longrightarrow \gamma\phi^\Delta$ is also a continuous map from $\mathcal{E}(\mathbb{R}^{2n})$ into $\mathcal{D}(\mathbb{R}^{2n})$, the continuity of $S * T$ on $\mathcal{D}(\mathbb{R}^n)$ is established since $S \otimes T \in \mathcal{D}'(\mathbb{R}^{2n})$.

<u>CONCLUSIONS</u>:
(A) Under the conditions of Definition 2 and Sect. 1.6.2, Proposition 1 we have

$$< S * T, \phi >=< S \otimes T, \phi^\Delta >=< S(x), < T(y), \phi(x + y) >>=< T(y), < S(x), \phi(x + y) >> .$$

(B) If one of the distributions S and T has compact support, then the convolution $S * T$ is defined. It is because this property implies the condition Γ. We can see this as follows : Suppose that $supp S = K$ (compact) and $supp T = F$ (closed). First we show that $K - F \equiv \{x - y : x \in K, y \in F\}$ is closed in $\mathbb{R}^n$. Let $z \in \overline{K - F}$. Then there exists a sequence of points $\{(x_n, y_n)\} \in K \times F$ such that $x_n - y_n$ tends to z. Since the sequence $\{x_n\}$ belongs to the compact set K it has a sub-sequence $\{x_{n_k}\}$ converging to some point $x_0 \in K$. Then as $-y_{n_k} = (x_{n_k} - y_{n_k}) - x_{n_k}$ is convergent and F is closed , $-y_{n_k}$ converges to some point $-y_0 \in F$. In this way $x_{n_k} - y_{n_k}$ converges to $z = x_0 - y_0 \in K - F$. Hence $M = K \cap (K - F)$ is compact as a closed subset of K.

Now if $(x, y) \in (K \times F) \cap K^\Delta$, then by the definition of K^Δ, $x + y \in K$, hence $x \in K - F$, but also $x \in K$ thus $x \in M$. On the other hand $y \in K - x \subset K - M$ which is compact (it is closed as shown above and clearly bounded). Calling $N = K - M$ we find that $(x, y) \in M \times N$, thus $(K \times F) \cap K^\Delta$ being compact, the condition Γ is verified.

<u>EXAMPLES</u>
(1) $T * \delta = T$, $(T \in \mathcal{D}'(\mathbb{R}^n))$. The Dirac measure δ obviously have compact support : If ω is an open subset of $\mathbb{R}^n$ not containing the origin, for all $\phi \in \mathcal{D}(\omega)$ we have $< \delta, \phi >= 0$. Hence $supp\delta = \{0\}$ and by Conclusion B) the above convolution exists.
$$< T * \delta, \phi >=< T(y), < \delta(x), \phi(x + y) >>=< T(y), \phi(y) > \quad \text{shows the}$$
above result. Thus δ acts as the unit element of the convolution product.

Similarly $< \delta * T, \phi >=< T, \phi >$, (see ahead Theorem 3a).
(2) Convolution by the derivatives of δ, (case $n = 1$):
$< \delta' * T, \phi >=< \delta' \otimes T, \phi^\Delta >=< \delta'(x), < T(y), \phi(x + y) >>=< T(y), <$
$\delta'(x), \phi(x + y) >>= - < T(y), \phi'(y) >=< T', \phi >$. By induction we find $\delta^{(m)} * T = T^{(m)}$. Thus in order to differentiate a distribution m times , it suffices to form a convolution by the m<u>th</u> order derivative of δ. Ahead this property will be generalized to $D^\alpha T$, (see Theorem 4a).

<u>Convolution of a distribution with a test function</u>
Based on the formula for the convolution of two functions it is natural to define

Definition 3 (a) If $S \in \mathcal{D}'(\mathbb{R}^n)$ and $\phi \in \mathcal{D}(\mathbb{R}^n)$, $(S * \phi)(x) =< S(y), \phi(x - y) >$.
REMARK. (a) $S * \phi$ yields a function of x and should not be confused with $S * T_\phi$ which is a distribution.

(b) If $T \in \mathcal{S}'_n$ and $\phi \in \mathcal{S}_n$, then $(T * \phi)(x) =< T(y), \phi(x - y) >$ still makes sense since $\phi(x - y)$ is obviously a rapidly decreasing function for fixed x.

<u>Translation operator</u>
If f is a function defined on $\mathbb{R}^n$ and $h \in \mathbb{R}^n$, then define the function $\tau_h f$ by $(\tau_h f)(x) = f(x - h)$ for all $x \in \mathbb{R}^n$. Let $f \in \mathcal{C}(-\infty, \infty)$ and $g = \tau_h f$, then the associated regular distribution T_g will satisfy

$$< T_g, \phi >= \int_{\mathbb{R}^n} f(x - h)\, \phi(x)\, dx = \int_{\mathbb{R}^n} f(x)\, \phi(x + h)\, dx =< T_f, \tau_{-h}\, \phi >;$$

this motivates the definition

Definition 4 Let $T \in \mathcal{D}'$. For $h \in \mathbb{R}^n$ the translate of T by h is the distribution $\tau_h T$ defined by

$$< \tau_h T, \phi >=< T, \tau_{-h}\phi > \quad \text{for all} \quad \phi \in \mathcal{D}(\mathbb{R}^n).$$

(Symbolically $(\tau_h T)(x) \equiv T(x - h)$).

Remarks (a) The map $\tau_h : \mathcal{D} \to \mathcal{D}$ is an isomorphism since considering the topology of $\mathcal{D} : \sup_x |D^\alpha \phi(x)| = \sup_x |D^\alpha (\tau_h \phi)(x)|$ for all multi-indices $\alpha \in \mathbb{N}^n$. Also the map $\tau_h : \mathcal{D}' \to \mathcal{D}'$, which is the transpose of the map $\tau_{-h} : \mathcal{D} \to \mathcal{D}$, is an isomorphism $\mathcal{D}' \to \mathcal{D}'$. Clearly $\tau_h^{-1} = \tau_{-h}$ and we have $< \tau_h T, \tau_h \phi >=< T, \phi >$ for all $T \in \mathcal{D}'$ and $\phi \in \mathcal{D}$.

(b) If $S * \phi = T * \phi$ for all test functions ϕ, this implies that $S = T$. This is because in this case we shall have by Definitions 3 and 4 : $\tau_{-h} S = \tau_{-h} T$, then apply $\tau_{-h}^{-1} = \tau_h$ to both sides.

<u>EXAMPLES</u> (a) $\tau_a \delta = \delta_a$, $a \in \mathbb{R}^n$. This is because $< \tau_a, \phi >=< \delta, \tau_{-a}\phi >=< \delta(x), \phi(x + a) >= \phi(a) =< \delta_a, \phi >$.

(b) To translate a distribution by $h \in \mathbb{R}^n$ we use the commutativity of the tensor product given in Sect. 1.6.2, Proposition 1 to find

$$< \delta(x - h) * T, \phi >=< \delta(x - h) \bigotimes T, \phi >=< T \bigotimes \delta(x - h), \phi >$$

$$=< T(y), < \delta(x - h), \phi(x + y) >>=< T(y), \phi(h + y) >=< \tau_h T, \phi >;$$

i.e. $\delta(x - h) * T(x) \equiv T(x - h)$. (For the translation of convolution product of distributions see Theorem 5 ahead).

Theorem 2 *Let $T \in \mathcal{D}'(\mathbb{R}^n)$, $\phi \in \mathcal{D}(\mathbb{R}^n)$. Then*

(a) $\tau_h(T * \phi) = (\tau_h T) * \phi = T * (\tau_h \phi), \quad \forall h \in \mathbb{R}^n$,
(b) $D^\alpha(T * \phi) = (D^\alpha T) * \phi = T * (D^\alpha \phi)$ *for every multi-index α*
(c) $T * \phi \in \mathcal{C}^\infty$.

Proof (a) For any $x \in \mathbb{R}^n$: $[\tau_h(T * \phi)](x) = (T * \phi)(x - h) =< T(y), \phi(x - h - y) >$ (Definition 3a) also $[(\tau_h T) * \phi](x) =< (\tau_h T)(y), \phi(x - y) >=< T(y), \phi[x - (x + y)] >=< T(y), \phi(x - h - y) >$.

Similarly $[T * (\tau_h \phi)](x) =< T(y), \phi(x - y - h) >$ shows part a).

(b) $[T * (D^\alpha \phi)](x) =< T(y), D^\alpha \phi(x - y) >$, if $\xi = x - y$
$=< T(y), (-1)^{|\alpha|} D_\xi^\alpha \phi(x - y) >=< D^\alpha T(y), \phi(x - y) >= [(D^\alpha T) * \phi](x)$
proves the second equality.

For the equality of the first and the last members : let e be a unit vector in $\mathbb{R}^n$ and let $\frac{\tau_0 - \tau_{re}}{r}$, $(r > 0)$. Regarding q_r as a translation operator, part a) gives $q_r(T * \phi) = T * (q_r \phi)$. As $r \to 0$, $q_r \phi \to D_e \phi$ in $\mathcal{D}(\mathbb{R}^n)$, so that $T * (q_r \phi))(x) =< T(y), q_r \phi(x - y) >\longrightarrow< T(y), D_e \phi(x - y) >= (T * D_e \phi)(x)$. Hence

$$q_r(T * \phi)(x) \longrightarrow (T * D_e \phi)(x) \qquad (\checkmark)$$

But at the same time

$$[q_r(T * \phi)](x) \longrightarrow [D_e(T * \phi)](x) \qquad (\checkmark\checkmark)$$

By ($\checkmark$) and ($\checkmark\checkmark$) we have $D_e(T * \phi) = T * D_e\phi$ and iterating on differentiation we get the result.

(c) $D^\alpha[(T * \phi)](x) =< T(y), D^\alpha\phi(x - y) >$ by (b) , then the infinite continuous differentiability of ϕ and continuity of T will yield the continuity of partial derivatives of $T * \phi$ of any order.

Remarks (1) If $T \in \mathcal{S}'_n$ and $\phi \in \mathcal{S}_n$, then $T * \phi$ is defined again by Definition 3 (a) since $\phi(x - y) \in \mathcal{S}_n$ and (c) still holds. This is because (b) can be proved exactly the same way for this case.

(2) The following associativity is also established without much difficulty: $T * (\phi * \psi) = (T * \phi) * \psi = (T * \psi) * \phi$.

Some Important Properties of Convolution Products
Convolution of More Than Two Distributions
Consider the distributions $T_1, T_2, \ldots, T_m$ on $\mathbb{R}^n$ and sets $F_1 = suppT_1, \ldots, F_m = suppT_m$. For a compact set K of $\mathbb{R}^n$, K^Δ denotes the subset of $\mathbb{R}^{nm}$ formed by all points $(x_1, \ldots, x_m)$ such that $x_j \in \mathbb{R}^n$, $(j = 1, 2, \ldots, m)$ and $x_1 + x_2 + \cdots + x_m \in K$. Then the generalization of the condition Γ would be:

(Γ_m) : For every compact subset K of $\mathbb{R}^n$, the set $\left(\prod_{j=1}^m F_j\right) \cap K^\Delta$ is compact in $\mathbb{R}^{nm}$.

Definition 5 The convolution product $\prod_{j=1}^{m*} T_j$ of m distributions on $\mathbb{R}^n$ is defined by

$$< \prod_{j=1}^{m*} T_j, \phi > =< T_1 * \cdots * T_m, \phi > \doteq < \bigotimes_{j=1}^{m} T_j, \phi^\Delta >$$

for all $\phi \in \mathcal{D}(\mathbb{R}^n)$ and $\phi^\Delta \in \mathcal{D}(\mathbb{R}^{nm})$.
(Note that $\bigotimes_{j=1}^m T_j$ is the unique distribution on $\mathcal{D}(\mathbb{R}^{nm})$ satisfying $< \bigotimes_{j=1}^m T_j, \otimes_{j=1}^m \psi_j >$, $\psi_j \in \mathcal{D}(\mathbb{R}^n)$. Compare with Sect. 1.6.2, Definition 2.

The condition Γ_m is satisfied if all the sets F_j except possibly one are compact.

Theorem 3 *(a) Let S and T be two distributions on $\mathbb{R}^n$ such that $F_1 = suppS$ and $F_2 = suppT$ satisfy condition Γ. Then $S * T = T * S$.*

(b) Let R, S and T be three distributions on $\mathbb{R}^n$ whose supports satisfy the condition Γ_3. Then the associativity holds: $(R * S) * T = R * (S * T) = R * S * T$.

Note. The conclusions of the theorem hold under less stringent conditions, i.e. if at least one of S and T has compact support in (a) and if at least two of the supports of R, S and T are compact in (b).

Proof (a) If $\phi \in \mathcal{D}(\mathbb{R}^n)$ and $K = supp\phi$, following the idea of the Lemma ,choose $\zeta \in \mathcal{D}(\mathbb{R}^{2n})$ such that $\zeta(x, y) = 1$ on a neighborhood of the set $(F_1 \times F_2) \cap K^\Delta$. Noticing that K^Δ is the support of ϕ^Δ, we have $< S * T, \phi > = < S \otimes T, \zeta\phi^\Delta > = < S(x), < T(y), \zeta(x, y)\phi(x + y) >> = < T(y), < S(x), \zeta(x, y)\phi(x + y) >> = < T \otimes S, \zeta\phi^\Delta > = < T * S, \phi >$, (Sect. 1.6.2, Proposition 1 was used).

   (b) Let $F_1 = suppR$, $F_2 = suppS$, $F_3 = suppT$, then for any compact set $K \subset\subset \mathbb{R}^n$, $(F_1 \times F_2 \times F_3 \cap K^\Delta)$ is compact in $\mathbb{R}^{3n}$ by the assumption, so there exist compact sets M, N and L in $\mathbb{R}^n$ such that $D = (F_1 \times F_2 \times F_3) \cap K^\Delta \subset M \times N \times \times L$. If $x \in \kappa_1 \doteq F_1 \cap (K - F_2 - F_3)$, $\exists y \in F_2$, $\exists z \in F_3$, so that $x + y + z \in K$, i.e. $(x, y, z) \in K^\Delta$, also $(x, y, z) \in F_1 \times F_2 \times F_3$, thus $(x, y, z) \in M \times N \times L$, in particular $x \in M$ which implies $\kappa_1 \subset \mathbb{R}^n$ is compact, (the closedness of $K - F_2 - F_3$ is shown like in Conclusion B above). κ_1 can be looked upon as the first projection of the compact set D. Similarly the compact $\kappa_2 \doteq F_2 \cap (K - F_1 - F_3)$ is its second projection. Let now K be the support of a test function in $\mathcal{D}$. Choose functions $\xi \in \mathcal{D}$ and $\eta \in \mathcal{D}$ such that $\xi(x) = 1$ on a neighborhood of κ_1 and $\eta(y) = 1$ on a neighborhood of κ_2. Let ζ be the function defined on $\mathbb{R}^{3n}$ as $(x, y, z) \longrightarrow \xi(x)\eta(y)$. If $(x, y, z) \in D$, then x and y belong to the first and second projections of D respectively, hence $\xi(x) = \eta(y) = 1$ so that $\zeta(x, y, z) = 1$ on a neighborhood of D. Also $\zeta\phi^\Delta$ has compact support in $\mathbb{R}^{3n}$. Furthermore for every $x \in F_1$ we have $\eta(y) = 1$ on a neighborhood of $F_2 \cap (supp(\phi(. + x) - F_3)$ and $\xi(x) = 1$ on a neighborhood of $F_1 \cap (K - supp(S * T)$. Then we have $R * S * T, \phi > = < R \otimes S \otimes T, \zeta\phi^\Delta > = < R(x), \xi(x) < S(y), \eta(y) < T(z), \phi(x + y + z) >>> = < R(x)\xi(x), < (S * T)(y), \phi(x + y) >> = < R * (S * T), \phi >$, i.e. $R * S * T = R * (S * T)$. Similarly choose $\psi, \chi \in \mathcal{D}$ such that $\psi(y) = 1$ on a neighborhood of $F_2 \cap (K - F_1 - F_3)$ and $\chi(z) = 1$ on a neighborhood of $\kappa_3 \doteq F_3 \cap (K - F_1 - F_2)$. Then $< R * S * T, \phi > = < R(x), < S(y), \psi(y) < T(z), \chi(z)\phi(x + y + z >>> = < (R * S)(x), < T(z), \chi(z)(x + z) >> = < (R * S) * T, \phi >$. $\qquad\square$

Remarks (1) For any distribution $T \in \mathcal{D}'(\mathbb{R}^n)$ we have by part a) $\delta * T = T * \delta = T$, (see also Example 1 before Definition 3).

   (2) $\mathcal{E}'$ is a commutative ring if the convolution is taken for multiplication with the Dirac measure δ as the unit element. From an algebraic point of view , $\mathcal{D}'$ is an $\mathcal{E}'$-module if the multiplication between the elements of $\mathcal{E}'$ and $\mathcal{D}'$ is again convolution.

   (3) By induction and using easy consequences of the definition of the condition Γ_m, the following generalization of associativity is demonstrated:

$$(\prod_{j=1}^{l*} T_j) * (\prod_{j=l+1}^{m*} T_j) = \prod_{j=1}^{m*} T_j$$

for $1 \leq l < m$, provided T_j are distributions whose supports satisfy the condition Γ_m.

Theorem 4 *Let $S \in \mathcal{D}'(\mathbb{R}^n)$, $T \in \mathcal{D}'(\mathbb{R}^n)$ satisfying the condition Γ (or at least one of them has compact support) and let α be a multi-index. Then* *(a)* $D^\alpha T = (D^\alpha \delta) * T$;

(b) $D^\alpha(S * T) = (D^\alpha S) * T = S * (D^\alpha T)$.

<u>Note</u>. For part (a), $n = 1$, see also Example 2 before Definition 3.

Proof (a) If $\phi \in \mathcal{D}$, then $(\delta * \phi)(x) = < \delta(y), \phi(x - y) > = \phi(x)$. Then by Theorem 2 (b) and associativity $(D^\alpha T) * \phi = T * D^\alpha \phi = T * D^\alpha(\delta * \phi) = T * (D^\alpha \delta) * \phi$. But then by Remark (b) following the Definition 4 this implies $D^\alpha T = T * D^\alpha \delta = (D^\alpha \delta) * T$.

(b) By Theorem 3 and part (a) we have $D^\alpha(S * T) = (D^\alpha \delta) * (S * T) = ((D^\alpha \delta) * S) * T = (D^\alpha S) * T$ and $((D^\alpha \delta) * S) * T = (S * D^\alpha \delta) * T = S * (D^\alpha \delta * T) = S * (D^\alpha T)$.

Translation of the Convolution Product

In Theorem 2 we investigated the commutativity of the translation operator and the convolution product as applied to $T * \phi$, ($T \in \mathcal{D}'$, $\phi \in \mathcal{D}$. Now a similar result for $S * T$ whenever $S, T \in \mathcal{D}'$.

Theorem 5 *If the supports of S and T satisfy the condition Γ and $h \in \mathbb{R}^n$, then*

$$\tau_h(S * T) = (\tau_h S) * T = S * (\tau_h T).$$

Proof First we show the result under the restrictive assumption that one of the distributions, say S, compact support. If $\phi \in \mathcal{D}(\Omega)$ we have:
$< \tau_h(S * T), \phi > = < S * T, \tau_{-h}\phi > = < S(x), < T(y), \phi(x + y + h) >> = < S(x), < (\tau_h T)(y), \phi(x + y) >> = < s * \tau_h T, \phi >$; i.e. $\tau_h(S * T) = S * (\tau_h T)$. Similarly by commutativity $\tau_h(S * T) = (\tau_h S) * T$. In the general case : as δ has compact support we have by the Remark following Theorem 3 and the first part: $\tau(S * T) = \tau_h(\delta * (S * T)) = \tau_h \delta * (S * T)$ and we further have by associativity: $\tau_h \delta * (S * T) = (\tau_h \delta * S) * T = \tau_h(S * \delta) * T = (\tau_h S) * T$. The other relation follows from commutativity.   □

Regularization

By this we mean approximation of a distribution T by infinitely differentiable functions. An example is provided by Example II, Sect. 1.2 following Proposition 2.

Theorem 6 *Let ρ be a continuous function defined on $\mathbb{R}^n$ satisfying*

(i) $supp\rho \subset B_1$,

(ii) $\rho(x) \geq 0$ *for all* $x \in \mathbb{R}^n$,

(iii) $\int_{\mathbb{R}^n} \rho(x)\,dx = 1$.

where $B_1 = \{x : |x| \leq 1\}$ is the unit ball in $\mathbb{R}^n$,. For every $\epsilon > 0$ define $\rho_\epsilon(x) = \frac{1}{\epsilon^n}\rho(\frac{x}{\epsilon})$. Then $\rho_\epsilon \xrightarrow{\mathcal{D}'} \delta$ (Dirac measure) as $\epsilon \to 0$.

<u>Note</u>. $supp\rho_\epsilon \subset B_\epsilon$ and $\int_{\mathbb{R}^n} \rho_\epsilon(x)\,dx = 1$.

Proof (c.f. Example II, Sect. 1.2).

Theorem 7 *If ρ_ϵ is as in the previous theorem and $\phi \in \mathcal{D}(\mathbb{R}^n)$, $T \in \mathcal{D}'(\mathbb{R}^n)$, then*

(a) $\lim_{\epsilon \to 0} \phi * \rho_\epsilon = \phi$ *in* $\mathcal{D}$,
(b) $\lim_{\epsilon \to 0} T * \rho_\epsilon = T$ *in* $\mathcal{D}'(\mathbb{R}^n)$.

Proof (a) Let f be any continuous function on $\mathbb{R}^n$. Then $\lim_{\epsilon \to 0} \int_{\mathbb{R}^n} \rho_\epsilon(y) f(x - y)\, dx = \lim_{\epsilon \to 0} \int_{\mathbb{R}^n} \frac{1}{\epsilon^n}\rho(\frac{y}{\epsilon}) f(x - y)\, dx = \lim_{\epsilon \to 0} \int_{\mathbb{R}^n} \rho(u) f(x - \epsilon u)\, du = \int_{\mathbb{R}^n} \rho(u) \lim_{\epsilon \to 0} f(x - \epsilon u)\, du = f(x)$ uniformly on compact sets. If we replace f by $D^\alpha \phi$ we have $D^\alpha \phi * \rho_\epsilon \longrightarrow D^\alpha \phi$ uniformly. Also the supports of all $\phi * \rho_\epsilon$ lie in some compact set since $supp \rho_\epsilon$ shrink to $\{0\}$. This gives (a) by Sect. 1.2 , Remark to Theorem 1.

(b) Using Remark 2 to Theorem 2 and part (a): $< T(y), \phi(-y) > = < T(y), \phi(x - y) > |_{x=0} = (T * \phi)(0) = \lim_{\epsilon \to 0}(T * (\rho_\epsilon * \phi)) = \lim_{\epsilon \to 0}((T * \rho_\epsilon) * \phi) = \lim_{\epsilon \to 0} < (T * \rho_\epsilon(y), \phi(x - y) > |_{x=0} = < \lim_{\epsilon \to 0}(T * \rho_\epsilon)(y), \phi(-y) > .$ $\square$

By these theorems every distribution is the limit of a filter of an infinitely differentiable functions.

Theorem 7 suggests an effective method of approximating a distribution by a sequence of infinitely differentiable functions. For this we may let ϵ run through the values $1/k$ $(k \geq 1)$. Then we get $\lim_{k \to \infty} T * \rho_{1/k} = T$. As $T * \rho_{1/k} \in C^\infty(\mathbb{R}^n)$ by Theorem 2c , for large k, $T * \rho_{1/k}$ approximates T in $\mathcal{D}'$.

J. Mikusinski also relates the concept of distributions to the unifor limits of sequeces of smooth functions and defines stable operations on such sequences.

1.8   Fourier Transform of Distributions

1.8.1   Fourier Transform in Classical Analysis

It was stated in Sect. 1.6 that the Fourier transform is one of the monic operations on distributions that deserved a separate treatment. As we will see in the next chapter , the Fourier transform of distributions are important in the study of partial differential equations. By large the Fourier methods are among the most powerful tools in both classical and modern analysis, also in probability theory.

<u>Characters.</u> For each $\xi \in \mathbb{R}^n$, the character e_ξ is the function defined by

$$e_\xi(x) = e^{2\pi i <\xi, x>} = e^{2\pi i(\xi_1 x_1 + \cdots + \xi_n x_n)}, \quad (x \in \mathbb{R}^n, \, i^2 = -1).$$

e_ξ satisfies the equation $e_\xi(x + y) = e_\xi(x)e_\xi(y)$. Hence for each $\xi \in \mathbb{R}^n$, e_ξ is a homomorphism of the additive group $\mathbb{R}^n$ onto the multiplicative group of the complex numbers of modulus 1; i.e. the circle group $S_1 = \mathbb{R}/\mathbb{Z}$. Regard X and Y as two copies of $\mathbb{R}^n$ in duality with respect to the bilinear form

$$< x, \xi > = x.\xi = x_1\xi_1 + \cdots + x_n\xi_n, \quad x \in X, \xi \in Y.$$

Definition 1 The <u>Fourier transform</u> of a function $f \in L^1(\mathbb{R}^n)$ is the function $\hat{f}$ defined by

$$\hat{f}(\xi) = \int_X e^{-2\pi i <\xi, x>} f(x)\, dx \equiv \int_X e_{-\xi}(x)\, f(x)\, dx.$$

<u>Notes</u>. (1) The term Fourier transformation will refer to the mapping $f \longrightarrow \hat{f}$, occassionally this transformation will be indicated by $\mathcal{F}$ for the convenience of notation.

(2) An alternative expression for the Fourier transform is $\hat{f}(\xi) = (f * e_\xi)(0)$.

(3) $\hat{f}$ is a bounded function since $|\hat{f}| \leq \int_X |f(x)|\, dx < \infty$.

Definition 2 The <u>inverse Fourier transform</u> $\tilde{g}$ of $g \in L^1(\mathbb{R}^n)$ is defined by

$$\tilde{g}(x) = \int_Y e^{2\pi i <x, \xi>} g(\xi)\, d\xi \equiv \int_Y e_x(\xi)\, g(\xi)\, d\xi.$$

Remarks (1) More precise specification of the domains of Fourier and inverse Fourier transforms will be given in the following paragraphs. When this is done, the original function f will be retrieved from its transform by $f = (\tilde{\hat{f}})$ as the term 'inverse' suggests.

<u>Physical motivation</u>. In view of the above remark we have a pair of transforms for the case $n = 1$, $X = Y = \mathbb{R}$:

$$\hat{f}(\xi) = \int_{\mathbb{R}} e^{-2\pi i (\xi . x)} f(x)\, dx \quad (1)$$

$$f(x) = \int_{\mathbb{R}} e^{2\pi i (x . \xi)} \hat{f}(\xi)\, d\xi \quad (2)$$

<u>Interpretation</u>. By an analogy from optics, given a light beam (1) indicates the spectral analysis to determine the monochromatic beams it contains, while (2) represents the spectral synthesis ; the reconstruction of the beam from its monochromatic components (harmonic oscillations) where $\hat{f}$ is the amplitude corresponding to the frequency ξ. x is usually called the time variable and ξ is the frequency variable. Also X is the time domain and Y is the frequency domain.

(2) X and Y are isomorphic to $\mathbb{R}^n$. Whenever there is an emphasis on the duality the domains of the integrals will be indicated by X and Y i.e. whether the integration is in the time domain or frequency domain, otherwise $\mathbb{R}^n$ will be sustained.

<u>Some Standard One-dimensional Examples</u>.

(i) $f(x) = e^{-\pi x^2}$, $\hat{f}(\xi) = e^{-\pi \xi^2}$,

(ii) $f(x) = \Pi(x)$ (c.f. Sect. 1.2, Example II following Propostion 2)
$\hat{f} = \dfrac{\sin \pi \xi}{\pi \xi}$,

(iii) $f(x) = e^{-a|x|}$, $\hat{f}(\xi) = \dfrac{2a}{a^2 + 4\pi^2 \xi^2}$.

Fig. 1.5 The integral of an entire function around an oriented contour vanishes

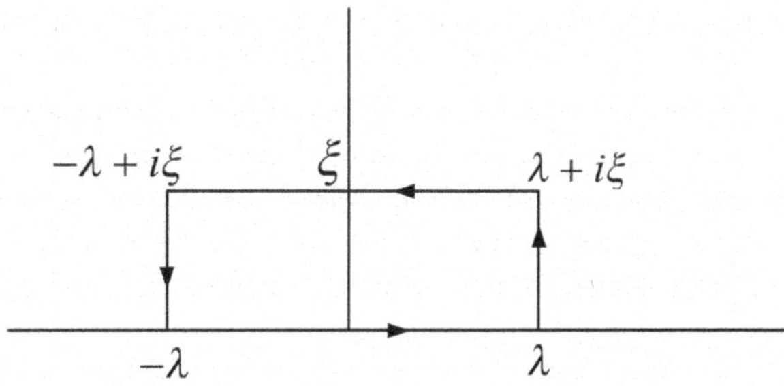

These transforms are obtained using the standard complex integration techniques . To illustrate let us demonstrate the transform formula of (i):

$\hat{f}(\xi) = \int_{-\infty}^{\infty} e^{-\pi x^2 - 2\pi i \xi x}\, dx = e^{-\pi \xi^2} \int_{-\infty}^{\infty} e^{-\pi (x + i\xi)^2}\, dx$. The formula will be proved if we show that the last integral, i.e. $I = \int_{-\infty}^{\infty} e^{-\pi (x + i\xi)^2} dx = \lim_{\lambda \to \infty} \int_{-\lambda + i\xi}^{\lambda + i\xi} e^{-\pi z^2} dz = 1$ (Fig. 1.5).

As $e^{-\pi z^2}$ is an entire function of the complex variable z, its integral around an oriented contour $\gamma(\lambda)$ as in the figure below vanishes:
Hence

$$\lim_{\lambda \to \infty} \left[\int_{-\lambda}^{\lambda} e^{-\pi z^2} dz + \int_{\lambda}^{\lambda + i\xi} e^{-\pi z^2} dz + \int_{\lambda + i\xi}^{-\lambda + i\xi} e^{-\pi z^2} dz + \int_{-\lambda + i\xi}^{\lambda} e^{-\pi z^2} dz \right] = 0 \quad (*)$$

On the vertical segments $(\lambda, \overrightarrow{\lambda + i\xi})$ and $(\overrightarrow{-\lambda + i\xi}, -\lambda)$ the integrals tend to zero as $\lambda \to \infty$, since e.g. on $(\lambda, \overrightarrow{\lambda + i\xi})$ the variable is $z = \lambda + iy$, $dz = i\,dy$, thus $|\int_{\lambda}^{\lambda + i\xi} e^{-\pi z^2} dz| = |i \int_0^{\xi} e^{-\pi(\lambda^2 - y^2)} e^{-2\pi i \lambda y}\, dy| \le |\int_0^{\xi} e^{-\pi \lambda^2 + \pi y^2}\, dy| \le |\xi| e^{-\pi \lambda^2 + \pi \xi^2}| \downarrow 0$ as $\lambda \to \infty$.

On the oriented segment $(-\lambda, \lambda)$, ξ is zero and $\lim_{\lambda \to \infty} \int_{\lambda}^{\lambda} e^{-\pi x^2}\, dx = \int_{-\infty}^{\infty} \frac{1}{\sqrt{2\pi}} e^{-y^2/2}\, dy = 1$ $(y = \sqrt{2\pi} x)$ (Gauss integral). Substituting into (*) we find $\lim_{\lambda \to \infty} \int_{\lambda + i\xi}^{-\lambda + i\xi} e^{-\pi z^2} dz + 1 = -I + 1 = 0$ yields the result.

<u>Note</u>. Example (i) is the case of a 'self-dual' function, i.e. a function which is equal to its Fourier transform.

<u>Important Remark.</u>

There are several other conventions in the literature for the character $e_\xi(x)$, accordingly different definitions of the Fourier transform:

(1) $e_\xi(x) = e^{2\pi i <\xi, x>}$; $\quad \hat{f}(\xi) = \int_X e^{-2\pi i <\xi, x>} f(x)\, dx$;

(2) $e_\xi(x) = e^{i <\xi, x>}$; $\quad \hat{f}(\xi) = \int_X e^{-i <\xi, x>} f(x)\, dx$;

(3) $e_\xi(x) = (2\pi)^{-n/2} e^{i <\xi, x>}$; $\quad \hat{f}(\xi) = \frac{1}{\sqrt{2\pi}^n} \int_X e^{-i <\xi, x>} f(x)\, dx$.

<u>Note</u>. For (1) and (2), the inverse transforms are $\int_Y e_x(\xi) \hat{f}(\xi) d\xi$ while in (3) there comes a factor $(2\pi)^{-n/2}$

Each definition has its pros and cons as far as the simplification of expressions in different results. The one we have chosen , namely (1) has the merit of having $\hat{\delta} = 1$ and $\hat{1} = \delta$ as we will see in Sect. 1.8.2. Nevertheless we shall have a factor of 2π popping up here and there while with definition (3) this is not the case. However in that case then $\hat{\delta} = (2\pi)^{-\frac{n}{2}}$ and $\hat{1} = (2\pi)^{\frac{n}{2}}\delta$ would be the price for it. Besides many transform formulae look clumsier with the Definition (3). We also note that with Definition (3) $f(x) = e^{-\pi x^2}$ is no more self-dual. In fact its transform will be $\hat{f}(\xi) = \dfrac{1}{\sqrt{2\pi}} e^{-\frac{\xi^2}{4}}$.

Theorem 1 (Basic properties of the Fourier transform) *Let $f, g \in L^1(\mathbb{R}^n)$, $h \in \mathbb{R}^n$, then*

(a) $\widehat{f(-x)} = \hat{f}(-\xi)$ *(transposition) ;*

(b) $\widehat{\overline{f(x)}} = \overline{\hat{f}(-\xi)}$ *(conjugation) ;*

(c) $\widehat{\lambda f + \mu g} = \lambda \hat{f} + \mu \hat{g}, \quad (\lambda, \mu \in \mathbb{R}) \quad$ *(linearity) ;*

(d) $\widehat{\tau_h f)} = e_{-h}\hat{f}$ *(translation) ;*

(e) $\widehat{e_h f} = \tau_h \hat{f}$ *(modulation)*

(f) *For $a > 0$,* $\widehat{f(ax)} = \frac{1}{a^n} \hat{f}(\frac{\xi}{a})$ *(change of the unit) ;*

(g) $\widehat{f * g} = \hat{f}\hat{g}$ *(Fourier transform converts convolution product into ordinary product).*

Proof (a), (b) and (c) follows directly from the definitions.

(d) $\widehat{\tau f}(\xi) = \int_X e^{-2\pi i <\xi,x>} f(x-h)\, dx = \int_X e^{-2\pi i <\xi,y+h>} f(y)\, dy = e^{-2\pi i <\xi,h>} \hat{f})(\xi) = (e_{-h}\hat{f})(\xi)$.

(e) $\widehat{e_h f}(\xi) = \int_X e^{2\pi <h,x>} e^{-2\pi i <\xi,x>} f(x)\, dx = \int_X e^{-2\pi i <\xi-h,x>} f(x)\, dx = \hat{f}(\xi-h) \equiv (\tau_h \hat{f})(\xi)$.

(f) Posing $y = ax$ and recalling that $dx = dx_1 dx_2 \cdots dx_n$ we have:
$\int_X e^{-2\pi i <\xi,x>} f(ax)\, dx = \int_X \frac{1}{a^n} e^{-2\pi \frac{1}{a} <\xi,y>} f(y)\, dy = \frac{1}{a^n} \hat{f}(\frac{\xi}{a})$.

(g) $\widehat{f * g}(\xi) = \int_X e^{-2\pi i <\xi,x>} \int_X f(t)g(x-t)\, dt dx$, by the Fubini' theorem:
$\widehat{f * g}(\xi) = \int_X f(t) \int_X e^{-2\pi i <\xi,x>} g(x-t)\, dx\, dt = \int_X f(t) e^{-2\pi i <\xi,t>} (\int_X e^{-2\pi i <\xi,y>} g(y)\, dy)\, dt = \hat{f}(\xi)\hat{g}(\xi)$:
$\qquad\qquad\qquad\qquad\qquad\qquad\qquad\qquad\qquad\qquad\qquad\qquad\qquad\qquad\qquad\qquad\quad\Box$

$$|(2\pi i\xi)^m| |\hat{f}(\xi)| = |\widehat{f^{(m)}}(\xi)| = |\int_{\mathbb{R}} e^{-2\pi i\xi x} f^{(m)}(x)\, dx| \quad \text{or} \quad |\xi^m| |\hat{f}(\xi)| \le \frac{1}{(2\pi)^m} \int_{\mathbb{R}} |f^{(m)}(x)|\, dx.$$

Remarks Thus the higher order derivatives of f is integrable, the more rapidly $\hat{f}$ decreases at infinity, (see also the next Proposition 1).

The Fourier transform theory in L^1 has some pathological features. For instance for L^1, $(\hat{\cdot})$ is a 1-1 mapping $L^1(\mathbb{R}^n)$ into the algebra of functions of class $\mathcal{C}_0(\mathbb{R}^n)$ of complex continuous functions which vanish at infinity. However the actual image is a very complicated subalgebra.

Also $f = (\hat{\tilde{f}})$ works out in the almost everywhere sense. A more regular theory is based on S_n, (c.f. [Rud], Chap. 7).

Let P be a polynomial of n variables with complex coefficients of the form $P(\xi) = \sum_\alpha c_\alpha \xi^\alpha = \sum c_\alpha \xi_1^{\alpha_1} \cdots \xi^{\alpha_n}$.
Then differential operators $P(D)$ and $P(-D)$ are defined by

$$P(D) = \sum c_\alpha D_\alpha, \quad P(-D) = \sum [\frac{1}{2\pi}(-1)]^{|\alpha|} c_\alpha D_\alpha$$

where D_α is defined by Definition 1, Sect. 1.2.

We note that $D_\alpha e_\xi = D_\alpha e^{2\pi i(\xi_1 x_1 + \cdots + \xi_n x_n)} = (2\pi \xi_1)^{\alpha_1} \cdots (2\pi \xi_n)^{\alpha_n} e^{2\pi i <\xi, x>}$ implies $P(D)e_\xi = P(2\pi\xi)e_\xi$.

Proposition 1 *With the notation given above, if $f \in S_n$ and α is any multi-index*

(a) $\widehat{P(D)f}(\xi) = P(2\pi\xi)\hat{f}(\xi)$ and $\widehat{(Pf)} = P(-D)\hat{f}$,
(b) The Fourier transform is a continuous linear mapping of S_n into S_n, (or $S_n(X)$ into $S_n(Y)$).
(c) If $f \in L^1(\mathbb{R}^n)$, then $\hat{f} \in C_0(\mathbb{R}^n)$ and $\|\hat{f}\|_\infty \leq \|f\|_1$, ($C_0(\mathbb{R}^n)$ is a sup-normed B-space of complex continuous functions that vanish at infinity).

<u>Note</u>. Part (b) will be extended to the inverse transform in Proposition 2.

Proof (a) If $f \in S_n$, then clearly $D_\alpha f \in S_n$, so is $D_\alpha f$ and $P(D)f$. By Theorem 2b, Sect. 1.7 we have

$$P(Df) * e_\xi = f * P(D)e_\xi = f * P(2\pi\xi)e_\xi = P(2\pi\xi)(f * e_\xi) \quad (\dagger)$$

now in view of $\hat{f} = (f * e_\xi)(0)$, (Definition 1, Note 2) $\dagger$ yields the first part.
If $\xi = (\xi_1, \ldots, \xi_n)$, let $\xi^+ = (\xi_1 + \epsilon, \xi_2, \ldots \xi_n \epsilon \neq 0$. Then
$$\frac{\hat{f}(\xi^+) - \hat{f}(\xi)}{i\epsilon} = \int_{\mathbb{R}^n} x_1 f(x) \frac{e^{-2\pi i x_1 \epsilon} - 1}{i x_1 \epsilon} e^{-2\pi <x, \xi>} dx.$$
As $\epsilon \to 0$, the dominated convergence theorem can be applied since $x_1 f \in L^1$ and by L'Hopital's rule, we obtain $-\frac{1}{i}\frac{\partial}{\partial \xi_1} = \int_{\mathbb{R}^n} 2\pi x_1 f(x) e^{-2\pi i <x, \xi>} dx$. This is the case $P(x) = x_1$ of the second part and the general case follows by iteration.

(b) Suppose $f \in S_n$ and $g(x) = [\frac{-1}{2\pi}]^{|\alpha|} x^\alpha f(x)$, then $g \in S_n$ and the second part of (a) implies $\hat{g} = D_\alpha \hat{f}$. This time by the first part of (a) $P(2\pi\xi)D_\alpha \hat{f} = P(2\pi\xi)\hat{g} = (P(D)g)$ which is a bounded function since $P(D)g \in L^1$. Then recalling the definition of S_n (Definition 2, Sect. 1.2) we find that $\hat{f} \in S_n$.

For the continuity assertion if $f_i \to f$ in S_n, then $\hat{f_i} \to \hat{f}$ in $L^1(\mathbb{R}^n)$, therefore $\hat{f_i}(\xi) \to \hat{\xi}$ for all $\xi \in \mathbb{R}^n$. Hence $(f_i, \hat{f_i}) \to (f, \hat{f}$ in L^1 and the continuity follows from the closed graph theorem.

(c) Since $|e_\xi| = 1$, it is clear that $|\hat{f}(\xi)| \leq \|f\|_1$, $(f \in L^1, \xi \in \mathbb{R}^n$ ($\ddagger$). Since $\mathcal{D}(\mathbb{R}^n \subset S_n$, S_n is dense in L^1. To each $f \in L^1(\mathbb{R}^n)$ correspond functions $f_i \in S_n$) such that $\|f - f_i\|_1 \to 0$. Since $\hat{f_i} \in S_n \subset C_0(\mathbb{R}^n)$ and since ($\ddagger$) implies that $\hat{f_i} \to \hat{f}$ uniformly on $\mathbb{R}^n$, part (c) is proved. $\square$

Remark If $f \in L^1 \setminus S_n$, then as it was pointed out before $\hat{f}$ may not be integrable, thus the inverse transform may not be defined. More precisely (see also Note 3) following the Definition 1) if $f \in L^1(\mathbb{R}^n)$, then $\hat{f} \in C_0(\mathbb{R}^n)$ and $\|f\|_\infty \leq \|f\|_1$.

Proof Since $|e_\xi(x)| = 1$, clearly $\hat{f}(\xi)| \leq \|f\|_1$ $(\xi \in \mathbb{R}^n)$. As $\mathcal{D}(\mathbb{R}^n) \subset \mathcal{S}_n$ and $\mathcal{S}_n$ is dense in $L^1(\mathbb{R}^n)$; to each $f \in L^1(\mathbb{R}^n)$ there corresponds a sequence $f_j \in \mathcal{S}_n$ such that $\|f - f_j\|_1 \longrightarrow 0$. But $\hat{f}_j \in \mathcal{S}_n \subset \mathcal{C}_0(\mathbb{R}^n)$ (by Proposition 1b) and in view of the above inequality $\hat{f}_j \longrightarrow \hat{f}$ uniformly on $\mathbb{R}^n$, hence $\hat{f}$ is continuous and vanishes at infinity.

EXAMPLE In parallel to Example (i) following Definition 2, we calculate the Fourier transform of $f(x) = e^{-\frac{1}{2}\|x\|^2}$, $(x \in \mathbb{R}^n)$:
$\hat{f}(\xi) = \int_{\mathbb{R}^n} e^{-2\pi i <\xi,x>} e^{-\frac{1}{2}\|x\|^2} = \int_{-\infty}^{\infty} \cdots \int_{-\infty}^{\infty} e^{-2\pi i(\xi_1 x_1 + \cdots + \xi_n x_n)} e^{-\frac{1}{2}(x_1^2 + \cdots + x_n^2)} dx_1 \cdots$
$dx_n = \prod_{j=1}^{n} \int_{-\infty}^{\infty} e^{-\frac{1}{2}(x_j + 2\pi i \xi_j)^2} dx_j (e^{-2\pi^2(\xi_1^2 + \cdots + \xi_n^2)})$. Again using Cauchy's theorem for the entire function $e^{-\frac{1}{2}(x_j + 2\pi i \xi_j)^2}$ along the lines in Example (i) we find

$$\hat{f}(\xi) = \prod \sqrt{2\pi} e^{-2\pi^2 \xi_j^2} = (2\pi)^{n/2} e^{-2\pi^2 \|\xi\|^2}.$$

If the Fourier transform is defined on $\mathcal{S}_n$, then the inverse transform given in Definition 2 is actually the inverse of the mapping $f \longrightarrow \hat{f}$.

The Inversion of Fourier Transform

Proposition 2 *(a) If $f \in \mathcal{S}_n$, then $f(x) = \int_Y e^{2\pi i <x,\xi>} \hat{f}(\xi) d\xi = \int_Y e_x \hat{f} d\xi$, thus $\tilde{\hat{f}} = f$ and $\hat{\tilde{f}} = f$ (for $(\tilde{.})$ c.f. Definition (2).*
   (b) Fourier transform is a continuous linear one-to-one mapping of $\mathcal{S}_n$ onto $\mathcal{S}_n$, (or $S(X)$ onto $S(Y)$) whose inverse is also continuous.

Note. The expression in (a) is the inversion formula for the Fourier transform.

Proof (a) (X and Y are isomorphic to $\mathbb{R}^n$). Consider:
$\int_Y g(\xi) \hat{f}(\xi) e^{2\pi i <x,\xi>} d\xi = \int_Y g(\xi)[\int_X e^{-2\pi <\xi,y>} f(y) dy] e^{2\pi i <x,\xi>} d\xi$.
By Fubini's theorem:
$= \int_X [\int_Y e^{-2\pi i <\xi,y-x>} g(\xi) d\xi] f(y) dy = \int_X \hat{g}(y - x) f(y) dy = \int_X \hat{g}(y) f(x + y) dy$ ($\sharp$).
For $\epsilon > 0$, $\widehat{g(\epsilon y)} = \epsilon^{-n} \hat{g}(\frac{y}{\epsilon})$, (Theorem 1f). Replacing $g(\xi)$ by $g(\epsilon \xi)$ in ($\sharp$) we have $\int_Y g(\epsilon \xi) \hat{f}(\xi) e^{2\pi i <X,\xi>} d\xi = \int_X \epsilon^{-n} \hat{g}(\frac{y}{\epsilon}) f(x + y) dy = \int_X \hat{g}(u) f(x + \epsilon u) du$ $(u = \frac{y}{\epsilon})$.

   Now taking $g = e^{-\frac{1}{2}\|x\|^2}$ as in the Example before Proposition 2 and letting $\epsilon \downarrow 0$ we find

$$g(0) \int_Y \hat{f}(\xi) e^{2\pi i <x,\xi>} d\xi = f(x) \int_X (2\pi)^{n/2} e^{-2\pi^2 \|u\|^2} du = f(x) \prod_{j=1}^{n} \int_{\mathbb{R}} (2\pi)^{\frac{1}{2}} e^{-2\pi^2 u_j^2} du_j$$

$$= f(x) \prod_{j=1}^{n} \int_{\mathbb{R}} \frac{1}{(2\pi)^{1/2}} e^{-\frac{v_j^2}{2}} dv_j \quad \text{(by } u_j = \frac{v_j}{2\pi})$$

$= f(x)$. Since $g(0) = 1$ this proves $\tilde{\hat{f}} = f$. Similarly repeating the same arguments by replacements $f \leftrightarrow \tilde{f}, e_\xi \leftrightarrow e_{-\xi}, X \leftrightarrow Y$ we find

$g(0) \int_Y \tilde{f}(x) e^{-2\pi i > \xi, x>} \, dx = f(x)$ completes the proof of the first part, (b) The linearity of the mapping $f \to \hat{f}$ is obvious.

One-to-oneness : If $\hat{f}_1 - \hat{f}_2 = 0$, by the inversion formula of part (a) we have $f_1 - f_2 = 0$.

For being onto: given $f \in \mathcal{S}_n$, it is the inverse Fourier transform of $\tilde{f}$.

The continuity of $f \to \hat{f}$ was already proved in Proposition 1. The proof of the continuity of the inverse transform is parallel to this proof. $\square$

Theorem 2 *If $f \in \mathcal{S}_n$, $g \in \mathcal{S}_n$:*

(a) $\widehat{fg} = \hat{f} * \hat{g}$,

(b) $f * g \in \mathcal{S}_n$.

Proof By Theorem 1g: $\widehat{f * g} = \hat{f}\hat{g}$. If we replace f and g by $\hat{f}$, $\hat{g}$

$$\widehat{\hat{f} * \hat{g}} = \hat{\hat{f}}\hat{\hat{g}} \quad (\sqrt{}).$$

But $\hat{\hat{f}}(x) = \int_{\mathbb{R}^n} e^{-2\pi i <x,\xi>} \hat{f}(\xi) \, d\xi = \int_{\mathbb{R}^n} e^{2\pi i <-x,\xi>} \hat{f}(\xi) \, d\xi$, then by the inversion formula of Proposition 2a

$$\hat{\hat{f}}(x) = f(-x) \doteq \check{f}(x). \text{ Substituting in } (\sqrt{}):$$

$\widehat{\hat{f} * \hat{g}} = \check{f}\check{g} = (fg)^{\vee} = \widehat{\widehat{fg}}$. Now apply the inverse transform of Proposition 2a to get (a).

(b) By Theorem 1g and the inversion formula $f * g = \widetilde{\hat{f}\hat{g}}$, then as $\hat{f}\hat{g} \in \mathcal{S}_n$, we also have $f * g \in \mathcal{S}_n$. $\square$

Theorem 3 (Parseval Formula). *For $f \in \mathcal{S}_n$ there is the isometry $\|f\|_2 = \|\hat{f}\|_2$.*

<u>Note.</u> As $\mathcal{S}_n \subset L^2$, $\| . \|$ is well-defined. If the field is complex , then the inner product in L^2 is given by $< f, g >= \int f\bar{g} \, dx$ where $(\bar{.})$ indicates the conjugation).

Proof By Proposition 2a $\int_{\mathbb{R}^n} f(x) \bar{g}(x) \, dx = \int_{\mathbb{R}^n} \bar{g} \, dx \int_{\mathbb{R}^n} e^{2\pi i <x,\xi>} \hat{f}(\xi) \, d\xi$

$= \int_{\mathbb{R}^n} \hat{f}(\xi) \, d\xi \int_{\mathbb{R}^n} \bar{g}(x) e^{2\pi i <x,\xi>} \, dx$, (Fubini).

Noticing that the last inner integral is the complex conjugate of $\hat{g}(\xi)$:

$\int_{\mathbb{R}^n} f(x) \bar{g}(x) \, dx = \int_{\mathbb{R}^n} \hat{f}(\xi) \bar{\hat{g}}(\xi) \, d\xi, \quad (f, g \in \mathcal{S}_n)$

or $< f, g >=< \hat{f}, \hat{g} >$ which yields $\|f\|_2 = \|\hat{f}\|_2$.

(equivalently $\int |f(x)|^2 \, dx = \int |\hat{f}(\xi)|^2 \, d\xi$). $\square$

<u>Fourier–Plancherel transform.</u>

Using the isometry given in the last theorem we state:

Theorem 4 *There is a linear isometry Φ of $L^2(\mathbb{R}^n)$ onto itself which is uniquely determined by the requirement that $\Phi f = \hat{f}$ for every $f \in \mathcal{S}_n$.*

Proof $\mathcal{S}_n$ is dense in $L^2(\mathbb{R}^n)$. Thus $f \longrightarrow \hat{f}$ is an isometry (relative to the L^2-metric) of the dense subspace $\mathcal{S}_n$ of $L^2(\mathbb{R}^n)$ to $\mathcal{S}_n$. This mapping is also onto by the inversion given in Proposition 2. In this way by a well-known result of analysis $f \to \hat{f}$ has a unique continuous extension $\Phi : L^2(\mathbb{R}^n) \longrightarrow L^2(\mathbb{R}^n)$ and that this Φ is a linear isometry onto $L^2(\mathbb{R}^n)$.

Remark This extension Φ is still called the Fourier transform (sometimes Fourier-Plancherel transform) and $\hat{f}$ denotes Φf when $f \in L^2$ as well for $f \in L^2(\mathbb{R}^n)$ or $f \in \mathcal{S}_n$.

1.8.2   Fourier Transform of Distributions

In the light of analyses of Sect. 1.8.1, it is natural to define the Fourier transform of a distribution as a map from $\mathcal{S}'(X)$ into $\mathcal{S}'(Y)$, $(X \cong \mathbb{R}^n, Y \cong \mathbb{R}^n)$. To motivate the definition consider $f \in \mathcal{S}(X)$ and the regular distribution T_f. Then for all $\phi \in \mathcal{S}(Y)$:

$$< T_f, \hat{\phi} > = \int_X f(\xi)(\int_Y \phi(x)e^{-2\pi i <x,\xi>}\,dx)\,d\xi = \int_X (\int_Y f(\xi)e^{-2\pi i <x,\xi>}\,d\xi)\,\phi(x)\,dx = < \hat{f}, \phi >$$

suggesting the definition

Definition 1 If $T \in \mathcal{S}'(X)$, then the Fourier transform $\hat{T}$ is defined as

$$< \hat{T}, \phi > = < T, \hat{\phi} > \quad \text{for all} \quad \phi \in \mathcal{S}(Y).$$

Since $\phi \longrightarrow \hat{\phi}$ is a continuous, linear mapping from $\mathcal{S}(X)$ onto $\mathcal{S}(Y)$ and $T \in \mathcal{S}'(X)$, then the Definition yields $\hat{T}$ as a tempered distribution in $\mathcal{S}'(Y)$, (c.f. 1.5 , 0 1).

Remarks (1) If the Fourier transform is denoted by $\mathcal{F}$, then $\hat{T}$ or $\mathcal{F}T$ is the image of T under the transpose of the mapping $\mathcal{F} : \mathcal{S}(Y) \longrightarrow \mathcal{S}(X)$. This is continuous and $\mathcal{S}$ is dense in $\mathcal{S}'$, (c.f. Sect. 1.5, Example d) then $\mathcal{F} : \mathcal{S}'(X) \longrightarrow \mathcal{S}'(Y)$ is the unique extension of $\mathcal{F} : \mathcal{S}(Y) \longrightarrow \mathcal{S}(X)$.

(2) Consistency question arises when $f \in L^1(\mathbb{R}^n)$. On one hand the Fourier transform $\hat{f}$ is given by Sect. 1.8.1, Definition 1, on the other hand T_f is a tempered distribution , (Sect. 1.5, Example b) and $\hat{T}_f$ is given by Definition 1. Are the two transforms same ? The answer is affirmative. Consider $\int_{\mathbb{R}^n} \int_{\mathbb{R}^n} e^{-2\pi i <\xi,x>}\,dx\,d\xi$, $\phi \in \mathcal{S}_n$. Using Fubini's theorem we can write this double integral in two ways to obtain

$$\int_{\mathbb{R}^n} \hat{f}(\xi)\,\phi(\xi)\,d\xi = \int_{\mathbb{R}^n} f(x)\,\hat{\phi}(x)\,dx,$$

then

$$< \hat{T}_f, \phi > = < T_f, \hat{\phi} > = \int f\hat{\phi} = \int \hat{f}\phi = < T_{\hat{f}}, \phi >$$

i.e. $\hat{T}_f = T_{\hat{f}}$ showing the consistency.

The counter-part of Sect. 1.8.1, Propositions 1 and 2 will be

Proposition 1 *(a) The Fourier transform is a continuous, one-to-one mapping of $\mathcal{S}'_n$ onto $\mathcal{S}'_n$ of period 4, whose inverse is also continuous.*
(b) If $T \in \mathcal{S}'_n$ and P is a polynomial, then
$$(\widehat{P(D)T})(\xi) = P(2\pi\xi)\hat{T}(\xi) \text{ and } [P(2\pi x)T(x)]\hat{}(\xi) = P(-D)\hat{T}(\xi).$$
($P(D)$ and $P(-D)$ are defined in terms of D_α, not D^α, c.f. Definition 1, Sect. 1.2).

<u>Note</u>. The continuity in (a) refers to the weak topology that $\mathcal{S}_n$ induces on $\mathcal{S}'_n$.

Proof (a) Let U be a weak neighborhood of 0 in $\mathcal{S}'_n$. Then there exist functions $\phi_1, \ldots, \phi_k \in \mathcal{S}_n$ such that $\{T \in \mathcal{S}'_n : | < T, \phi_i > | < 1 \text{ for } 1 \leq i \leq k\} \subset U$.

Define $V = \{T \in \mathcal{S}'_n : |T, \hat{\phi}_i| < 1 \text{ for } 1 \leq i \leq k\}$. Then $T \in V \Rightarrow \hat{T} \in U$ since $< \hat{T}, \phi_i >=< T, \hat{\phi}_i >$. This shows that $\mathcal{F}$ is continuous.

As it was found in the proof of Sect. 1.8.1, Theorem 2a:

If $\phi \in \mathcal{S}_n$, then $\mathcal{F}\phi = \hat{\phi}$, $\mathcal{F}^2\phi = \check{\phi}$, recall $\check{\phi}(x) = \phi(-x)$). Thus $\mathcal{F}^4\phi = \phi$ (i.e. the period of $\mathcal{F}$ is 4). This also implies that $\mathcal{F}^4 T = T$, $(T \in \mathcal{S}'_n)$ since $< \mathcal{F}^2 T, \phi >=< \mathcal{F}(\mathcal{F}T), \phi >=< \mathcal{F}T, \mathcal{F}\phi >=< T, \mathcal{F}^2\phi >=< T, \check{\phi} >$; so that $< \mathcal{F}^4, \phi >=< \mathcal{F}^2(\mathcal{F}^2)\phi >=< \mathcal{F}^2 T, \check{\phi} >=< T, \check{\check{\phi}} >=< T, \phi >$.

Thus $\mathcal{F}^4 T = T$. One-to-oneness of the mapping $\mathcal{F}$ follows : $\mathcal{F}w = 0 \Rightarrow \mathcal{F}^4 w = w = 0$. Onto : given $T \in \mathcal{S}'_n$, $T = \mathcal{F}(\mathcal{F}^3 T)$ and $\mathcal{F}^3 \in \mathcal{S}'_n$.

On the other hand $\mathcal{F}^{-1} = \mathcal{F}^3$ shows the continuity of $\mathcal{F}^{-1}$.

(b) This is the distribution counter-part of Sect. 1.8.1, Proposition 1a. First note that as $f \to Pf$, $f \to gf$, $f \to D^\alpha f$ are continuous linear mappings of $\mathcal{S}_n$, if T is a tempered distributions, so are $D^\alpha T$, PT and gT, $(g: \text{smooth})$. Using these we have the following computations:

$\widehat{P(D)T}(\xi), \phi(\xi) >=< (P(D)T)(x), \hat{\phi}(x) >=< T(x), P(-D)\hat{\phi}(x) >$
$=< T(x), P(\widehat{2\pi\xi})\phi(\xi) >$ (by 1.8.1, Proposition 1a) $=< \hat{T}(\xi), P(2\pi\xi)\phi(\xi) >=< P(2\pi\xi)\hat{T}(\xi), \phi(\xi) >$ gives the first part of (b). For the second part:
$< P(-D)\hat{T}(\xi), \phi(\xi) >=< \hat{T}(\xi), P(D)\phi(\xi) >=< T(x), \widehat{P(D)\phi}(x) >$
$=< T(x), P(2\pi x)\hat{\phi}(x) >$
$=< (P(2\pi x)T)(x), \hat{\phi}(x) >=< \widehat{P(2\pi x)T}(\xi), \phi(\xi) >$.   □

Remark Denote $\mathcal{F}^{-1} \equiv\, \sim T$, then $<\sim T, \phi >=< T, \sim \phi >$ and considering the inversion formula of 1.8.1, Proposition 2 we have $\sim T(x) = T(-x)$.

<u>Fourier Transform and Convolution of Distributions</u>

Theorem 1 *Let $\phi \in \mathcal{S}_n$ and T be a tempered distribution in $\mathcal{S}'_n$. Then*

(i) $\widehat{(T * \phi)} = \hat{\phi}\hat{T}$,
(ii) $\widehat{(\hat{T} * \hat{\phi})} = \hat{\phi}\hat{T}$.

<u>Note</u>. $T * \phi$ was defined in Sect. 1.7, Definition 3b

Proof First let us see that $T * \phi$ has polynomial growth, therefore it defines a tempered distribution. Consider the easily verified inequality

$$(1 + \|x + y\|^2)^N \leq 2(1 + \|x\|^2)^N (1 + \|y\|)^N, \quad (x, y \in \mathbb{R}^n) \quad (1)$$

$p_N(f) = \sup_{|\alpha| \leq N} \sup_{x \in \mathbb{R}^n} |(D^\alpha f)(x)|\}, \quad N = 0, 1, 2, \ldots$
multiply both sides of (1) by $|(D^\alpha)(y)|$ and take supremums on α and y to find

$$p_N(\tau_x f) \leq 2^N (1 + \|x\|^2)^N p_N(f) \quad (2)$$

As T is a continuous linear functional on $\mathcal{S}_n$ and $\{p_N\}$ define the topology of $\mathcal{S}_n$, there exist N and constant $c > 0$ such that

$$| < T, f > | \leq C p_N(f) , \quad (f \in \mathcal{S}_n), \quad (3)$$

By (2) and (3) : $|(T * \phi)(x)| = | < T(y), \phi(x - y) > | = | < T, \tau_x \check{\phi} > | \leq 2^N C(1 + \|x\|^2)^N p_N(\phi)$, showing that $T * \phi$ has polynomial growth, therefore it has a Fourier transform in $\mathcal{S}_n'$. If $\psi \in \mathcal{D}(\mathbb{R}^n)$ with support K, then $\widehat{T * \phi}(\hat{\psi})$
$= < T * \phi, \hat{\hat{\psi}} > = < T * \phi, \check{\psi} > \quad (\check{\psi} = \psi(-x))$
$= \int_{\mathbb{R}^n} (T * \phi)(x)\psi(-x)\, dx = \int_{-K} < T(y), \psi(-x)\phi(x - y) > dx = \int_{-K} < T(y), \psi(-x)\phi(x - y) > dx = T(y), \int_{-K} \psi(-x)\phi(x - y)\, dx > = < T, (\phi * \psi)^{\check{}} > = < \hat{T}, \widehat{\phi * \psi} > = < \hat{T}, \hat{\phi}\hat{\psi} > \quad$ (by Sect. 1.8.1, Theorem 1g) so that $\widehat{T * \phi}(\hat{\psi}) = < \hat{\phi}\hat{T}, \hat{\psi} > \quad (4)$
As $\mathcal{D}(\mathbb{R}^n)$ is dense in $\mathcal{S}_n$, the Fourier transform of members of $\mathcal{D}(\mathbb{R}^n)$ is also dense in $\mathcal{S}_n'$ by Sect. 1.8.1, Proposition 1b. Thus (4) holds for every $\hat{\psi} \in \mathcal{S}_n$, showing (i).

(ii) By (i) $\widehat{\hat{T} * \hat{\phi}} = \hat{\hat{\phi}}\hat{\hat{T}} = \check{\phi}\check{T} = (\phi T)^{\check{}}$. This proves (ii) since $(\phi T)^{\check{}} = (\widehat{(\widehat{\phi T})})$.

More Properties of Fourier Transform

Theorem 2 *(a) Change of the unit :* $\widehat{T(ax)}(\xi) = \frac{1}{|a|^n}\hat{T}(\frac{\xi}{a})$, $\quad a \in \mathbb{R}$.
(b) Translation : $\widehat{\tau_h T}(\xi) = e^{-2\pi i <h,\xi>}\hat{T}(\xi) = e_{-h}(\xi)\hat{T}(\xi)$.
(c) $\widehat{e_h T}(\xi) = (e^{2\pi i <h,x>}T)\hat{}(\xi) = \tau_h(\xi)\hat{T}(\xi)$.

<u>Note</u>. Based on regular distributions $T(ax)$ is defined as $< T(ax), \phi > = < T(x), \frac{1}{|a|^n}\phi(\frac{x}{a}) >$ and $< \tau_h T, \phi > = < T, \tau_{-h}\phi > \quad$ (Sect. 1.7, Definition 4i), equivalent notation : $\tau_h T(x) \equiv T(x - h)$.

Proof (a) $\quad\quad < \widehat{T(a\xi)}(\xi), \phi(\xi) > = < T(ax), \hat{\phi}(x) > = < T(x), \frac{1}{|a|^n}\hat{\phi}(\frac{x}{a}) > ,$
(Sect. 1.8.1, Theorem 1d)
$= < T(x), \widehat{\phi(a\xi)}(x) > = < \hat{T}(\xi), \phi(a\xi) > = < \frac{1}{|a|^n}\hat{T}(\frac{\xi}{a}), \phi(\xi) >.$

(b) $< \widehat{\tau_h T}, \phi > = < \tau_h T, \hat{\phi} > = < T, \tau_{-h}\hat{\phi} > = < T, \widehat{e_{-h}\phi} > \quad$ (Sect. 1.8.1, Theorem 1e)
$= < \hat{T}, e_{-h}\phi > = < e_{-h}\hat{T}, \phi > \quad$ (Sect. 1.6, operation V).
Thus $\widehat{\tau_h T} = e_{-h}\hat{T}$.

(c) $\quad < \widehat{e_h T}, \phi > = < e_h T, \hat{\phi} > = < T, e_h\hat{\phi} > = < T, \widehat{\tau_h\phi} > = < \hat{T}, \tau_{-h}\phi > = < \tau_h\hat{T}, \phi >$. Hence $\widehat{e_h T} = \tau_h\hat{T}$. $\quad\quad\quad\square$

The Case of Distributions with Compact Support

Consider $V(\xi) = < T(x), e^{-2\pi i \xi} > ,$ $\quad (T \in \mathcal{E}'(\mathbb{R}^n))$ which is well-defined and infinitely differentiable (ξ real or complex). Therefore $V(\xi)$ is a holomorphic entire function. Let us show that $V(\xi)$ is the Fourier transform of T:
For $\quad \phi \in \mathcal{C}^\infty(\mathbb{R}^n),$ $\quad < \hat{T}, \phi > = < T, \hat{\phi} > = < T(x), \int_{\mathbb{R}^n} e^{-2\pi i <x,\xi>} \phi(\xi)\, d\xi > = <$

$$T(x), < \phi(\xi), e^{-2\pi i <x,\xi>} >>=< \phi(\xi)T(x), e^{-2\pi i <x,\xi>} >=<$$
$$\phi(\xi), T(x)e^{-2\pi i <x,\xi>} >>= \int_{\mathbb{R}^n} \phi(\xi) < T(x), e^{-2\pi i <x,\xi>} d\xi =$$
$$\int_{\mathbb{R}^n} \phi(\xi) V(\xi) d\xi =< V, \phi >, \text{ thus } \hat{T} = V.$$

Bochner Theorem for tempered distributions.
We need some more definitions:

Definition 2 For $\phi \in \mathcal{S}_n$, the involution ϕ^* is defined as $\overline{\phi(-x)}$.

Definition 3 Let T be a tempered distribution. We say that T is positive if $< T, \phi >\geq 0$ for every test function $\phi \in \mathcal{S}_n$ with $\phi \geq 0$. (Provided that ≥ 0 makes sense for complex-valued functions T and ϕ).

Definition 4 Let T be a tempered distribution. We say that T is positive definite if $< T, \phi * \phi^* >\geq 0$ for every test function $\phi \in \mathcal{S}_n$.

Theorem 4 (Bochner). *Let T be a tempered distribution. Then T is positive definite $\Longleftrightarrow$ if its Fourier transform $\hat{T}$ is positive.*

Proof ($\Longleftarrow$) Suppose $\hat{T}$ is positive. Then

$$< T, \phi * \phi^* >=< \tilde{\hat{T}}, \phi * \phi^* >=< \hat{T}, (\phi * \phi^*)^{\sim} >=< \hat{T}, |\tilde{\phi}|^2 > \quad (\sqrt{})$$

Since $\hat{T}$ is positive and $|\tilde{\phi}|^2 \geq 0$, we have $< T, \phi * \phi^* >\geq$. So that T is positive.

($\Longrightarrow$) Let T be positive definite. If we want to use the same equalities ($\sqrt{}$) for showing this implication, we should note that the positivity of $|\tilde{\phi}|^2 = |\phi|^2$ will not imply in general that its square root ϕ be a C^∞ function, (consider for instance a neighborhood of a point where $|\phi|^2 = 0$). But we can overcome this difficulty by noting that given $\phi \in \mathcal{S}_n$, $\phi \geq 0$, there exists a sequence $\Theta_n \in \mathcal{D}(\mathbb{R}^n)$, $\Theta_n \geq 0$ such that $\Theta_n^2 \to \phi$ in $\mathcal{S}_n$. To show this we can use a mesa function ψ (c.f. Note to the Lemma of Sect. 1.7), i.e. $\psi \in \mathcal{D}$, $0 \leq \psi \leq 1$ and $\psi(x) \equiv 1$ on some neighborhood of $x = 0$. Let $\psi_a(x) = \psi(x/a)$. Then $\psi_a^2 \phi \to \phi$ in $\mathcal{S}_n$ as $a \to \infty$. For any ϵ, since $\phi(x) \geq 0$, $\phi(x) + \epsilon$ is bounded from below by ϵ. Hence $[\phi(x) + \epsilon]^{\frac{1}{2}} \in C^\infty$. Therefore $\Theta_{a,\epsilon} = \psi_a(x)[\phi(x) + \epsilon]^{\frac{1}{2}} \in \mathcal{D}$. Hence $\Theta_{a,\epsilon}^2 = \psi_a^2(x)[\phi(x) + \epsilon] \longrightarrow \phi(x) + \epsilon$ as $a \to \infty$. It means that we can create a sequence Θ_n by choosing any sequence of a-values approaching infinity, e.g. by setting $a_n = n$, $\epsilon_n = \epsilon(a_n)$.

Consequently it suffices to show that $< \hat{T}, \Theta^2 >\geq 0$. In fact we can replace $\Theta^2 \leftrightarrow |\phi|^2$ for any $\phi \in \mathcal{S} \supset \mathcal{D}$ and show the stronger inequality $< \hat{T}, |\phi|^2 >\geq 0$. This follows from ($\sqrt{}$) upon substituting $\hat{\phi}$ for ϕ and noticing that $\tilde{\hat{\phi}} = \phi$. If then T is positive definite we have $< \hat{T}, |\phi|^2 >=< T, \hat{\phi} * \hat{\phi}^* >\geq 0$ for all $\phi \in \mathcal{S}$, whence $< \hat{T}, |\phi|^2 >\geq 0$, as desired., (c.f. [Rich-Youn], p. 76). $\qquad \square$

1.8.3 *Examples of Fourier Transforms of Distributions*

(I) Polynomials are tempered distributions. To obtain their Fourier transforms we start by the polynomial 1 regarded as a regular distribution. If $\phi \in \mathcal{S}_n$ we may note that as $\phi(x) = \int_Y \hat{\phi}(\xi)\, e_x(\xi)\, d\xi$ by the inversion formula Sect. 1.8.1, Proposition 2a, we have $\phi(0) = \int_Y \hat{\phi}(\xi)\, d\xi$. Then $< \hat{1}, \phi > = < 1, \hat\phi > = \int_Y 1.\hat{\phi}(\xi)\, d\xi = \phi(0) = < \delta, \phi >$. Thus

$$\hat{1} = \delta \qquad (*)$$

Also $< \hat{\delta}, \phi > = < \delta, \hat\phi > = \hat{\phi}(0) = \int_X \phi(x)\, dx = < 1, \phi >$, whence

$$\hat{\delta} = 1 \qquad (**)$$

If P is an arbitrary polynomial on $\mathbb{R}^n$, then by Sect. 1.8.2, Proposition 1b as applied to $T = \delta$ and then to $T = 1$ yields:

$$\widehat{P(D)\delta} = P(2\pi\xi)\hat{\delta} = P(2\pi\xi)$$

$$\widehat{(2\pi x)}(\xi) = P(-D)\hat{1}(\xi) = P(-D)\delta(\xi).$$

(II) $\widehat{e_{\xi_0}}(\xi) = \widehat{e^{2\pi i \xi_0 x}}(\xi) = \delta(\xi - \xi_0)$.

Proof $\widehat{e_{\xi_0}}(\xi) = \widehat{e_{\xi_0}.1}(\xi) = \tau_{\xi_0}\hat{1}(\xi)$, (1.8.2, Theorem 2 (c)) $\overset{(*)}{=} \tau_{\xi_0}\delta(\xi) = \delta(\xi - \xi_0)$.

(III) $[\cos(2\pi < x, \xi_0 >)]\hat{}(\xi) = \frac{1}{2}[\delta(\xi - \xi_0) + \delta(\xi + \xi_0)]$,
$[\sin(2\pi < x, \xi_0 >)]\hat{} = \frac{1}{2i}[\delta(\xi - \xi_0) + \delta(\xi + \xi_0)]$.

Proof By (II) the right-side of the first equation is $\frac{1}{2}(\widehat{e_{\xi_0}}(\xi) + \widehat{e_{-\xi_0}}(\xi)) = [\frac{1}{2}(e_{\xi_0}(x) + e_{-\xi_0}(x))]\hat{} = [\cosh(2\pi i < x, \xi_0 >)]\hat{} = [\cos(2\pi < x, \xi_0 >)]\hat{}$.
The second part is similar.

(IV) $(v.p.\frac{1}{x})\hat{}(\xi) = \pi i(1 - 2H(\xi))$
(For the 'principal value $\frac{1}{x}$' distribution and the Heaviside distribution see Sect. 1.2 Example I following Proposition 2 and Example 4).

Proof First we note that in the distributional sense $x(v.p.\frac{1}{x}) = 1$. For by Sect. 1.6, Operation V) : $< x(v.p.\frac{1}{x}), \phi > = < v.p.\frac{1}{x}, x\phi > = v.p. \int_\mathbb{R} x\frac{\phi(x)}{x}\, dx = < 1, \phi >$. Then for $P(x) = x$, $P(-D_c) = -\frac{1}{i}\frac{d}{dx}$; taking $T = v.p.\frac{1}{x}$ and using Sect. 1.8.2, Proposition 1b:

$[2\pi x(v.p.\frac{1}{x})]\hat{}(\xi) = -\frac{1}{i}\frac{d}{d\xi}\widehat{v.p.\frac{1}{x}}(\xi)$ or

$2\pi\hat{1}(\xi) = -\frac{1}{i}\frac{d}{d\xi}\widehat{v.p.\frac{1}{x}}(\xi)$ yielding by Example (I) (*):

$\frac{d}{d\xi}\widehat{v.p.\frac{1}{x}}(\xi) = -2\pi i\delta(\xi)$ and this means in view of Sect. 1.3, Example 1):

$$v.p.\widehat{\frac{1}{x}}(\xi) = -2\pi i\, H(\xi) + C \qquad (\dagger)$$

$v.p.\frac{1}{x}$ and its Fourier transforms are odd distributions , thus
$0 = -2\pi i\,(H(\xi) + H(-\xi) + 2C = -2\pi i + 2C$ giving $C = \pi i$ and substituting in
$(\dagger):\ \ v.p.\widehat{\frac{1}{x}}(\xi) = -2\pi i\, H(\xi) + \pi i = \pi i(1 - 2H(\xi))$.
Recall that the 'sign function' $sgn\xi$ is defined by

$$sgn\xi = \begin{cases} +1 & \text{if } \xi > 0 \\ -1 & \text{if } \xi < 0 \end{cases}$$

We have $sgn\xi = 2H(\xi) - 1$ yielding
$v.p.\widehat{\frac{1}{x}} = -\pi i\, sgn\,\xi.$
   (V) Fourier transforms of derivatives and translates of the delta function:
(i) $\widehat{\delta'}(\xi) = 2\pi i\xi,$ (Sect. 1.8.2, Proposition 1b), $P(x) = x$).
   $\widehat{\delta^{(m)}}(\xi) = (2\pi i\xi)^m,$ $(P(x) = x^m)$.
(ii) $[\delta(x - a)]\widehat{}(\xi) = e^{-2\pi i\xi a},$ (Sect. 1.8.2, Theorem 2a).
   (VI) Transforms of $H(x)$, $\Pi(x)$ and $\Lambda(x)$:
$[\frac{1}{i}\frac{d}{dx}H(x)]\widehat{} = 2\pi\xi\hat{H}(\xi)$ yields $\hat{H}(\xi) = \frac{1}{2\pi i\xi}\widehat{\delta(x)}(\xi == -\frac{i}{2\pi\xi}.$
   The 'gate function' $\Pi(x)$ is defined by

$$\Pi(x) = \begin{cases} 0 & |x| > 1/2 \\ 1/2 & |x| = 1/2 \\ 1 & |x| < 1/2 \end{cases}$$

Clearly $\Pi(x) = H(x + \frac{1}{2}) - H(x - \frac{1}{2})$, except $x = \pm 1/2$, then
$[H(x - \frac{1}{2})]\widehat{}(\xi) = e^{\pi i\xi}(-\frac{i}{2\pi\xi})$, (Sect. 1.8.2, Theorem 2b) and
$[H(x + \frac{1}{2})]\widehat{}(\xi) = e^{\pi i\xi}(-\frac{i}{2\pi\xi})$,   therefore
$\hat{\Pi}(\xi) = -\frac{i}{2\pi\xi}(e^{\pi i\xi} - e^{-\pi i\xi}) = \frac{\sin \pi\xi}{\pi\xi}.$
   The 'lambda distribution' $\Lambda(x)$ is defined as $\Lambda(x) = \Pi(x) * \Pi(x)$. Evaluating the
integral $\int \Pi(y)\,\Pi(x - y)\,dy$, we easily find $\Lambda(x) = \begin{cases} 1 + x & \text{if } -1 \le x \le 0 \\ 1 - x & \text{if } 0 \le x \le 1 \\ 0 & \text{otherwise} \end{cases}$.

   Thus $\hat{\Lambda}(\xi) = [\hat{\Pi}(\xi)]^2 = \dfrac{\sin^2 \pi\xi}{\pi^2\xi^2}.$
(VI) Fourier transform of tensor products of distributions:
   Consider r Euclidean spaces $\mathbb{R}^n$ $(1 \le j \le r)$ and of each $\mathbb{R}^{n_j}$ we shall have two
copies X_j and Y_j which will form a dual system. Then $\prod_{j=1}^{r} X_j$ and $\prod_{j=1}^{r} Y_j$ are
both isomorphic to $\mathbb{R}^n$ $(n = n_1 + n_2 + \cdots + n_r)$ and they form a dual system with
respect to the bilinear form $< x, \xi > = \sum_{j=1}^{r} < x_j, \xi_j >$ $(x_j \in X_j,\ \xi_j \in Y_j)$.
If $T_j \in \mathcal{S}'(X_j)$ $(1 \le j \le r)$, then

$$\bigotimes_{j=1}^{r} T_j \in \mathcal{S}'\left(\prod_{j=1}^{r} X_j\right),$$

$$\widehat{\bigotimes_{j=1}^{r} T_j} = \bigotimes_{j=1}^{r} \hat{T}_j.$$

(For the tensor product of more than two distributions see Sect. 1.7, Definition 4).

Proof Consider a family $\{\phi_j\}_{1 \le j \le r}$ of functions such that $\phi_j \in \mathcal{S}(X_j)$. If x_j denotes the vector variable in X_j and D^{α_j} a partial derivative with respect to the components of x_j, then we have for $\alpha = \alpha_1 + \cdots + \alpha_r$ $\quad D^{\alpha}(\phi_1 \otimes \cdots \otimes \phi_r) = D^{\alpha_1}\phi_1 \otimes \cdots \otimes D^{\alpha_r}\phi_r$ and $(1 + \|x\|^2)^k |D^{\alpha} \otimes_{j=1}^{r} \phi_j(x)| \le \prod_{j=1}^{r}(1 + \|x_j\|^2)^k |D^{\alpha_j}\phi_j(x)|$. Hence $\otimes_{j=1}^{r}\phi_j \in \mathcal{S}(\prod_{j=1}^{r} X_j)$, i.e. $\otimes_{j=1}^{r} \mathcal{S}(X) \subset \mathcal{S}(\prod_{j=1}^{r} X_j)$.

Also it can be shown that $\otimes_{j=1}^{r} \mathcal{S}(X_j)$ is dense in $\mathcal{S}(\prod_{j=1}^{r} X_j)$ since $\otimes_{j=1}^{r} \mathcal{D}(X_j)$ is dense in $\mathcal{D}(\prod_{j=1}^{r} X_j)$ and in turn it is dense in $\mathcal{S}(\prod_{j=1}^{r} X_j)$,   (c.f. [Hor], 4.11, Proposition 4).

For $\phi_j \in \mathcal{S}(X_j)$ we have

$$\widehat{\otimes_{j=1}^{r}\phi_j}(\xi) = \int_{\prod X_j}(\otimes_{j=1}^{r}\phi_j)(x)\, e^{-2\pi i <\xi, x>}\, dx = \prod_{j=1}^{r} \int_{X_j} \phi_j(x_j)\, e^{-2\pi i <\xi_j, x_j>}$$
$$dx_j = (\otimes_{j=1}^{r}\hat{\phi}_j)(\xi).$$

We use the identity $\quad < \otimes_{j=1}^{r} T_j, \otimes_{j=1}^{r}\phi_j > = \prod_{j=1}^{r} < T_j, \phi_j > \quad$ to show $\otimes_{j=1}^{r} T_j \in \mathcal{S}'(\prod_{j=1}^{r}(X_j)$. In conclusion

$$< \widehat{\bigotimes_{j=1}^{r} T_j}, \otimes_{j=1}^{r}\phi_j > = < \bigotimes_{j=1}^{r} T_j, \widehat{\otimes_{j=1}^{r}\phi_j} > = < \bigotimes_{j=1}^{r} T_j, \otimes_{j=1}^{r}\hat{\phi}_j > =$$
$$\prod_{j=1}^{r} < T_j i\hat{\phi}_j > = \prod_{j=1}^{r} < \hat{T}_j, \phi_j > = < \bigotimes_{j=1}^{r} \hat{T}_j, \otimes_{j=1}^{r}\phi_j >.$$

1.9   Other Theories of Linear Generalized Functions: Hyperfunctions and Ultradistributions

1.9.1   Hyperfunctions

The hyperfunctions are a class of generalized functions introduced by M. Sato in 1958 [Sato] only ten years later than Schwartz' distributions. However both hyperfunctions and ultradistributions are not so commonly used as distributions. One reason seems to be that their mere definitions need a lot of preparations. The theory behind them is quite involved with an extensive supporting literature.

For these reasons our coverage will be neither comprehensive nor fully technical. Many proofs will be omitted but properly referenced.

However according to Komatsu ([Komat-1]) in the one-dimensional case the definition of hyperfunctions is elementary.

Let Ω be an open set in $\mathbb{R}$ and V be an open set in $\mathbb{C}$ containing Ω as a closed set (Fig. 1.6).

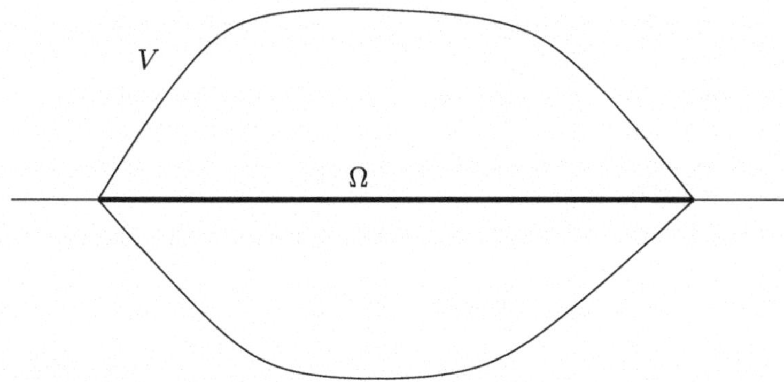

Fig. 1.6   A closed subset Omega of R, contained in an open set V

Let also $\mathcal{O}(V \setminus \Omega)(\text{resp.}\mathcal{O}(V))$ be the space of all holomorphic functions on $V \setminus \Omega$ (resp V). Then the space $\mathcal{B}(\Omega)$ of hyperfunctions on Ω is defined to be the quotient-space

$$\mathcal{B}(\Omega) = \mathcal{O}(V \setminus \Omega)/\rho_{V \setminus \Omega}^{V}\mathcal{O}(V)$$

where $\rho_{V \setminus \Omega}^{V} : \mathcal{O}(V) \to \mathcal{O}(V \setminus \Omega)$ is the restriction mapping. (We omit $\rho_{V \setminus \Omega}^{V}$ when there is no ambiguity). Every element of $\mathcal{O}(V \setminus \Omega)$ that is extended to a holomorphic function is identified with 0. Thus we simply write

$$\mathcal{B}(\Omega) = \mathcal{O}(V \setminus \Omega)/\mathcal{O}(V) \quad (1)$$

Each element $f(x)$ of $\mathcal{B}(\Omega)$ is a hyperfunction. If it is represented by $F(z) \in \mathcal{O}(V \setminus \Omega)$, its class is denoted by $[F(z)]$. Then hyperfunction $f(x)$ is written as

$$f(x) = F(x + i0) - F(x - i0) \equiv F_{+}(x + i0) - F_{-}(x - i0) \quad (2)$$

and has the intuitive meaning of the difference of the "boundary values" of $F(x)$

The main properties of these functions are the following:

(1) $\mathcal{B}(\Omega)$, $\Omega \subset \mathbb{R}$ forms a <u>sheaf</u> over $\mathbb{R}$. That means for every open set Ω in $\mathbb{R}$ there is $\sigma(\Omega) \in \mathcal{B}(\Omega)$ called a section over Ω. For each Ω_1, Ω_2 open sets in $\mathbb{R}$ with $\Omega_1 \subset \Omega_2$ we have restriction mappings $\rho_{\Omega_1,\Omega_2} : \sigma(\Omega_2) \to \sigma(\Omega_1)$. (Also $\mathcal{B}(\Omega_2) \to \mathcal{B}(\Omega_1)$). Then $(\sigma, \rho_{\Omega_1,\Omega_2})$, is called a sheaf of sections over $\mathbb{R}$ if four axioms given in Sect. 3.4.7 are satisfied. (With the notation of Sect. 3.4.7 $X = \mathbb{R}$, $\mathfrak{S} = \mathcal{B}(\Omega)$). Then the following conditions follow for any open covering $\Omega = \bigcup_{\alpha} \Omega_{\alpha}$

$(S.1)$: If $f \in \mathcal{B}(\Omega)$ satisfies $f|_{\Omega_{\alpha}} = 0$, then $f = 0$. (This follows from the locality principle of sheaves given in Sect. 3.4.7).

$(S.2)$: If $f_{\alpha} \in \mathcal{B}(\Omega_{\alpha})$ satisfy $f_{\alpha}|_{\Omega_{\alpha} \cap \Omega_{\beta}} = f_{\beta}|_{\Omega_{\alpha} \cap \Omega_{\beta}}$ for all $\Omega_{\alpha} \cap \Omega_{\beta} \neq 0$, then there is an $f \in \mathcal{B}(\Omega)$ such that $f_{\alpha} = f|_{\Omega_{\alpha}}$. (And this follows from 'glueing property' of Sect. 3.4.7)

(2) The sheaf $\mathcal{B}$ is flabby. That is if $\Omega_1 \subset \Omega$, the restriction mappings $\rho_{\Omega_1,\Omega}$ are always surjective.

(3) If K is a compact set in Ω, then $\mathcal{B}_K(\Omega) = \mathcal{A}(K)'$. For the symbols:

By property $(S.1)$ each $f \in \mathcal{B}(\Omega)$ has the maximal open subset of Ω on which it vanishes. Its complement is called the support of f denoted by $suppf$. Then $\mathcal{B}_K(\Omega) = \{f :\in \mathcal{B}(\Omega), suppf \subset K\}$.

$\mathcal{A}(K)'$ denotes the space of all continuous linear functionals on the locally convex space

$$\mathcal{A}(K) = \varinjlim_{V \supset K} \mathcal{O}(V)$$

of the germs of real-analytic functions defined on a neighborhood of K, (the limit is the inductive limit).

EXAMPLES. The inclusion of distributions in the space of hyperfunctions are discussed ahead. But as some elementary examples we consider the following (the branch of a multi-valued function is to be understood as the principal one):

(I) Heaviside function $Y(x) = [-(1/2\pi i)\log(-z)]$,

(II) Dirac's Delta function : $\delta(x) = [-\frac{1}{2\pi i z}]$,

(III) $(x + i0)^\lambda$, with defining function $F_+(z) = z^\lambda$, $F_-(z) = 0$,
$(x - i0)^\lambda$, with defining function $F_+(z) = 0$, $F_-(z) = -z^\lambda$.

Some elementary operations on hyperfunctions are:

 (i) $\mathbb{C}$-linear operations,
(ii) Multiplication by real analytic functions
(iii) Differentiation : We differentiate the defining function $F(z)$ of $f(x) \in \mathcal{B}(\Omega)$:
$f'(x) = [F(z)]' = [F'(z)]$ or symbolically as $\frac{d}{dx}(F_+(x + i0) - F_-(x - i0)) = \frac{dF_+}{dz}(x + i0) - \frac{dF_-}{dz}(x - i0)$.

For the higher dimensional case a different outlook at the construction of hyperfunctions is necessary. In other words the use of some notions from the algebraic topology is essential. For this purpose we re-read (1) as

$$\mathcal{B}(\Omega) = H^1_\Omega(V; \mathcal{O}) \quad (3)$$

where V is an open set in $\mathbb{C}$ and Ω is a closed set in V. The right side is the first cohomology group of V with support in Ω and the coefficients in the sheaf $\mathcal{O}$

[for the reader unfamiliar with the algebraic topology terms we include a rudimentary introduction: A sequence of homomorphisms of Abelian groups

$$0 \to A^0 \xrightarrow{h_1} A^1 \xrightarrow{h_2} A^2 \xrightarrow{h_3} A^3 \xrightarrow{h_4} \cdots \quad (*)$$

is called a complex if it satisfies the condition $Imh_j \subset Kerh_{j+1}$ $j = 1, 2, \ldots$. Usually we denote by $A*$ the complex $(*)$. We define the pth cohomology group of the complex $A*$ by

$$H^p(A*) = Kerh_{p+1}/Imh_p, \quad p = 0, 1, 2, \ldots$$

In our case we take $A^0 = 0$ and

$$0 \xrightarrow{h_1} \mathcal{O}(V) \xrightarrow{h_2} \mathcal{O}(V \setminus \Omega)$$

yields for $p = 1$

$$H^1_\Omega(V; \mathcal{O}) = \mathcal{O}(V \setminus \Omega)/\mathcal{O}(V).$$

(Note that the condition for the cohomology group is satisfied since as pointed out before, Imh_2 consists of holomorphic functions in $\mathcal{O}(V \setminus \Omega)$ which can be extended to a holomorphic function in $\mathcal{O}(V)$, therefore in the zero of the quotient).]

If every connected component of V intersects $V \setminus \Omega$, then the sequence

$$0 \to \mathcal{O} \xrightarrow{rest} \mathcal{O}(V \setminus \Omega) \to H^1_\Omega(V; \mathcal{O}) \to 0$$

is exact, i.e. the image of each component is the kernel of the next component.

Let V' be another open set in $\mathbb{C}$ and let Ω' be a closed set in V'. If we have $V' \subset V$ and $\Omega' \supset \Omega$ then we have the commutative diagram (Fig. 1.7):

Therefore there is a canonical homomorphism,

$$H^1_\Omega(V; \mathcal{O}) \longrightarrow H^1_{\Omega'}(V'; \mathcal{O})$$

This canonical homomorphism is also an isomorphism (c.f. [Morim], Theorem 3.1.1). Then take inductive limit with respect to canonical homomorphism to get

$$H^1(\Omega; \mathcal{O}) \doteq \varinjlim_{\supset \Omega} H^1_\Omega(V; \mathcal{O}). \quad (4)$$

$H^1(\Omega; \mathcal{O})$ is naturally an $\mathcal{O}(\Omega)$ module. Let us replace Ω by a locally closed set S in $\mathbb{R}$.

Definition Let S be a locally closed set in $\mathbb{R}$. An element of $\mathcal{B}[S] \doteq H^1(S; \mathcal{O})$ is a hyperfunction on S. $\mathcal{B}[S]$ is an $\mathcal{A}(S)$ module. ($\mathcal{A}$ is analytic functions on S).

$$\mathcal{O}(V) \xrightarrow{rest} \mathcal{O}(V \setminus \Omega)$$
$$\downarrow rest \qquad\qquad\qquad \downarrow rest$$
$$\mathcal{O}(V') \xrightarrow{rest} \mathcal{O}(V' \setminus \Omega')$$

Fig. 1.7   Commutativity relation between $\mathcal{O}(V)$ and $\mathcal{O}(V' \setminus \Omega')$ under restriction mappings

A locally closed set in $\mathbb{R}$ is characterized as $F \cap U$, where F is a closed set in $\mathbb{R}$ and U is an open set in $\mathbb{R}$. By taking $F = \mathbb{R}$ and $U = u$ an open set in $\mathbb{R}$ we write $\mathcal{B}(u) = \mathcal{B}[u] = H^1(u; \mathcal{O})$

Analytic Functions and Hyperfunctions

Let S be a locally closed set in $\mathbb{R}$ and let u be a real open neighborhood of S, (i.e. $S \subset u$). Then the natural imbedding

$$i : \mathcal{B}[S] \to \mathcal{B}(u)$$

is injective. Thus i is called the canonical imbedding and we consider $\mathcal{B}[S] \subset \mathcal{B}$.

Put $\mathbb{C}^+ = \{z \in \mathbb{C}; Imz > 0\}$, $\mathbb{C}^- = \{z \in \mathbb{C}; Imz < 0\}$. For an open set V in $\mathbb{C}$ we put $V^+ = V \cap \mathbb{C}^+$, $V^- = V \cap \mathbb{C}^-$.

If u is an open set in $\mathbb{R}$ and a neighborhood $\hat{u}$ satisfies $\hat{u} \cap \mathbb{R} = u$, then $\phi \in \mathcal{O}(\hat{u} \setminus u)$ is composed of $\phi_+ \in \mathcal{O}(\hat{u}^+)$ and $\phi_- \in \mathcal{O}(\hat{u}^-)$:

$$\mathcal{O}(\hat{u} \setminus u) = \mathcal{O}((\hat{u})^+) \oplus \mathcal{O}(\hat{u}^-)$$

ϕ is a defining function of the hyperfunction 0 if and only if ϕ_+ and ϕ_- are analytic continuation of each other across u. See the Fig. 1.8.

Define $\epsilon, \bar{\epsilon} \in \mathcal{O}(\mathbb{C} \setminus \mathbb{R})$ by
$$\epsilon(z) = \begin{cases} 1 & z \in \mathbb{C}^+ \\ 0 & z \in \mathbb{C}^- \end{cases} \quad \bar{\epsilon}(z) = \begin{cases} 0 & z \in \mathbb{C}^+ \\ 1 & z \in \mathbb{C}^- \end{cases}$$

We have $\epsilon(z) + \bar{\epsilon}(z) = 1 \in \mathcal{O}(\mathbb{C})$, selecting $\hat{u}$ as above so that $\hat{u} \cap \mathbb{R} = u$. Given $\phi \in \mathcal{O}(\hat{u} \setminus u)$, we define the boundary values $\phi(x + i0)$ and $\phi(x - i0) \in \mathcal{B}(u)$ by

$$\phi(x + i0) = [\epsilon(z)\phi(z)]_{z=x}, \quad \phi(x - i0) = -[\bar{\epsilon}\phi(z)]_{z=x}.$$

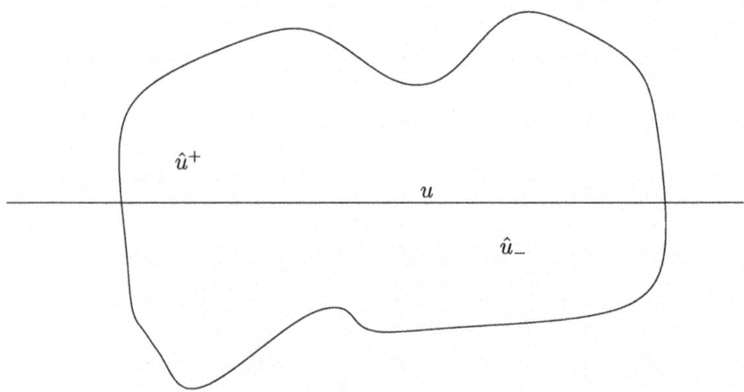

Fig. 1.8 $\hat{u}_+$ and $\hat{u}_-$ are analytic continuations of each other across the real line

Then like (2) the hyperfunction $[\phi]$ is represented as the difference of two boundary values

$$g(x) \equiv [\phi(z)]_{z=x} = \phi(x + i0) - \phi(x - i0).$$

We define the hyperfunction 1 by $1 = [\epsilon(z)]_{z=x} = -[\bar{\epsilon}(z)_{z=x} \in \mathcal{B}(\mathbb{R})$

Proposition 1 *Associating* $f \in \mathcal{A}(u)$ *with the hyperfunction*

$$f(x).1 = f(x + i0) = f(x - i0) \in \mathcal{B}(u),$$

we define an injection $\mathcal{A}(u) \to \mathcal{B}(u)$. *With this we consider* $\mathcal{A}(u) \subset \mathcal{B}(u)$, *that is we identify* f *with* $f.1$. $\mathcal{A}(u)$ *is an* $\mathcal{A}(u)$-*submodule of* $\mathcal{B}(u)$. *([Morim], Proposition 3.1.8.*

If $g \in \mathcal{B}(u)$ *is represented in the form* $g = f.1$ *with* $f \in \mathcal{A}(u)$, *then* g *is said to be a real analytic hyperfunction.* $\mathcal{A}(u)$ *is a subsheaf of* $\mathcal{B}(u)$.

Hyperfunctions as Boundary Values of Harmonic Functions

Hyperfunctions are closed under taking non-characteristic boundary values of solutions of linear partial differential equations. That means let

$$P(x, D) = \sum_{|\alpha| \leq m} a_\alpha(x) D^\alpha$$

be a partial differential equation with real-analytic coefficients $a_\alpha(x)$ on an open set V in $\mathbb{R}^{n+1}$, and let $\Omega = V \cap \mathbb{R}^n$ be a non-characteristic hypersurface. Then we have $H^1_\Omega(V, \mathcal{B}^P) \cong \mathcal{B}(\Omega)^m$, ([Komat-2]). If P has constant coefficients, then

$$H^1_\Omega(V, \mathcal{B}^P) \cong (\mathcal{B}^P(V \setminus \Omega)/\mathcal{B}^P(V)$$

Here $\mathcal{B}^P$ denotes the sheaf of hyperfunctions solution of $Pu = 0$. If $P(D)$ is elliptic, then the hyperfunction solutions $\mathcal{B}^P$ are real analytic and hence we may replace $\mathcal{B}^P$ by the sheaf $\mathcal{A}^p$ of real-analytic solutions. In case $n > 1$, there are no single elliptic operators of first order. The Laplacian Δ would be the simplest elliptic operator in that case.

We denote points in $\mathbb{R}^{n+1}$ by $v = (x, t), \omega = (y, s)$ with $x, y \in \mathbb{R}^n$ and $t, s \in \mathbb{R}$, and the sheaf of harmonic functions on $\mathbb{R}^{n+1}$ by $\mathcal{P}$ Thus we have for any open set V in $\mathbb{R}^{n+1}$

$$\mathcal{P}(V) = \{H(v) \in C^2; \Delta H = (\Delta_x + D_t^2)H = 0\}.$$

Let $\Omega = V \cap \mathbb{R}^n = \{x \in \mathbb{R}^n; (x, 0) \in V\}$. Then with any $H(v) \in \mathcal{P}(V \setminus \Omega)$, two boundary values $H(x, +0) - H(x, -0) \in \mathcal{B}(\Omega)$, $D_t H(x, +0) - D_t H(x, -0)$ are associated and $H(v)$ can be continued to a harmonic function on V if and only if these boundary values vanish. Moreover all pairs of hyperfunctions on Ω appear as boundary values.

Distributions and Hyperfunctions

Generally speaking distributions are generalized functions in the C^∞ category and hyperfunctions in the real analytic category. Distributions can be defined on C^∞ manifolds (c.f. Sect. 2.4) and hyperfunctions on real analytic manifolds.

We discuss only the one variable case, but a similar discussion is possible in the case of several variables. Consider the property (3) of the sheaf of hyperfunctions ,i.e. $\mathcal{B}_K(\Omega) = \mathcal{A}(K)'$. By inductive limit this gives $\mathcal{B}_c(u) = \mathcal{A}'(u)$, where $\mathcal{B}_c$ is the hyperfunctions with compact support in u.

The canonical injection ρ from $\mathcal{A}(\mathbb{R})$ into $\mathcal{E}(\mathbb{R}) \equiv C^\infty(\mathbb{R})$ has a dense image. In fact the totality of polynomials are dense in $\mathcal{E}(\mathbb{R})$. Therefore the dual mapping $\rho^* : \mathcal{E}'(\mathbb{R}) \to \mathcal{A}'(\mathbb{R}) = \mathcal{B}_c(\mathbb{R})$ is injective and conserves the support. In fact we have the following proposition.

Proposition 2 *let K be a compact set in $\mathbb{R}$. The support of $T \in \mathcal{E}'(\mathbb{R})$ is contained in K if and only if $\rho^* T \in \mathcal{A}'(\mathbb{R})$ can be extended to a linear functional on $\mathcal{A}(K)$.*

Proof (c.f. [Morim], Proposition 3.9.1).

<u>Note.</u> Recall that $\mathcal{E}'(\mathbb{R})$ contains the distributions of compact support.

Theorem *Let u be an open set in $\mathbb{R}$. Then there is a linear injection i from the space $\mathcal{D}'(u)$ of distributions on u into the space $\mathcal{B}(u)$ of hyperfunctions on u:*

$$i : \ \mathcal{D}' \to \mathcal{B}(u).$$

i conserves the support; that is $supp T = supp i(T)$ for $T \in \mathcal{D}'(u)$.

Proof By the partition of unity we can express $T \in \mathcal{D}'(u)$ as a locally finite sum of distributions T_i with compact support : $T = \sum_i T_i$. By Proposition 2 $\rho^*(T_i)$ is a hyperfunction with compact support n we have $supp T = supp \rho^*(T_i)$. Therefore $\sum_i \rho^*(T_i)$ is a locally finite sum of hyperfunctions. We define $i(T) = \sum_i \rho^* T_i) \in \mathcal{B}(u)$. Thus i has the properties claimed in the Theorem, ([Morim], Theorem 3.9.2). $\qquad\qquad \Box$

Higher Dimensional Case

Hyperfunctions in higher dimensional case have the same properties (1)–(3) and are characterized by them. However it was not easy to define the spaces $\mathcal{B}(\Omega)$, $\Omega \subset \mathbb{R}^n$ of hyperfunctons having these properties. Sato had to spend two years before he succeeded in giving a definition for the higher dimensional hyperfunctions, ([Komat-2]). His definition is

$$\mathcal{B}(\Omega) = H^n(V, V \setminus \Omega, \mathcal{O})$$

where the right hand side is the n-th cohomology group of the open pair $(V, V \setminus \Omega)$ with coefficients in the sheaf $\mathcal{O}$ of holomorphic functions, which he invented for this purpose.

The same concept was independently introduced by Grothendieck, ([Groth]) under the name of local cohomology group with support in Ω. In Grothendieck's notation it is written $H^n_\Omega(V, \mathcal{O})$.

Micro Analyticity and Singular Spectrum

Any real analytic function is the one which can be analytically continued from the real axis to a complex neighborhood. Let us consider the boundary-representation of a hyperfunction $f(x) = F_+(x + i0) - F_-(x - i0)$ at the real axis. This leads to the observation that a hyperfunction and hence any function in particular, is the composite of two components $F_+(x + i0)$ and $F_-(x - i0)$ in which the former can be continued analytically only in the region above the real axis $(y > 0)$ and the latter only to $(y < 0)$. This interpretation of hyperfunctions leads to a more refined concept of analyticity of functions breaking them into two parts and considering them separately.

Let $S^0 = \pm 1$ and denote a point (x, ξ) of $\mathbb{R} \times S^0$ by a widely used notation ([Kanek], Def. 1.6.1) $(x, \xi/i)dx\infty$ for convenience. A hyperfunction $f(x)$ is said to be microanalytic at the point $(x, +(1/i)dx\infty)$ if, for suitable complex neighborhood V of x a pair of defining functions $F_\pm$ of f can be both analytically continued to the complex upper half-neighborhood $V_+ = V \cap (Imz) > 0$ of x. (This means in effect that F_- can be analytically continued to a neighborhood of Ω). Micro analyticity at the point $(x, -(1/i)dx\infty)$ is similarly defined using the lower half-neighborhood $V \cap (Imz < 0)$.

The set of all points where f is not microanalytic is called the singular spectrum of x and denoted by $S.S.f$.

EXAMPLE. We have $S.S.\delta(x) = \{0\}$.

1.9.2   Ultradistributions

Ultradistributions considered in this Subsection will mean elements in the dual of the space of non-quasi analytic class of infinitely differentiable functions equipped with a natural locally convex topology as defined by Roumieu and Beurling. Every ultradistribution is a hyperfunction , so like hyperfunctions its definition takes a preparation. The ultradistributions in a class are characterized by the behaviour of the defining functions of hyperfunctions.

Roughly the basic idea is to consider the class of functions which are between the analytic functions $\mathcal{A}(\Omega)$ and the space of of infinitely differentiable functions $\mathcal{C}^\infty$, (e.g. like functions of Gevrey class). Thus we have class of test functions narrower than the test functions of the Schwartz theory, accordingly a larger class of dual spaces.

Let M_p, $p = 0, 1, 2, \ldots$ be a sequence of positive numbers. An infinitely differentiable function on an open set $\Omega \subset \mathbb{R}^n$ will be called an ultradifferentiable function of class M_p if on each compact set $K \subset\subset \Omega$ its derivatives are estimated in the form

$$\|D^\alpha f\|_{\mathcal{C}(K)} \le Ch^{|\alpha|}M_{|\alpha|}, |\alpha| = 0, 1, 2, \ldots \quad (1)$$

with multi-index notation and $|\alpha| = \alpha_1 + \cdots + \alpha_n$.

To make the class of functions invariant under affine transformations, there are two possibilities of the choice of constant h. We call $f \in \mathcal{E}(\Omega) \equiv C^\infty(\Omega)$ an ultradifferentiable function of class (M_p) (respectively of class $\{M_p\}$) if (1) holds for every $h > 0$ (respectively for some $h > 0$).

Following conditions are imposed:

(M.1)   (Logarithmic convexity)

$$M_p^2 \leq M_{p-1} M_{p+1}, \; ; \; p = 1, 2, \ldots$$

(M.2)   (Stability under ultradifferentiable operator)
There are constants A and H such that

$$M_p \leq A H^p \min_{0 \leq p \leq q} M_q M_{p-q}, \; p = 0, 1, 2, \ldots$$

(M.3)   (Strong non quasi-analyticity)
there is a constant $A > 0$ such that

$$\sum_{q=p+1}^{\infty} \frac{M_{q-1}}{M_q} \leq A p \frac{M_p}{M_{p+1}}, \quad (p = 1, 2, \ldots)$$

Some results remain valid however, when $(M.2)$ and (M_3) are replaced by the following weaker conditions

(M.2)'   (Stability under differential operators)
There are constants $A > 0$ and $H > 0$ such that $M_{p+1} \leq A H^p M_p$, $(p = 0, 1, \ldots)$
(M.3)'   (Non quasi-analyticity)

$$\sum_{p=1}^{\infty} \frac{M_{p-1}}{M_p} < \infty$$

The <u>associated function</u> of M_p is defined as

$$M(t) = \sup_{p \in \aleph} \log(\frac{t^p}{M_p}) \; t \geq 0$$

Note that if $s > 1$ the Gevrey sequences $M_p = (p!)^{ps}$ or p^s or $\Gamma(1 + ps)$ satisfy the above conditions.

For two sequences M_p and N_p of positive numbers we define their orders

Definition 1 Let M_p and N_p be the sequences of positive numbers,

(i) $M_p \subset N_p$ if there exist constants $L > 0$ and $C > 0$ such that $M_p \leq C L^p N_p$ for any p.

(ii) $M_p \preceq N_p$ if for any $L > 0$ there exist a constant $C > 0$ such that $M_p \le CL^p N_p$
for any p

In order to define quasi-analytic classes , we impose the following conditions, (QA)
and (NA), instead of $(M.3)$ or $(M.3)'$:
(QA) (quasi-analyticity)

$$p! \subset M_p, \quad \sum_{p=1}^{\infty} \frac{M_{p-1}}{M_p} = \infty.$$

(NA) (non-analyticity)

$$\lim_{p \to \infty} \sqrt[p]{\frac{p!}{M_p}} = 0.$$

Takiguchii [Takig] Let $\mathcal{E}^{(M_p)}(\Omega)$ $(\mathcal{E}^{\{M_p\}}(\Omega))$ be the spaces of all ultra differentiable
functions of class (M_p) (of class $\{M_p\}$) on Ω and let $\mathcal{D}^{(M_p)}(\Omega)$ $(\mathcal{D}^{\{M_p\}}(\Omega))$ be the
subspace of $\mathcal{E}^{(M_p)}(\Omega)$ (of $\mathcal{E}^{\{M_p\}}(\Omega)$) composed of all functions with compact support.
We can introduce natural locally convex topologies in these spaces.

If M_p satisfies $(M.1)$ and $(M.3)'$, then there are sufficiently many functions in
them. If we assume these conditions and denote by $\mathcal{D}^{(M_p)'}(\Omega)$ (by $\mathcal{D}^{\{M_p\}'}(\Omega)$), the
strong dual of $\mathcal{D}^{(M_p)}(\Omega)$ (of $\mathcal{D}^{\{M_p\}}(\Omega)$).

Definition 2 The elements of $\mathcal{D}^{(M_p)'}(\Omega)$ (of $\mathcal{D}^{\{M_p\}'}(\Omega)$) will be called
<u>ultradistributions</u> of class M_p of Beurling type (of Roumieu type).

With a widely used notation $* = (M_p)$ or $\{M_p\}$.

Thus we have symbols like $\mathcal{E}^*(\Omega)$, $\mathcal{D}^*(\Omega)$, $\mathcal{D}^{*'}(\Omega)$ etc. Since for any open covering of Ω there is a partition of unity subordinate to it in $\mathcal{D}^{(M_p)}(\Omega)$, the theory
of ultradistributions can be constructed in a way quite similar to that of Schwartz
distributions.

In particular, the ultradistributions $\mathcal{D}^{*'}(\Omega)$, $\Omega \subset \mathbb{R}^n$ form a soft sheaf (c.f. GLOS-
SARY). In this sheaf the multiplication by a function in $\mathcal{E}^*(\Omega)$ operates as a sheaf
homomorphism, e.g. if $g \in \mathcal{E}^{(M_p)}(\Omega)$, $U \subset V$ and ϕ is a function $g \times \mathcal{D}^{(M_p)'}(\Omega) \to$
$\mathcal{D}^{M_p'}(\Omega)$, we have the following commutative diagram (Fig. 1.9) (where $\rho_{U,V}$ is the
restriction mapping)

If M_p satisfies $(M.2)'$, any differential operator of finite order with coefficients in
$\mathcal{E}^*(\Omega)$ operates also as a sheaf homomorphism. If, moreover , M_p satisfies $(M.2)$,
then any differential operator

$$
\begin{array}{ccc}
g \times \mathcal{D}^{(M_p)'}(\Omega)(U) & \xrightarrow{\phi_{U,V}} & \mathcal{D}^{(M_p)'}(\Omega)(U) \\
\downarrow{\scriptstyle \rho_{U,V}} & & \downarrow{\scriptstyle \rho_{U,V}} \\
g \times \mathcal{D}^{(M_p)'}(\Omega)(V) & \xrightarrow{\phi_{U,V}} & \mathcal{D}^{(M_p)'}(\Omega)(V)
\end{array}
$$

Fig. 1.9 Sheaf homomorphism of ultradistributions of Beurling type under multiplication by
smooth functions

$$P(D) = \sum_{|\alpha|}^{\infty} a_\alpha D^\alpha$$

of infinite order operates in the sheaf as a sheaf homomorphism whenever the coefficients satisfy the estimate $|\alpha| \leq CL^{|\alpha|}/M_{|\alpha|}$, $|\alpha| = 0, 1, \ldots$. for some L and C. Such a differential operator will be called an ultradifferential operator of class (M_p).

The dual $\mathcal{E}^{*'}(\Omega)$ of $\mathcal{E}^*(\Omega)$ is identified with the subspace of $\mathcal{D}^{*'}$ composed of all ultradistributions with compact support in Ω.

The spaces $\mathcal{D}^*(\Omega)$, $\mathcal{E}^*(\Omega)$, $\mathcal{D}^{*'}$ and $\mathcal{E}^{*'}(\Omega)$ are all complete, barreled, bornological nuclear spaces, (see the GLOSSARY). We have

$$\mathcal{A}(\Omega) \subset \mathcal{E}^*(\Omega) \subset \mathcal{E}(\Omega)$$

and the inclusion mappings are continuous and of dense range. Hence we may consider

$$\mathcal{E}'(\Omega) \subset \mathcal{E}^{*'} \subset \mathcal{A}'(\Omega).$$

Since these inclusion mappings preserve the support, they can be extended to the inclusions

$$\mathcal{D}'(\Omega) \subset \mathcal{D}^{*'}(\Omega) \subset \mathcal{B}(\Omega)$$

showing that the distributions, ultradistributions are hyperfunctions.

Like structure theorems for Schwartz distributions and tempered distributions given in Sects. 1.4 (Theorem 2) and Sect. 1.5, there are structure theorems for ultradistributions.

Komatsu states two structure theorems.

Theorem 1 *(First structure theorem)*
$f \in \mathcal{D}^{(M_p)'}(\Omega)$ $(\in \mathcal{D}^{\{M_p\}'}(\Omega))$ if and only if on every relatively compact open subset G of Ω (on the open set Ω),

$$f = \sum_{|\alpha|=0}^{\infty} D^\alpha f_\alpha$$

with measures f_α on G (on Ω) such that the estimate holds

$$\|f_\alpha\|_{C'(\bar{G}} \leq CL^{|\alpha|}/M_{|\alpha|}, \quad |\alpha| = 0, 1, \ldots$$

for some L and C (for every relatively compact open set G, every $L > 0$ and some C).

Theorem 2 (Second structure theorem) *$f \in \mathcal{D}^{*'}$ if and only if for every relatively compact convex open set G in Ω there are ultradifferential operator $P(D)$ of class $*$ and a measure g on G such that $f = P(D)g$, (the complicated proofs can be found in [Komat-1, Komat-3].*

Examples leading to the need to go beyond the generalized functions and even nonlinear distributions like Colombeau, (covered in Chap. 3) are provided by Colombini and Spagnoli [Col-1, Col-2]. They deal with the Cauchy problem

$$u_{tt} - a(t)u_{xx} = 0, \quad u(x, 0) = \phi(x), u_t(x, 0) = \psi(x) \quad (\sqrt{})$$

where $0 \leq t \leq T$, $x \in \mathbb{R}$ and $a(t)$ is a C^∞ function on the interval $[0, T]$ satisfying the assumption $a(t) \geq 0$.

They show that $(\sqrt{})$ may not be well-posed in the class $\mathcal{E}(\mathbb{R})$, of the C^∞ functions contrary to what occurs when $a(t) \geq \lambda > 0$.

More precisely they construct a C^∞ function $a(t)$, strictly positive on $[0, \rho]$ and identically zero on $[\rho, +\infty]$ where ρ is a given positive number and two C^∞ functions $\phi(x)$ and $\psi(x)$ in such a way that $(\sqrt{})$ has no solution in the class of distributions on $\mathbb{R} \times [0, T]$ as $T > \rho$.

By virtue of the strict positivity of the coefficient $a(t)$ for $t < \rho$, this problem has a C^∞ solution on $[0, \rho] \times \mathbb{R}$, which is the unique solution in the class $C^1([0, \rho), \mathcal{D}'(\mathbb{R}))$. But this solution can not be continued as a distribution on any strip $(0, \rho + \epsilon) \times \mathbb{R} \; \forall \epsilon > 0$. In particular it does not belong to $\mathcal{C}([0, \rho], \mathcal{D}'(\mathbb{R}))$.

Later they showed well-posedness of this and other weakly hyperbolic equations in Gevrey classes and therefore within the ultradistributions theory.

Chapter 2
Some Applications of Linear Distributions

2.1 Applications in Calculus and Analysis

2.1.1 Application to Integration Techniques

Some improper integrals can be evaluated by the linear distributional methods in an ingenious way. In the following example as $e^{-\pi x^2}$ has no anti-derivative, evaluation by the standard techniques would be hard.

Example 1 $I = \int_{-\infty}^{\infty} \cos(2\pi ax)\, e^{-\pi x^2}\, dx$.

Evaluation : By Sect. 1.8.1, Example 1 $I = \int_{-\infty}^{\infty} \cos 2\pi ax\, \widehat{e^{-\pi \xi^2}}(x)\, dx$
Regarding $\cos 2\pi ax$ as a regular distribution
$I = <(\cos 2\pi ax)\hat{}(\xi), e^{-\pi \xi^2}> = < \frac{1}{2}[\delta(\xi - a) + \delta(\xi + a)],\ e^{-\pi \xi^2} >,$ Sect. 1.8.3,
(III) $= \frac{1}{2}(e^{-\pi a^2} + e^{-\pi(-a)^2}) = e^{-\pi a^2}$.

Example 2 $I = \int_{-\infty}^{\infty} \dfrac{\sin^2 \pi x \cos \pi x}{\pi^2 x^2}\, dx$.

Evaluation : By Sect. 1.8.3, (VI) we have $I = \int_{-\infty}^{\infty} \widehat{\Lambda(\xi)}(x) \cos \pi x\, dx = <$
$\Lambda(\xi), (\cos \pi x)\hat{}(\xi) >$
$= < \Lambda(\xi), \frac{1}{2}[\delta(\xi - \frac{1}{2} + \delta(\xi + \frac{1}{2}] > = \frac{1}{2}$.
Note that the evaluation of this integral otherwise would have been quite difficult.

2.1.2 Periodic Distributions, Fourier Series

Periodicity of distributions is defined by the translation operation, i.e. a distribution $T \in \mathcal{D}'(\mathbb{R})$ is said to be periodicofperioda> 0, if it satisfies

$$\tau_a T = T.$$

© The Author(s), under exclusive license to Springer Nature Switzerland AG 2026   67
U. Çapar, *A Guide to Generalized Functions*,
https://doi.org/10.1007/978-3-032-09184-0_2

Equivalently for every $\phi \in \mathcal{D}(\mathbb{R})$

$$< T(x-a), \phi(x) > = < T(x), \phi(x+a) > = < T(x), \phi(x) > .$$

<u>Dirac Comb.</u> This is also known as an "impulse train" or the "shah function". since its graph resembles the shape of the Cyrillic letter 'sha'. Dirac comb is a Schwartz distribution defined by

$$\text{Ш}_\Theta (x) = \sum_{n=-\infty}^{\infty} \delta(x - n\Theta), \quad \Theta > 0.$$

It is a tempered distribution . For if $\phi \in \mathcal{S}(\mathbb{R})$, then $< \text{Ш}_\Theta, \phi > = \sum_{-\infty}^{\infty} \phi(n\Theta)) < \infty$. This is because $\frac{1}{\Theta} \sum_{-\infty}^{\infty} \Theta|\phi(n\Theta)| < \infty$ since $\mathcal{S} \subset L^1$. The continuity is also established easily. With the subsequent analysis Ш_Θ will turn out to be a periodic distribution with period Θ. We also note that considering the action of delta functions on test functions :

$$\text{Ш}_\Theta (x) = \frac{1}{\Theta} \text{Ш} (\frac{x}{\Theta}) = \frac{1}{\Theta} \sum_{-\infty}^{\infty} \delta(\frac{x}{\Theta} - n)$$

where for simplicity Ш_1 is denoted by Ш.

The Dirac comb is an essential tool in the Fourier analysis of distributions, also it has many interesting applications in the signal engineering and sampling.

But first a Lemma:

Lemma 1 *(a) If $T \in \mathcal{D}'(\mathbb{R})$ is an unknown distribution satisfying $xT(x) = 0$ for all x, then $T(x) = C\delta(x)$.*

(b) Suppose $g(x)T(x) = 0$ for all x and $g(x)$ is analytic except at the origin where it has a series expansion. If $g^{(n)}(0)$ is the first non-zero derivative at the origin, then $T(x) = \sum_{k=0}^{n-1} C_k \delta^{(k)}(x)$.

Proof By Sect. 1.8.2, Proposition 1b $[2\pi x T(x)]\hat{}(\xi) = -\frac{1}{i} \frac{d}{d\xi} \hat{T}(\xi)$. As $xT(x) = 0$ we have $\frac{d\hat{T}}{d\xi} = 0$ or $\hat{T}(\xi) = C$. Considering $\hat{\delta} = 1$ we find $T(x) = C\delta(x)$.

(b) By the assumptions $g(x) = \sum_{k=n}^{\infty} \frac{x^k}{k!} g^{(k)}(0)$ and $g(x)T(x) = \sum_{k=n}^{\infty} \frac{g^{(k)}(o)}{k!} x^k$

$T(x) \overset{\mathcal{F}}{=} \sum_{k=n}^{\infty} \frac{(1)^k g^{(k)}(0)}{(2i\pi)^k k!} \frac{d^k}{d\xi^k} \hat{T}(\xi) = 0,$ (again by Sect. 1.8.2, Proposition 1b). This yields $\frac{d^n}{d\xi^n} \hat{T}(\xi) = 0$ or $\hat{T}(\xi) = \sum_{k=0}^{n-1} a_k \xi^k$. Therefore by Sect. 1.8.3, Example (V) $T(x) = \sum_{k=0}^{n-1} a_k (2\pi i)^{-k} \delta^{(k)}(x)$, ($a_k$'s being arbitrary constants).

Remark The linear combination can not extend beyond the $(n-1)$th derivative of the delta function. Otherwise e.g. $< g(x)\delta^{(n)}(x), \phi(x) > = (-1)^n < \delta(x), (g(x)\phi(x))^{(n)} > = (-1)^n g^{(n)}(0)\phi(0)$ is not zero in general, contradicting the assumption.

As the Fourier transform is a continuous mapping we have

$$[\text{Ш}_\Theta(x)]\hat{}(\xi) = [\sum_{-\infty}^{\infty} \delta(x - n\Theta)]\hat{}(\xi) = \sum_{n=-\infty}^{\infty} [\delta(x - n\Theta)]\hat{}(\xi) = \sum_{-\infty}^{\infty} e^{-2\pi i n\Theta\xi}$$

Section 1.8.2, Theorem 2b and Sect. 1.8.3, Example (I).

For $\Theta = 1$: $\hat{\text{Ш}}(\xi) = \sum_{n=-\infty}^{\infty} e^{-2\pi i n\xi}$.

Calling $\hat{\text{Ш}}_\Theta(\xi) = S(\xi)$, we see that $S(\xi)$ has period $\frac{1}{\Theta}$. Also

$$e^{2\pi i \Theta\xi} S(\xi) = \sum_{-\infty}^{\infty} e^{-2\pi i (n-1)\Theta\xi} = S(\xi),$$

(Note that $S(\xi)$ is not the space of rapidly decreasing functions).

Hence $(e^{2\pi i \Theta\xi} - 1)S(\xi) = 0$.

Comparing with the Lemma 1b: $g(\xi) = e^{2\pi i \Theta\xi} - 1$, $g'(0) = 2\pi i \Theta \neq 0, n - 1 = 0$, so that $S(\xi)$ should be represented by linear combinations of delta functions. Considering the periodicity of $S(\xi)$ then we have

$$S(\xi) = a \sum_{-\infty}^{\infty} \delta(\xi - \frac{n}{\Theta}). \qquad (*)$$

$(S(\xi + \frac{1}{\Theta}) = a \sum_{n=-\infty}^{\infty} \delta(\xi - \frac{n-1}{\Theta}) = S(\xi))$.

To find the constant a , consider $\tilde{\Pi}$ defined by

$$\tilde{\Pi}(x) = \begin{cases} 0 & \text{if } |x| > \frac{1}{2\Theta} \\ 1/2 & \text{if } |x| = \frac{1}{2\Theta} \ ; \\ 1 & \text{if } |x| < \frac{1}{2\Theta} \end{cases}$$

Then $< S(\xi), \tilde{\Pi}(\xi) > = \int_{-\frac{1}{2\Theta}}^{\frac{1}{2\Theta}} [\sum_{-\infty}^{\infty} e^{-2\pi i n\Theta\xi}] d\xi = \sum_{-\infty}^{\infty} \int_{-\frac{1}{2\Theta}}^{\frac{1}{2\Theta}} e^{-2\pi i n\Theta\xi} d\xi$, the integral is zero for all $n \neq 0$ and equals $\frac{1}{\Theta}$ for $n = 0$. Therefore by $(*)$ $\frac{1}{\Theta} = < a \sum_{-\infty}^{\infty} \delta(\xi - \frac{n}{\Theta}), \tilde{\Pi}(\xi) > = a\tilde{\Pi}(0)$. Thus $a = \frac{1}{\Theta}$ yielding

$$S(\xi) = \hat{\text{Ш}}^\Theta(\xi) = \frac{1}{\Theta} \sum_{n=-\infty}^{\infty} \delta(\xi - \frac{n}{\Theta}), \qquad (1)$$

$$\hat{\text{Ш}}(\xi) = \sum_{-\infty}^{\infty} \delta(\xi - n). \qquad (2)$$

So that $\text{Ш}(\xi) = \hat{\text{Ш}}(\xi)$, i.e. Dirac comb period 1 is self-dual. Using this fact and also Sect. 1.8.2, Theorem 2a we also have

$$\frac{1}{\Theta} \sum_{-\infty}^{\infty} \delta(\frac{x}{\Theta} - n) \xrightarrow{\mathcal{F}} \sum_{-\infty}^{\infty} \delta(\Theta\xi - n) \quad (3)$$

Let $f(x)$ be any function, $f(x) * \delta(x - n\Theta) =< \delta(y - n\Theta), f(x - y) >= f(x - n\Theta)$ and we have

$$f(x) * \amalg_{\theta}(x) >= f(x) * \sum_{-\infty}^{\infty} \delta(x - n\Theta) = \sum_{-\infty}^{\infty} f(x - n\Theta).$$

This is a periodic function with period Θ since $\sum_{-\infty}^{\infty} f(x - (n - 1)\Theta) = \sum_{-\infty}^{\infty} f(x - n\Theta)$. If f has a Fourier transform, (e.g. $f \in \mathcal{S}(\mathbb{R})$) , then using (1)

$$[\sum_{-\infty}^{\infty} f(x - n\Theta)]\hat{}(\xi) = \frac{1}{\Theta} \hat{f}(\xi) \sum_{-\infty}^{\infty} \delta(\xi - \frac{n}{\Theta}) = \frac{1}{\Theta} \sum_{-\infty}^{\infty} \hat{f}(\frac{n}{\Theta}) \, \delta(\xi - \frac{n}{\Theta}). \quad (4)$$

The last equality follows since $\hat{f} \in \mathcal{C}^{\infty}(\mathbb{R})$, thus for $\phi \in \mathcal{S}(\mathbb{R})$ we have $< \hat{f}(\xi) \, \delta(\xi - \frac{n}{\Theta}), \phi(\xi) >=< \delta(\xi - \frac{n}{\Theta}), \hat{f}(\xi)\phi(\xi) >= \hat{f}(\frac{n}{\Theta})\phi(\frac{n}{\Theta})$, i.e. as a distribution $\hat{f}(\xi)\delta(\xi - \frac{n}{\Theta}) = \hat{f}(\frac{n}{\Theta}))\delta(\xi - \frac{n}{\Theta})$.

To invert (4), we note that by Sect. 1.8.2, Theorem 2c $(e^{2\pi i x \frac{n}{\Theta}}.1)\hat{} = \tau_{\frac{n}{\Theta}}.\hat{1} = \delta(\xi - \frac{n}{\Theta})$.

Applying then the inverse Fourier transform to (4):

$$\sum_{-\infty}^{\infty} f(x - n\Theta) = \frac{1}{\Theta} \sum_{-\infty}^{\infty} \hat{f}(\frac{n}{\Theta}) \, e^{2\pi i x \frac{n}{\Theta}}. \quad (5)$$

(Poisson summation formula)

For $x = 0$

$$\sum_{n-\infty}^{\infty} f(n\Theta) = \frac{1}{\Theta} \sum_{-\infty}^{\infty} \hat{f}(\frac{n}{\Theta}) \quad (6)$$

In turn for $\Theta = 1$

$$\sum_{-\infty}^{\infty} f(n) = \sum_{-\infty}^{\infty} \hat{f}(n) \quad (7)$$

If $f \in \mathcal{S}(\mathbb{R})$, then the series on the left converges since $\sum_{-\infty}^{\infty} f(n) =< \amalg(x), f(x) >$ and $\amalg \in \mathcal{S}'$. Similarly $\hat{f} \in \mathcal{S}$ and the series on the right converges too.

Fourier Series of Functions

The Dirac comb can also be favorably used in developing the Fourier series of functions.

Following [Rod] , let $f(x)$ be a periodical function integrable on a period of length Θ. Define

$$f_0(x) = \begin{cases} f(x) & \text{if } a \le x < a + \Theta \ \ (\text{for some } a) \\ 0 & \text{otherwise} \end{cases}$$

Then noticing that $f_0(x) * \delta(x - n\Theta) = f_0(x - n\Theta)$ shifts $f_0(x)$ $n\Theta$ units to the right, the periodic function $f(x)$ can be represented as $f(x) = f_0(x) * \sum_{n=-\infty}^{\infty} \delta(x - n\Theta)$ and then

$$\hat{f}(\xi) = \mathcal{F}[f_0(x) * \sum_{n=-\infty}^{\infty} \delta(x - n\Theta)] = [\sum_{n=-\infty}^{\infty} f_0(x - n\Theta)]\hat{\ }$$

and by (4)

$$\hat{f}(\xi) = \frac{1}{\Theta} \sum_{n=-\infty}^{\infty} \hat{f}_0(\frac{n}{\Theta})\delta(\xi - \frac{n}{\Theta}). \quad (8)$$

Note. $f_0 \in L^1(\Theta)$ and has bounded support, then $\hat{f}_0$ exists and is infinitely differentiable. This follows from the inequality

$$|\hat{f}_0^{(m)}(\xi)| \le (2\pi)^m \int_{\mathbb{R}} |x|^m |f_0(x)| dx$$

given in Sect. 1.8.1 before Proposition 1.

Thus any function f integrable over a period of length Θ has a Fourier transform given in terms of the impulses of the Dirac comb. The coefficients of the Dirac impulses are found by taking the inverse Fourier transform in (8) using Sect. 1.8.2, Theorem 2c:
$f(x) = \frac{1}{\Theta} \sum_{n=-\infty}^{\infty} \hat{f}_0(\frac{n}{\Theta}) e^{2\pi i n \frac{x}{\Theta}}$, or

$$f(x) = \sum_{n=-\infty}^{\infty} c_n e^{2\pi i n \frac{x}{\Theta}} \quad (9)$$

where

$$c_n = \frac{1}{\Theta} \hat{f}_0(\frac{n}{\Theta}) = \frac{1}{\Theta} \int_{a}^{a+\Theta} f(x) e^{-2\pi i n \frac{x}{\Theta}} dx. \quad (10)$$

Remarks. (1) As (9), (10) were obtained by the distributional methods , the equality in (9) holds in distributional sense, i.e. the partial sums converge in $\mathcal{S}'(\mathbb{R})$ to the tempered distribution defined by $f \in L^1(\Theta)$. Every periodic function f integrable over a period Θ has a Fourier series development like in (9).

(2) The interval $[a, a + \Theta]$ having finite measure we have $L^2(a, a + \Theta) \subset L^1(a, a + \Theta)$ and the set of functions $\phi_n(x) = \{\frac{1}{\sqrt{\Theta}} e^{2\pi i n \frac{x}{\Theta}}\}_{n=-\infty}^{\infty}$ form an orthonormal basis in L^2. Thus for $f \in L^2(a, a + \Theta)$ the equality (9) makes also sense

in L^2. Let $\tilde{c}_n = \frac{1}{\sqrt{\Theta}} < f, \phi_n > = \frac{1}{\sqrt{\Theta}} \int_a^{a+\Theta} f \bar{\phi}_n dx = \frac{1}{\sqrt{\Theta}} f(x) e^{-2\pi i n \frac{x}{\Theta}} dx$, so that $\sum_{n=-\infty}^{\infty} \tilde{c}_n \frac{1}{\sqrt{\Theta}} e^{2\pi i n \frac{x}{\Theta}} = \sum_{n=-\infty}^{\infty} e^{2\pi i n \frac{x}{\Theta}}$ and we have the isometry $\| f \|_2 = \sum |c_n|^2$ as is well-known as Parseval's relation in classical Fourier analysis.

(3) If $f \in L^1 \backslash L^2$, then the convergence of (9) in L^1 is not guaranteed unless some extra conditions are verified. We therefore write rather

$$f(x) \sim \sum_{n=-\infty}^{\infty} c_n e^{2\pi i n \frac{x}{\Theta}}.$$

Pointwise convergence of Fourier series is even more complicated. Carlson [Carl] has shown that Fourier series of any function belonging to $L^2(-\pi, \pi)$ is convergent almost everywhere . In L^1 the situation is worse. Kolmogorov gave en example of the Fourier series which diverges at every point.

As $e^{2\pi i n \frac{x}{\Theta}} = \cos 2\pi n \frac{x}{\Theta} + i \sin 2\pi n \frac{x}{\Theta}$, (9) can be brought to the form

$$f(x) = a_0 + \sum_{n=1}^{\infty} (a_n \cos 2\pi n \frac{x}{\Theta} + b_n \sin 2\pi n \frac{x}{\Theta})$$

where $a_0 = c = 0$, $a_n = c_n + c_{-n}$, $b_n = i(c_n - c_{-n})$ and $a_0 = \frac{1}{\Theta} \int_a^{a+\Theta} f(x)\, dx$, $a_n = \frac{2}{\Theta} \int_a^{a+\Theta} f(x) \cos 2\pi n \frac{x}{\Theta}\, dx$, $b_n = \frac{2}{\Theta} \int_a^{a+\Theta} f(x) \sin 2\pi n \frac{x}{\Theta}\, dx$.

An even function has a cosine series development ($b_n = 0$) and an odd function has a sine series ($a_n = 0$). All such series converge in $\mathcal{S}'$ in the sense of distribution.

Fourier Series of Distributions

The results of the preceding section can not be immediately carried over to the periodic distributions. In fact f_0 was obtained as a multiplication of $f(x)$ by the gate function

$$\Pi(x) = \begin{cases} 1 & \text{if } a < x < a + \Theta \\ 1/2 & \text{if } x = a + \Theta \\ 0 \text{ otherwise} \end{cases}$$

As the gate function is not infinitely differentiable everywhere, the product of a distribution and a gate function does not always make sense. We go around this difficulty by defining a distribution on a circle.

Let S be the circle of circumference Θ in the complex plane centered at the origin, i.e. $S = \{z : \|z\| = \Theta/2\pi\}$. Points on the circle can be given arc length s, $0 \leq s \leq \Theta$. Thus to every periodic function $f(x)$, $x \in \mathbb{R}$ of period Θ, there corresponds in a one-to-one manner a function $f(s)$ on the circle. Let $\mathcal{D}(S)$ be the infinitely differentiable functions $\phi(s)$ on the circle S.

$\mathcal{D}'(S)$ will be all linear, continuous functionals on $\mathcal{D}(S)$. The space of linear distributions $\mathcal{D}'$ is then a locally convex topological vector space. In particular all locally integrable functions $f(s)$ on S define define regular distributions by

$$< f(s), \phi(s) >= \int_S f(s)\,\phi(s)\,ds.$$

It can be shown that there is a one-to-one correspondence between periodic distributions $\mathcal{F}(x)$ of period Θ and distributions $\mathcal{F}(s)$ on the circle S.

We may note that as S is a set of finite measure , all distributions on S have bounded support. For this reason the convolution product on S is well-defined : For $T, V \in \mathcal{D}'(S)$, $\phi \in \mathcal{D}(S)$ we have
$< T * V, \phi >=< T(s)V(t), \phi(s+t) >$, $s, t \in S$. In the case of two functions $f(s) * g(s) = \int_S f(t)\,g(s-t)\,dt$.

Fourier Series of Distributions on the Circle

According to (5) and also considering the fact that $\hat{\delta} = 1$ we found

$$\sum_{n=-\infty}^{\infty} \delta(x - n\Theta) = \frac{1}{\Theta}\delta\left(\frac{x}{\Theta} - n\right) = \frac{1}{\Theta}\sum_{n=-\infty}^{\infty} e^{2\pi i n\Theta\frac{x}{\Theta}}.$$

If we transpose these equalities onto the circle and notice that any $\phi \in \mathcal{D}(S)$ will be included in the support of only one delta function on the leftmost summation above, also substituting $s = x + n\Theta$ we get

$$\delta(s) = \frac{1}{\Theta}\sum_{n=-\infty}^{\infty} e^{2\pi i n\frac{s}{\Theta}}. \quad (11)$$

This gives the Fourier series development of the distribution $\delta(s)$ on the circle. It converges in $\mathcal{D}'(S)$ to $\delta(s)$. The Fourier series development of any distribution T on the circle is obtained by using (11) :
$T(s) = T(s) * \delta(s) = T(s) * \frac{1}{\Theta}\sum_{n=-\infty}^{\infty} e^{2\pi i n\frac{s}{\Theta}}.$
    Since the convolution is a continuous operation

$$T(s) = \sum_{n=-\infty}^{\infty} \frac{1}{\Theta}(T(s) * e^{2\pi i n\frac{s}{\Theta}}).$$

As the exponential is infinitely differentiable, this reads

$$T(s) = \sum_{n=-\infty}^{\infty} \frac{1}{\Theta} < T(t), e^{2\pi i n\frac{s-t}{\Theta}} = \sum_{n=-\infty}^{\infty} \frac{1}{\Theta} < T(t), e^{-2\pi i n\frac{t}{\Theta}} > e^{2\pi i n\frac{s}{\Theta}}$$

yielding

$$T(s) = \sum_{n=-\infty}^{\infty} c_n e^{2\pi i n\frac{s}{\Theta}}$$

where $c_n = \frac{1}{\Theta} < T(t), e^{-2\pi i n\frac{t}{\Theta}} >$.

We have found that every distribution in $\mathcal{D}'(S)$ can be expanded in a Fourier series. The series converges in $\mathcal{D}'(S)$.

Some Distributional Equations of Series

Some relations looking classically meaningless may be quite valid expressions in distribution theory.

Apply the Poisson summation formula (5) to the Dirac comb of period 2π and $f = \delta$ we have

$$\text{III}_{2\pi}(x) = \sum_{n=-\infty}^{\infty} \delta(x - 2n\pi) = \frac{1}{2\pi} \sum_{n=-\infty}^{\infty} e^{inx}$$

or in a more standard form

$$\text{III}_{2\pi}(x) \doteq D(x) = \frac{1}{\pi}(\frac{1}{2} + \sum_{n=1}^{\infty} \cos nx).$$

The partial sums $D_n(x) = \frac{1}{\pi}(\frac{1}{2} + \sum_{k=1}^{n} \cos kx)$ is called the Drichlet kernel and they converge in distribution. Differentiating two times we get

$$\sum_{n=-\infty}^{\infty} \delta'(x - 2n\pi) = -\frac{1}{\pi}(\sum_{n=1}^{\infty} n \sin nx),$$

$$\sum_{n=-\infty}^{\infty} \delta''(x - 2n\pi) = -\frac{1}{\pi}(\sum_{n=1}^{\infty} n^2 \cos nx).$$

Both of these formulas make sense in the theory of distributions.

Note that classically left hand sides are not defined as the derivatives of the delta function while the series on the right diverge.

2.2 Applications to Differential Equations

In this section we consider linear partial differential equations with constant coefficients. Namely the equations of the form

$$P(D)u = g \quad (1)$$

where P is a non-constant polynomial in n variables with complex coefficients. D can be D_α as defined in Sect. 1.8.1.

g is a given function, it can be a distribution. Then the function or distribution u is a solution of (1).

Definition 1 A distribution $E \in \mathcal{D}'(\mathbb{R}^n)$ is said to be a <u>fundamental solution</u> of the operator $P(D)$ if it satisfies (1) with $g = \delta$, i.e.

$$P(D)E = \delta \quad (2)$$

If there is such an E and g has compact support, then

$$u = E * g \quad (3)$$

is well-defined and a solution of (1) since by Sect. 1.7, Theorem 2b:
$P(D)(E * g) = [P(D)E] * g = \delta * g = g.$

Thus the existence of a fundamental solution implies the existence of a solution to (1) provided that the convolution $E * g$ is well-defined. Equation (3) also gives an idea about the solution u, for instance if $g \in \mathcal{D}(\mathbb{R}^n)$, then $u = E * g \in \mathcal{C}^\infty(\mathbb{R}^n)$, Sect. 1.7, Theorem 2c.

Moreover if $\tilde{u}$ is any solution of (1), then $P(D)(\tilde{u} - E * h) = P(D)\tilde{u} - P(D)(E * g) = g - \delta * g = 0.$ Thus any solution of (1) differs from $E * g$ by a solution of the homogeneous equation $P(D)u = 0$.

The convolution $E * g$ may exist even g does not have compact support, e.g. it may satisfy the condition Γ given in Sect. 1.7 preceding the Definition 2. We should not expect to find E as having a compact support. Otherwise as analyzed at the end of Sect. 1.8.2, $\hat{E}$ would be an entire function. On the other hand by Sect. 1.8.2, Proposition 1b $P(2\pi\xi)\hat{E} = \hat{\delta} = 1$. ($P$ is polynomial).But the product of an entire function and a polynomial can not be 1 unless both are constants. However if $\frac{1}{P(2\pi\xi)}$ is a tempered distribution, we can then find the fundamental solution by the inverse transform of,

$$\hat{E} = \frac{1}{P(2\pi\xi)}. \quad (4)$$

The main result of this section is that the fundamental solution always exist when the differential operator has constant coefficients. This is the Malgrange & Ehrenpreis theorem.

Obviously zero differential operator has no fundamental solution. A non-zero identically constant differential operator $P(D) \equiv c$ has a unique fundamental solution $E = \frac{1}{c}\delta$. Accordingly we should expect that a non-zero differential operator can not have a regular distribution (a locally integrable function) as a fundamental solution.

Theorem 1 (Malgrange & Ehrenpreis) *Evey non-zero linear partial differential operator with constant coefficients on $\mathbb{R}^n$ has a fundamental solution, i.e. a distribution E such that $P(D)E = \delta$.*

<u>Note</u>. To recapitulate the notation:
For $\alpha = (\alpha_1, \cdots, \alpha_n)$, $|\alpha| = \alpha_1 + \cdots + \alpha_n$

$$P(.) = \sum a_\alpha (\xi)^\alpha = \sum a_\alpha \xi_1^{\alpha_1} .... \xi_n^{\alpha_n}$$

$$D^\alpha = \frac{\partial^{\alpha_1 + \cdots + \alpha_n}}{\partial x_1^{\alpha_1} \cdots \partial x_n^{\alpha_n}} , \quad \text{(without the complex factor } \tfrac{1}{i} \text{).}$$

$P(D) = \sum a_\alpha D^\alpha$, the adjoint operator $P^*(D) = \sum (-1)^{|\alpha|} \bar{a}_\alpha D^\alpha$.

<u>Remark on Proofs.</u> There have been different approaches to prove this important theorem. One possible approach to show the existence of a fundamental solution is to obtain the inverse Fourier transform of (4), i.e. to find $\mathcal{F}^{-1}\left(\frac{1}{P(2\pi\xi)}\right)$. For this purpose Bernstein ([Bern]) used modules over a ring of differential operators and the concept of analytic continuation of distribution valued holomorphic functions.

The other approaches of proofs can be classified as non-constructive and constructive. In constructive proofs explicit formulae are developed for fundamental solutions.

In non-constructive proofs of the existence of fundamental solutions, the main tool is the Hörmander's inequality.

<u>HÖRMANDER's INEQUALITY</u>. Let $P(D)$ be a non-zero linear differential operator with constant coefficients of order m. Then for every bounded open set $\Omega \subset \mathbb{R}^n$ there exists a constant $C > 0$ such that for every $\phi \in \mathcal{D}(\Omega)$ we have $\|P(D)\phi\|_2 \geq C\|\phi\|_2$. A value of C is given by $C = |P_m| K_{m,\Omega}$ where $|P_m| = \max\{|\alpha| : |\alpha| = m\}$ and $K_{m,\Omega}$ depends only on m and the diameter of Ω.

(The proof of Hörmander's inequality can be found in [Hörm-1] and [Rosay]). <u>Outline of the original proof of Malgrange & Ehrenpreis</u>. They consider the linear functional

$$F : P^*(D)\mathcal{D} \longrightarrow \mathbb{C}, \quad F[P^*(D)]\phi = \phi(0)$$

defined on the subspace $P^*(D)\mathcal{D} = \{P^*(D)\phi : \phi \in \mathcal{D}\}$ of $\mathcal{D}(\Omega)$. The proof consists of two steps:

(1°) Show that F is a continuous with respect to the topology induced by $\mathcal{D}$. This is achieved by showing first the continuity in L^2 norm using the Hörmander's inequality. Then the Hahn-Banach theorem implies that F can be extended as a continuous linear extension $E \in \mathcal{D}'$. Now E is the desired fundamental solution since:

$$< \phi, P(D)E >=< P^*(D)\phi, E >=< P^*(D)\phi, F >= \phi(0) =< \delta, \phi > .$$

However E is not uniquely determined and thus an explicit representation is not obtained.

Rosay gives an elementary proof utilizing the fact that the Heaviside function is locally L^2 and its distributional derivative is the Dirac delta function. <u>Detailed account of ROSAY's proof.</u>

<u>L^2Solvability</u>. If Ω is a bounded set in $\mathbb{R}^n$, then for every $g \in L^2(\Omega)$ there exists $u \in L^2(\Omega)$ such that $P(D)u = g$.

Proof Apply Hörmander's inequality to the adjoint $P^*(D)$, thus $\|P^*(D)\phi\|_2 \geq C\|\phi\|_2, \phi \in \mathcal{D}(\Omega)$. $P(D)u = g$ means that for all $\phi \in \mathcal{D}(\Omega)$

$$< g, \phi >=< u, P^*(D)\phi > \qquad (*)$$

on the subspace $\mathcal{L} = P^*(D)\mathcal{D}(\Omega) = \{P^*(D)\phi : \phi \in \mathcal{D}\} \subset \mathcal{D}(\Omega) \subset L^2_{loc}(\Omega)$.

Given $g \in L^2(\Omega)$, define a linear operator $F : P^*(D)\mathcal{D}(\Omega) \longrightarrow \mathbb{C}$, $< F, P^*(D)\phi > = < g, \phi >$. F is continuous in the L^2 norm since $| < g, \phi > | \leq \|g\|_2 \|\phi\|_2 \leq \frac{1}{C}\|g\|_2 \|P^*(D)\phi\|_2$. Then F can be extended to $\bar{\mathcal{L}}$, the closure of $\mathcal{L}$ in $L^2(\Omega)$. As $\bar{\mathcal{L}}$ is complete, by the Riesz representation theorem there exists $u \in L^2(\Omega)$ representing F, i.e. $< F, P^*(D)\phi >$
$= < u, P^*(D)\phi > = < P(D)u, \phi > = < g, \phi >$, so that (*) holds.

Hörmander's inequality with weight and support

$P(D)$ and Ω as before, there exists $k > 0$ so that for all $\eta \in \mathbb{R}$ and $\phi \in \mathcal{D}(\Omega)$ we have

$$\int_{\Omega} e^{\eta x_1} |P(D)\phi|^2 \, dx \geq k \int_{\Omega} e^{\eta x_1} |\phi|^2 \, dx$$

$(dx = dx_1 \, dx_2 \cdots dx_n)$ and D does not depend on η.

Proof Apply Hörmander's inequality to $\psi = e^{(\frac{\eta}{2})x_1}\phi$ and the operator $Q(D)$ defined by

$$Q(D)(\psi) = e^{(\frac{\eta}{2})x_1} P(D)[e^{-(\frac{\eta}{2})x_1}\psi]$$

which is indeed a constant coefficient operator of the same order.

Corollary 1 *If $\phi \in \mathcal{D}(\mathbb{R}^n)$ and $P(D)\phi = 0$ in the half-space $\{x_1 > 0\}$, then $\phi = 0$ in this half-space.*

Proof Take Ω as an open set containing the support of ϕ. Apply then the above version of the Hörmander's inequality:

$$\int_{\Omega \setminus \{x_1 > 0\}} e^{-\eta|x_1|} |P(D)\phi|^2 \, dx + \int_{\{x_1 > 0\}} e^{\eta x_1} |P(D)\phi|^2 \, dx$$

$$\geq k \left[\int_{\Omega \setminus \{x_1 > 0\}} e^{-\eta|x_1|} |\phi|^2 \, dx + \int_{\{x_1 > 0\}} e^{\eta x_1} |\phi|^2 \, dx \right]$$

By the assumption the second integral on the left side is zero. Then the limit as $\eta \to \infty$ gives

$$0 \geq k \lim_{\eta \to \infty} \int_{\{x_1\}} e^{\eta x_1} |\phi|^2 \, dx$$

is possible only if $\phi = 0$ in the half-space $\{x_1 > 0\}$.

Corollary 2 *Let $\phi \in \mathcal{D}(\mathbb{R}^n)$ or more generally $\phi \in L^2(\mathbb{R}^n)$ and assume that ϕ has compact support. If $P(D)\phi$ is supported by the ball of radius r around 0, then so is ϕ.*

Proof If $\phi \in \mathcal{D}(\mathbb{R}^n)$, by using translations and rotations the result follows immediately from Corollary 1, by writing the ball as an intersection of half-spaces. For the case $\phi \in L^2(\mathbb{R}^n)$, consider $\chi \in \mathcal{D}(\mathbb{R}^n)$ supported by the unit ball around the origin satisfying $\int \chi = 1$. For $\epsilon > 0$, $\chi_\epsilon = \frac{1}{\epsilon^n}\chi(\frac{x}{\epsilon})$ set

$$\phi_\epsilon = \phi * \chi_\epsilon = \frac{1}{\epsilon^n} \int_{\mathbb{R}^n} \phi(x - t)\chi(\frac{t}{\epsilon})dt.$$

Then $\phi_\epsilon \in \mathcal{D}(\mathbb{R}^n)$, $P(D)\phi_\epsilon = P(D)(\phi * \chi_\epsilon) = [P(D)\phi] * \chi_\epsilon$ is supported by the ball of radius $r + \epsilon$, (this is by the assumption and the fact that $\chi(\frac{t}{\epsilon})$ is supported by the ball of radius ϵ , thus $|x| \leq |t| + \epsilon \leq r + \epsilon)$, also $\phi_\epsilon \to \phi$ in L^2 as $\epsilon \to 0$. This reduces the problem to the first case.

Approximation Theorem. For $\rho > 0$, let B_ρ be the ball of radius ρ around 0 in $\mathbb{R}^n$. Let $0 < r < r' < R$. If $v \in L^2(B_{r'})$ and satisfies $P(D)v = 0$ on $B_{r'}$, then there exists $\{v_j\}$, a sequence in $L^2(B_R)$ so that $P(D)v_j = 0$ on B_R and $v_j \to v$ in $L^2(B_r)$ as $j \to \infty$.

Proof By smoothing via convolution we can assure v to be smooth. We have to show that if $g \in L^2(B_R)$ and satisfies $< \alpha, g >_{B_r} = 0$ for all $\alpha \in L^2(B_R)$ such that $P(D)\alpha = 0$, then $< v, g >_{B_r} = 0$. This will suffice for proving the approximation theorem since then v will be in the null-space of the operator $P(D)$ with respect to $< , >_{B_r}$, thus such a series can be constructed .

Claim *There exists* $w \in L^2(B_R)$ *so that for all* $\phi \in \mathcal{D}(\mathbb{R}.)$, $< \phi, g >_{B_r} = < P(D)\phi, w >_{B_r}$.

Proof of the claim. By the Riesz representation theorem all we have to show is , (see also L^2 solvability)

$$| < \phi, g >_{B_r} | \leq C\|P(D)\phi\|_{B_R}.$$

If $P(D)\phi = 0$, we indeed have $< \phi, g > = 0$. If $P(D)\phi \neq 0$, then by L^2 solvability we can find $\psi \in L^2(B_R)$ so that $P(D)\psi = P(D)\phi$ and $\|\psi\|_{B_R} \leq C_1\|P(D)\phi\|_{B_R}$. Then $< \phi, g >_{B_r} = < \phi - \psi, g >_{B_r} = < \psi, g >_{B_r} = < \psi, g >_{B_r}$ since $P(D)(\phi - \psi) = 0$ and g is then orthogonal to $\phi - \psi$ by the assumption on g.

Now $| < \phi, g >_{B_r} | \leq \|\psi\|_{B_R}\|g\|_{B_R} \leq C_1\|g\|_{B_R}\|P(D)\phi\|_{B_R} = C\|P(D)\phi\|_{B_R}$ proves the claim.

Continue with the proof of Approximation Theorem

Pick w as given by the claim. Extend g and w on $\mathbb{R}^n$ to $\tilde{g}$ and $\tilde{w}$ by setting $\tilde{g} = 0$ on $\mathbb{R}^n \setminus B_r$ and $\tilde{w} = 0$ on $\mathbb{R}^n \setminus B_R$. We then have (by definition) $\tilde{g} = P^*(D)\tilde{w}$. Since $\tilde{g}$ has compact support and $P^*(D)\tilde{w}$ is supported by B_r, Corollary 2 implies $w = 0$ on $B_R \setminus B_r$.

Then take v as in the beginning of the proof and extend it to be a smooth function on $\mathbb{R}^n$ (but no longer satisfying $P(D)v = 0$ off B_r). Thus we have by the claim $< v, g >_{B_r} = < P(D)v, w >_{B_R} = < P(D)v, w >_{B_r}$ (as $P(D)v = 0$ on $B_{r'} \supset B_r$).

Solvability of $P(D)u = g$ in $L^2_{loc}(\mathbb{R}^n$

Let $P(D)$ be non-zero constant coefficient linear partial operator on $\mathbb{R}^n)$. Then for every $g \in L^2_{loc}(\mathbb{R}^n)$ there exists $u \in L^2_{loc}(\mathbb{R}^n)$ such that $P(D)u = g$.

Proof By the L^2 solvability there exists $u_1 \in L^2(B_2)$ so that $P(D)u_1 = g$ on B_2. Then, inductively assuming that u_r has been chosen in $L^2(B_{r+1})$ so that $P(D)u_r = g$, $u_{r+1} \in L^2(B_{r+2})$ is chosen in the following way : Let w be an arbitrary solution of $P(D)w = g$ in $L^2(B_{r+2})$ (which is again possible by L^2 solvability). On B_{r+2} we have $P(D)(u_r - w) = 0$. By the Approximation Theorem there exists $v \in L^2(B_{r+2})$ such that $P(D)v = 0$ and $\|v - (u_r - w)\|_{B_{r+1}} \leq \frac{1}{2^{r+1}}$. Set $u_{r+1} = v + w$. Then $P(D)u_{r+1} = g$ on B_{r+2} and $\|u_{r+1} - u_r\|_{B_{r+1}} = \|v + w - u_r\|_{B_{r+1}} \leq \frac{1}{2^{r+1}}$. Thus the sequence $\{u_r\}$ is convergent in $L^2_{loc}(\mathbb{R}^n)$ and its limit u satisfies $P(D)u = g$.

Proof of Malgrange–Ehrenpreis

Let H be the function (product of Heaviside functions) defined on $\mathbb{R}^n$ by $H(x_1, \cdots, x_n) = 1$ if all x_j are positive and 0 otherwise. Then $\dfrac{\partial^n H(x)}{\partial x_1 \cdots \partial x_n} = \delta(x)$. As $H \in L^2_{loc}$, by local solvability there exists $u \in L^2_{loc}(\mathbb{R}^n)$, so that $P(D)u = H$. Set the distribution E as $\dfrac{\partial^n u}{\partial x_1 \cdots \partial x_n}$. Then

$$P(D)E = P(D)\frac{\partial^n u}{\partial x_1 \cdots \partial x_n} = \partial^n \partial x_1 \cdots \partial x_n (P(D)u) = \delta$$

which ends the proof.

Constructive Proofs

Lars Hörmander was the first to work out explicit formulae for fundamental solutions of linear partial differential operators with constant coefficients. Main references are [Hörm-1], [Wag-1], [Wag-2].

The fundamental solution constructed by the so-called Hörmander's staircase is given by

$$E = \sum_{k=0}^{m} e^{kx\eta} \mathcal{F}^{-1}\left(\frac{\chi_k(\xi)}{P(i\xi + k\eta)}\right),$$

here $\{\chi_k\}$ is a partition of unity satisfying further

$$deg P = m, \ \eta \in \mathbb{R}^n \ \text{ with } \ P_m(\eta) \neq 0.$$

König proposed to replace the functions $\xi \longrightarrow \dfrac{\chi_k(\xi)}{P(i\xi + k\eta)}$ by $\xi \longrightarrow \dfrac{P(i\xi + \xi\eta)}{P(i\xi + \epsilon\eta)}$. Using König's approach [Wag-2] developed a simpler representation of a fundamental solution as

$$E = \frac{1}{P_m(\eta)} \int_T \lambda^m e^{\lambda\eta x} \mathcal{F}^{-1}\left(\frac{\overline{P(i\xi + \lambda\eta)}}{P(i\xi + \lambda\eta)}\right) \frac{d\lambda}{2\pi i \lambda}$$

$\lambda \in \mathbb{R}$ satisfying $\dfrac{E}{\cosh(\eta x)} \in \mathcal{S}'(\mathbb{R}^n)$ and T: one dimensional torus.

However the solutions defined by these explicit formulae are often difficult to work out. In some problems fundamental solutions can be constructed using physical considerations and then tested that they are actually distributions as required of a fundamental solution. We illustrate this approach in the following example of a 3-dimensional Laplace operator.

Example 1 The free-space fundamental solution in three-dimensional potential theory satisfies

$$-\left(\frac{\partial^2}{\partial x_1^2} + \frac{\partial^2}{\partial x_2^2} + \frac{\partial^2}{\partial x_3^2}\right) = \delta(x) \quad \text{or}$$

$$-\nabla^2 E = \delta(x) \quad \text{or} \quad \text{-div grad } E = \delta(x) \quad (5)$$

Here $E(x)$ can be interpreted as the electrostatic potential at an arbitrary point x due to a unit positive charge at $x = 0$. As $-\nabla^2$ operator is invariant under a rotation of coordinates , the solution could depend on $r \|x\|$. For $r > 0$ (off the origin) $E(r)$ satisfies the homogeneous equation $\nabla^2 E = 0$ which takes the form in the spherical coordinates

$$\frac{1}{r^2}\frac{\partial}{\partial r}\left(r^2 \frac{\partial E}{\partial r}\right) = 0.$$

that has the solution $E = \frac{C}{r} + D$. If we impose the condition that the potential should vanish at infinity , then $E = \frac{E}{r}$. To determine C we integrate (5) over a small sphere R_ϵ of radius $\epsilon > 0$ and centered at $x = 0$. Considering the unit source at the origin we find $-\int_{R_\epsilon}[\text{div grad } E]\,dx = 1$. By the divergence theorem then

$$-\int_{\sigma_\epsilon}\left(\frac{\partial E}{\partial r}\right)_{r=\epsilon} ds = 1 \quad (6)$$

where σ_ϵ is the surface of R_ϵ. Physical interpretation : the flux of the electric field through the closed surface σ_ϵ is equal to the charge created by the source in the interior.

As $\frac{\partial E}{\partial r}\big|_{r=\epsilon} = -\frac{C}{r^2}\big|_{r=\epsilon} = -\frac{C}{\epsilon^2}$ we find $C = \frac{1}{4\pi}$.

Hence

$$E(r) = \frac{1}{4\pi r} \quad (7)$$

To show that (7) satisfies $-\nabla^2 E = \delta$ distributionally it should be proved that for every test function ϕ

$$\int_{\mathbb{R}^3}\frac{1}{4\pi r}(\nabla^2)\,dx = -\phi(0).$$

Evaluate the left side by the integration by parts as :

$$\int_{\mathbb{R}^3} \frac{1}{4\pi r}(\nabla^2 \phi)\, dx = \lim_{\epsilon \to 0} \int_{\mathbb{R}^3 \setminus R_\epsilon} \frac{1}{4\pi r}(\nabla^2 \phi)\, dx$$

Applying the Green's identity

$$\int_\Omega (u\nabla^2 v - v\nabla^2 u)\, dx = \int_{\sigma_\Omega} (u\frac{\partial v}{\partial n} - v\frac{\partial u}{\partial n})\, du$$

to $\Omega = \mathbb{R}^3 \setminus R_\epsilon$, $u = \frac{1}{4\pi r}$, $v = \phi$ and $\frac{\partial \cdot}{\partial n} = \frac{\partial \cdot}{\partial r}$ for a sphere

$$\int_{\mathbb{R}^3 \setminus R_\epsilon} \frac{1}{4\pi r}(\nabla^2 \phi)\, dx = \int_{\mathbb{R}^3 \setminus R_\epsilon} \phi(\nabla^2 \frac{1}{4\pi r})\, dx + \int_{\sigma_\epsilon} [-\frac{1}{4\pi r}\frac{\partial \phi}{\partial r} + \phi\frac{\partial}{\partial r}(\frac{1}{4\pi r})]\, ds$$

where we have used the fact that the test functions have bounded support, so that the surface integrals at infinity are eliminated.

As $\frac{1}{4\pi r}$ satisfies the homogeneous equation $\nabla^2(\frac{1}{4\pi r}) = 0$ in $\mathbb{R}^3 \setminus R_\epsilon$ we therefore have

$$\int_{\mathbb{R}^3 \setminus R_\epsilon} \frac{1}{4\pi r}(\nabla^2 \phi)\, dx = -\int_{\sigma_\epsilon} [\frac{1}{4\pi r}\frac{\partial \phi}{\partial r} + \frac{1}{4\pi r^2}\phi]\, ds \quad (8)$$

As the derivatives of the test functions are again test functions we have $|\frac{\partial \phi}{\partial r}| < M$ for all x, hence

$$|\int_{\sigma_\epsilon} \frac{1}{4\pi r}\frac{\partial \phi}{\partial r}\, ds| \le \frac{M}{4\pi\epsilon}.4\pi\epsilon^2 = M\epsilon \quad (9)$$

$$\int_{\sigma_\epsilon} \frac{\phi(x)}{4\pi r^2}\, ds = \int_{\sigma_\epsilon} \frac{\phi(0)}{4\pi r^2}\, ds + \int_{\sigma_\epsilon} \frac{\phi(x) - \phi(0)}{4\pi r^2}\, ds = \phi(0) + \int_{\sigma_\epsilon} \frac{\phi(x) - \phi(0)}{4\pi r^2}\, ds$$

and $|\int_{\sigma_\epsilon} \frac{\phi(x)-\phi(0)}{4\pi r^2}\, ds| \le \max_{x \in \sigma_\epsilon} |\phi(x) - \phi(0)|$. Since $\phi(x)$ is continuous at $x = 0$ we have $\lim_{\epsilon \to 0}[\max_{x \in \sigma_\epsilon}] = 0$, thus by (8) and (9):

$$\lim_{\epsilon \to 0} \int_{\mathbb{R}^3 \setminus R_\epsilon} \frac{1}{4\pi r}(\nabla^2 \phi)\, dx = -\phi(0)$$

which completes the proof.

The fundamental solution $E_n(x)$ for the n-dimensional Laplace operator is obtained along the lines parallel to the case $n = 3$, although now we shall be far from physical intuition. It will satisfy:

$$-\nabla^2 E_n = -(\frac{\partial^2}{\partial x_1^2} + \cdots + \frac{\partial^2}{\partial x_n^2})E_n = \delta(x) \quad (10)$$

If we look for a solution that will depend only on $r = (x_1^2 + \cdots + x_n^2)^{\frac{1}{2}}$, the above can be reduced to an ordinary differential equation in r. If a function u depends only on r:

$$\frac{\partial u}{\partial x_i} = (\frac{\partial u}{\partial r})\frac{\partial r}{\partial x_i} = \frac{x_i}{r}(\frac{\partial u}{\partial r}),$$

$$\frac{\partial^2 u}{\partial x_i^2} = \frac{x_i^2}{r^2}\frac{\partial^2 u}{\partial r^2} - \frac{x_i^2}{r^3}\cdot\frac{\partial u}{\partial r} + \frac{1}{r}\frac{\partial u}{\partial r},$$

$$\nabla^2 u = \sum_{i=1}^{n} \frac{\partial^2 u}{\partial x_i^2} = \frac{\partial^2 u}{\partial r^2} - \frac{1}{r}\frac{\partial u}{\partial r} + \frac{n}{r}\frac{\partial u}{\partial r} = \frac{\partial^2 u}{\partial r^2} + \frac{n-1}{r}\frac{\partial u}{\partial r}.$$

Thus for $x \neq 0$ (10) becomes the homogeneous equation

$$\frac{\partial^2 E_n}{\partial r^2} + \frac{n-1}{r}\frac{\partial E_n}{\partial r} = \frac{1}{r^{n-1}}\frac{\partial}{\partial r}(r^{n-1}\frac{\partial E_n}{\partial r}) = 0.$$

The fundamental solution is found as $E_n = \frac{C}{r^{n-1}} + D$, $n > 2$. If we require the solution to vanish at infinities we must set $D = 0$. Again if we integrate over a small sphere R_ϵ and proceed along the same lines as it was in 3-dimensional case , we find $\int_{\sigma_\epsilon}(\frac{\partial E_n}{\partial r})|_{r=\epsilon}\, ds = -1$. Substituting $E_n = \frac{C}{r^{n-2}}$ and evaluating at $r = \epsilon$ yields

$$(-n + 2)C \int_{\sigma_\epsilon} \epsilon^{-n+1}\, ds = (-n + 2)C\epsilon^{-n+1} S_{n-1}(\epsilon) = -1 \quad (11)$$

The surface area of a sphere of radius r in n-dimensional space is given by $A_{n-1}(r) = \frac{2\pi^{n/2}r^{n-1}}{\Gamma(n/2)}$. Substituting into (11) yields $C = \frac{\Gamma(n/2)}{2\pi^{n/2}(n-2)}$ and hence $E_n(r) = \frac{\Gamma(n/2)}{2\pi^{n/2}(n-2)r^{n-2}}$ $(n > 2)$.

Verification that this is the fundamental solution satisfying (10) , follows the same lines as in 3-dimensional case, i..e. via the Green's identity.

2.3   Sobolev Spaces

A Sobolev space is a vector space of functions equipped with a norm that is a combination of L^p norms of the function itself as well as its derivatives up to a given order. The derivatives are understood as distributional (weak) derivatives. Intuitively a Sobolev space of functions possessing sufficiently high order derivatives. Such a set-up will be very suitable for the study of solutions to partial differential equations. The norms in Sobolev spaces will allow to measure both the size and the regularity of solutions. Stochastic counter-part of Sobolev spaces will play an important role in stochastic analysis , in particular in Malliavin calculus (c.f. Chap. 5).

Sobolev Spaces $W_k^p(\Omega)$.

Let Ω be an open set of $\mathbb{R}^n$ and k a positive integer. For $1 \le p < \infty$ we denote by $W_k^p(\Omega)$ the set of all complex-valued functions $f(x) = f(x_1, x_2, \cdots, x_n)$ defined in Ω such that its distributional derivative $D^\alpha f$ of order $|\alpha| = \sum_{j=1}^n |s_j| \le k$ all belong to $L^p(\Omega)$. Then $W_k^p(\Omega)$ is a normed linear space by $(f_1 + f_2)(x) = f_1(x) + f_2(x)$, $(cf)(x) = cf(x)$ and $\|f\|_{p,k} = \left(\sum_{|\alpha| \le k} \int_\Omega |D^\alpha f(x)|^p \, dx \right)^{\frac{1}{p}}$, $(dx = dx_1 \cdots dx_n)$. By the assumption f_1 and f_2 are identified if $f_1 = f_2$ almost everywhere (a.e.) in Ω.

For $p = +\infty$, $\|f\|_{\infty,k} = \max_{|\alpha| \le k} \|D^\alpha f\|_{L^\infty(\Omega)}$.

Note that the norms $\|f\|_{p,k}$ can be rewritten as $\|f\|_{p,k} = \left(\|f\|_{L^p}^p + \sum_{1 \le |\alpha| \le k} \|D^\alpha f\|_{L^p}^p \right)^{\frac{1}{p}}$ and W_k^p spaces can also be given by another set of equivalent norms :

$$
\|f\|_{k,p}^- = \begin{cases} \sum_{|\alpha| \le k} \|D^\alpha f\|_{L^p(\Omega)} & \text{if } 1 \le p < \infty \\ \sum_{|\alpha| \le k} \|D^\alpha f\|_{L^\infty(\Omega)} & \text{if } p = +\infty \end{cases}
$$

Proposition 1 *The space W_k^p is a Banach space . In particular W_k^2 is a Hilbert space by the inner product $(f, g)_{2,k} = \sum_{|\alpha| \le k} \int_\Omega D^\alpha f(x) \overline{D^\alpha g(x)} \, dx$.*

Proof Let $\{f_j\}$ be a Cauchy sequence in $W_k^p(\Omega)$. Then in view of the way that the norm is defined , for any differential operator D^α with $|\alpha| \le k$. The sequence $\{D^\alpha f_j\}$ is also a Cauchy sequence in $L^p(\Omega)$ and so by completeness of $L^p(\Omega)$, there exist functions $f^{(\alpha)} \in L^p(\Omega)$ such that

$$
\lim_{j \to \infty} \int_\Omega |D^\alpha f_j(x) - f^{(\alpha)}(x)|^p \, dx = 0, \quad (|\alpha| \le k) \qquad (*)
$$

We have for the compact subsets of Ω

$$
\int_K |f_j| . 1 \, dx \le (\int_K |f_j|^p \, dx)^{\frac{1}{p}} (\int_K 1 . dx)^{\frac{1}{p'}}
$$

(Hölder's inequality with $\frac{1}{p} + \frac{1}{p'} = 1$).

As $\int_K |f_j| \, dx \le \|f_j\|_{L^p} [m(K)]^{\frac{1}{p'}} < \infty$ (m being the Lebesgue measure) , f_j is locally integrable thus defining a regular distribution . Hence for any function $\phi \in \mathcal{D}(\Omega)$:

$$
T_{D^\alpha f_j}(\phi) = \int_\Omega D^\alpha f_j(x) \, \phi(x) \, dx = (-1)^{|\alpha|} \int_\Omega f_j(x) D^\alpha \phi(x) \, dx \qquad (\dagger)
$$

By $(*)$: $\lim_{j \to \infty} \int_\Omega |f_j(x) - f^{(0)}|^p \, dx = 0$. $\qquad (\sqrt{})$

Again by Hölder's inequality

$$\left| \int_\Omega (f_j(x) - f^{(0)}) \, D^\alpha \phi(x) \, dx \right| \le \left(\int_\Omega |f_j(x) - f^{(0)}|^p \, dx \right)^{\frac{1}{p}} \left(\int_\Omega |D^\alpha \phi(x)|^{p'} \, dx \right)^{\frac{1}{p'}}$$

which yields by ($\sqrt{\ }$) $\lim_{j\to\infty} \int_\Omega f_j(x) \, D^\alpha \phi(x) \, dx = \int_\Omega f^{(0)} \, D^\alpha \phi(x) \, dx$
$= < T_{f^{(0)}}(x), \, D^\alpha \phi(x) >$. Substituting in (†) we find

$$\lim_{j\to\infty} T_{D^\alpha f_j}(\phi) = (-1)^{|\alpha|} < T_{f^{[0]}}, \, D^\alpha \phi(x) >= D^\alpha T_{f^{(0)}}(\phi).$$

On the other hand starting by (*) a similar analysis gives

$$\lim_{j\to\infty} T_{D^\alpha f_j}(\phi) = T_{f^{(\alpha)}}(\phi).$$

Hence we must have $D^\alpha T_{f^{(0)}} = T_{f^{(\alpha)}}$. That is the distributional derivative $D^\alpha f^{(0)}$
equals $f^{(\alpha)}$. Then

$$\|f_j - f^{(0)}\|_{p,k} = \left(\sum_{|\alpha|\le k} \int_\Omega |D^\alpha f_j(x) - D^\alpha f^{(0)}(x)|^p \, dx \right)^{\frac{1}{p}}$$

$= \left(\sum_{|\alpha|\le k} \int_\Omega |D^\alpha f_j(x) - f^{(\alpha)}|^p \, dx \right)^{\frac{1}{p}} \to 0$ by (*) as $j \to \infty$. This shows the
completeness of W_k^p. The second half is an easy consequence of this. $\square$

<u>Note.</u> W_k^2 is commonly denoted by H^k.
For a positive integer k we have another characterization of $W_k^2(\mathbb{R}^n) \equiv H^k$ as given
in the following:

Theorem 1 *Let $f \in L^2(\mathbb{R}^n)$ and let $\hat{f}$ denote its Fourier-Plancherel transform.
Then the following two statements are equivalent :*

(a) $f \in H^k$,
(b) $\int_{\mathbb{R}^n} (1 + \|\xi\|^2)^k \, |\hat{f}|^2 \, d\xi < \infty$.

Proof Restricting to the case where $k = 1$, first we show that $a) \Rightarrow b)$. For $f \in H^1$,
by Sect. 1.8.2, Proposition 1-b

$$\left(\frac{\partial f}{\partial \alpha_j} \right)^\wedge(\xi) = 2\pi i \xi_j \hat{f}(\xi)$$

As by the definition of a Sobolev space $D^\alpha f \in L^2(\mathbb{R}^n)$ and $\alpha = (0, \cdots, 0, \alpha_j, 0,$
$\cdots, 0)$ ($\alpha_j = 0$ or 1) and considering Fourier-Plancherel transform we find that
$\xi_j \hat{f}(\xi_j) \in L^2$ or $\xi_j^2 [\hat{f}]^2 \in L^1$ and also $\hat{f} \in L^2$. Thus for $j = 1, \cdots, n$:
$[\hat{f}(\xi)]^2 (1 + \xi_1^2 + \xi_2^2 + \cdots + \xi_n^2) = [\hat{f}(\xi)]^2 (1 + \|\xi\|^2) < \infty$.

Proof of $b) \to a)$: Let $\phi \in \mathcal{D}$, then by the Plancherel theorem (Parseval formula)

$$\int_{\mathbb{R}^n} f(x) \overline{\frac{\partial \phi}{\partial \alpha_j}(x)} \, dx = \int_{\mathbb{R}^n} \hat{f}(\xi) \overline{\left(\frac{\partial \hat{\phi}}{\partial \alpha_j} \right)} \, d\xi \, ,$$

Again by Sect. 1.8.2, Proposition 1-b

$$\int_{\mathbb{R}^n} \overline{\frac{\partial \phi}{\partial \alpha_j}}(x_j) \, dx = - \int_{\mathbb{R}^n} 2\pi i \xi_j \, \hat{f}(\xi) \overline{\hat{\phi}(\xi)} \, d\xi. \quad (*)$$

By (b) $\xi_j \hat{f}(\xi) \in L^2$. There exists $u_j \in L^2$ such that $\hat{u}_j(\xi) = 2\pi i \xi_j \hat{f}(\xi)$. Substituting in (*) and recalling that ϕ is real-valued the Parseval's formula yields:

$$\int_{\mathbb{R}^n} f(x) \frac{\partial \phi(x)}{\partial \alpha_j}(x) \, dx = - \int_{\mathbb{R}^n} \hat{u}_j(\xi) \overline{\hat{\phi}(\xi)} \, d\xi = - \int_{\mathbb{R}^n} u_j(x) \, \phi(x) \, dx,$$

that is, the distributional derivative of f in the direction of α_j is $-u_j \in L^2$. Then $D^\alpha f \in L^2$, $(|\alpha| = 1)$, hence $f \in H^1$. This completes the proof since for $k > 1$ H^k is defined recursively as $H^k = \{ f \in H^1 : D^\alpha f \in H^{k-1}, |\alpha| = 1 \}$.

$\underline{\text{Sobolev Spaces } H^s \text{ with Non} - \text{integer } s}$

Let s be a positive real number that is not integer. Set, based on Theorem 1

$$H^s = \{ f \in L^2; \int_{\mathbb{R}^n} |\hat{f}(\xi)|^2 \, (1 + \|\xi\|^2)^s \, d\xi < \infty \}.$$

We define a norm in H^s by

$$\| f \|_{H^s} = \int_{\mathbb{R}^n} (1 + \|\xi\|^2)^s \, d\xi.$$

For $s = 1$ this norm is different from the W_1^2 norm, but the two norms are equivalent. The advantage of the present norm is that H^s becomes a Hilbert space with the scalar product

$$< f_1, f_2 >_{H^s} = \int_{\mathbb{R}^n} (\hat{f}_1 \overline{\hat{f}_2})(\xi) \, (1 + \|\xi\|^2) \, d\xi.$$

$\underline{\text{Notes.}}$ (i) Clearly for $s_1 < s_2$ we have $H^{s_2} \subset H^{s_1}$.
(ii) H^s is isometrically isomorphic to $L^2(\mu_s)$ where $d\mu_s(y) = (1 + \|y\|^2)^s \, dm(y)$; m : (Lebesgue measure). This easily follows from the definition of the norm in H^s and $\| f \|_{L^2} = \| \hat{f} \|_{L^2}$.

$\underline{H^s_{loc} \; : \; \text{Locally Sobolev Spaces}}$

Let Ω be an open subset of $\mathbb{R}^n$. A distribution $u \in \mathcal{D}'(\Omega)$ is said to be locally H^s if there corresponds to each point $x \in \Omega$ a $v \in H^s$ such that $u = v$ in some neighborhood $\omega \in \Omega$ of x.

Theorem 2 *If $u \in \mathcal{D}'(\Omega)$, the following two statements are equivalent*

(i) $u \in H^s_{loc}(\Omega)$,
(ii) $\psi u \in H^s$ *for every* $\psi \in \mathcal{D}(\omega)$.

<u>Note</u>. $< \psi u, \phi >=< u, \psi\phi >$ and $\psi\phi \in \mathcal{D}$ for $\phi \in \mathcal{D}(\mathbb{R}^n)$.

Proof $(i) \Rightarrow (ii)$: Assume that u is locally H^s. Let K be a compact set that is the support of some $\psi \in \mathcal{D}(\Omega)$. Then there are finitely many open sets $\omega_i \subset \Omega$ whose union covers K and in which u coincides with some $v_i \in H^s$. Then using the device of partition of unity (c.f. 1.4) there exists functions $\psi_i \in \mathcal{D}(\omega_i)$ such that $\sum \psi_i = 1$ on K and $\psi = \sum \psi_i \psi$. If $\phi \in \mathcal{D}(\mathbb{R}^n)$ it follows that $< \psi u, \phi >=< u, \psi\phi >= \sum < u, \psi_i\psi\phi >= \sum < v_i, \psi_i\psi\phi >$ where $\psi_i\psi\phi \in \mathcal{D}(\omega_i)$. Thus $\psi u = \sum \psi_i\psi v_i$. It is not difficult to see that $\psi_i\psi v_i \in H^s$. As there are finitely many terms in the summation we have $\psi u \in H^s$.

Conversely if (ii) holds and $x \in \Omega$ and if $\psi \in \mathcal{D}(\Omega)$ is unity in a neighborhood ω of x, then $u = \psi u$ in ω and $\psi u \in H^s$ by assumption. Hence (ii) implies (i).

2.4 Distributions on Manifolds

The locally convex vector spaces $\mathcal{E}(\Omega)$, $\mathcal{D}(\Omega)$, $\mathcal{D}'(\Omega)$ and $\mathcal{E}'(\Omega)$ can be extended from open sets $\Omega \subset \mathbb{R}^n$ to arbitrary manifolds. In this section only a concise outline of this rather involved subject is presented. Applications of the distributions on manifolds cover certain areas of partial differential equations, classical mechanics and the theory of relativity.

Although this section is intended to be fairly self-consistent , some background at the level of an introductory course on manifolds would be helpful. Otherwise the reader may refer one of the excellent texts on the subject like [Hicks], [Allen], [Munk], [Booth], [Martin].

A manifold is, roughly speaking, a topological space which is locally Euclidean, i.e. an observer at any point on it would conceive his immediate neighborhood as flat. More precisely we consider a Hausdorff, paracompact topological space M with a countable basis of open sets. Each point has a neighborhood homeomorphic to an open set of $\mathbb{R}^m$. Each pair (U, ϕ) where U is an open set of M and ϕ is a homeomorphism of U to an open subset of $\mathbb{R}^m$ is called a <u>coordinate chart</u>. U is known to be the domain of the chart. If (V, ψ) is another coordinate chart such that $U \cap V \neq \emptyset$, $\phi \circ \psi^{-1}$ and $\psi \circ \phi^-$ define homeomorphisms of the open sets $\phi(U \cap V)$ and $\psi(U \cap V)$. We shall say that (U, ϕ) and (V, ψ) are C^∞ − compatible if $\phi \circ \psi^{-1}$ and $\psi \circ \phi^{-1}$ are <u>diffeomorphisms</u> , i.e. mappings continuosly differentiable of any order (Fig. 2.1).

Similarly C^k-compatible coordinate charts can be defined. (A topological space is paracompact if every open cover has a sub-cover in which every point is covered finitely many times). For $i = 1, 2, \cdots , m$ let u_i be the slot functions of $\mathbb{R}^m$, i.e.

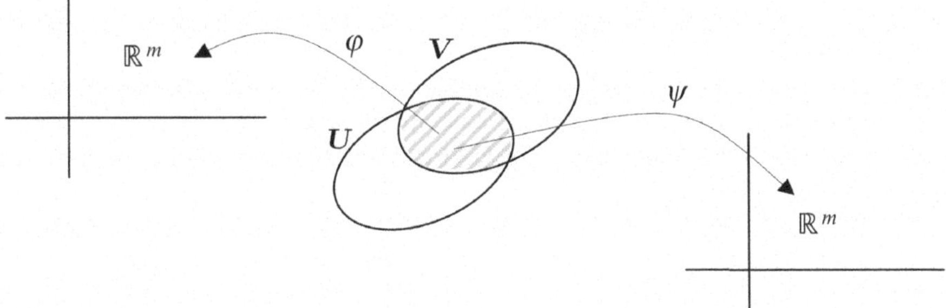

Fig. 2.1 In a manifold compatibility of two charts with non-empty intersection

$u_i : \mathbb{R}^m \to \mathbb{R}$ by $u_i(a_1, \cdots, a_m) = a_i$. Each coordinate chart (U, ϕ) on M induces coordinate functions $x_i = u_i \circ \phi$ for $i = 1, \cdots, m$, hence a local coordinate system. If a point q belongs to $U \cap V$, then it has two sets of coordinates $x_i = u_i \circ \phi(q)$ and $y_i = u_i \circ \psi(q)$ $(i = 1, \cdots, m)$. Thus $\phi \circ \psi^{-1}$ and $\psi \circ \phi^{-1}$ can be expressed in terms of coordinate functions by $x_i = g_i(y_1, \cdots, y_m)$ and $y_i = h_i(x_1, \cdots, x_m)$, $(i = 1, \cdots, m)$ where g_i, h_i are smooth functions inverses of each other.

Definition 1 A Hausdorff, paracompact topological space M is a differentiable (or C^∞) manifold if there is a family $\mathcal{U} = \{(U_\alpha, \phi_\alpha)\}$ of coordinate charts satisfying :

(a) U_α cover M,
(b) For any α, β (U_α, ϕ_α) and (U_β, ϕ_β) are C^∞-compatible,
(c) Any coordinate chart (V, ψ) compatible with every $(U_\alpha, \phi_\alpha) \in \mathcal{U}$ is itself in $\mathcal{U}$.

Notes. (1) $\mathcal{U}$ is called an atlas or a differentiable structure in M.

(2) If $\phi_\beta \circ \phi_\alpha^{-1}$ are only C^0 (for all α and β) mappings of $\phi_\alpha(U_\alpha \cap U_\beta)$ onto $\phi_\beta(U_\alpha \cap U_\beta)$, then M is called a topological manifold. Similarly atlases consisting of C^k-compatible coordinate charts can be defined ($k \geq 0$).

(3) (c) means that the atlas $\mathcal{U}$ is maximal (or complete). Any atlas (sub-atlas) $\mathcal{U}$ can be completed into a maximal one by taking the union of all coordinate charts which are C^∞-compatible to those in $\mathcal{U}$. Because of this completion property , to provide a reasonable set of coordinate charts satisfying (a) and (b) will be sufficient.

(4) m is called the dimension of the manifold. A sphere in $\mathbb{R}^3$ is 2-dimensional and in the theory of relativity space-time is taken to be a 4-dimensional differentiable manifold.

SOME MANIFOLD EXAMPLES.

I. Let M be $\mathbb{R}^m$ with a sub-atlas equal to the pair consisting of ϕ: identity and $U = \mathbb{R}^m$.

II. Let M be any open set of $\mathbb{R}^m$ and let ϕ be the identity map, $(U = M, \phi)$ being the sub-atlas.

III. The unit sphere S^2 in $\mathbb{R}^3$ can be covered by the sub-atlas of 6 coordinate charts, the domains U_i of which are $x_1 > 0$, $x_2 > 0$, $x_3 > 0$, $x_1 < 0$, $x_2 < 0$, and $x_3 < 0$ with homeomorphisms ϕ_i given by natural projections on the appropriate coordinate planes. On the intersection of two domains the coordinate systems are easily seen to be C^∞-compatible.

IV. M is $GL(n, \mathbb{R})$, the general linear group, i.e. non-singular linear transformations of $\mathbb{R}^n$ into itself. To establish that this is a manifold covered by a single coordinate chart we consider the representation of $GL(n, \mathbb{R})$ by non-singular $n \times n$ matrices $[a_{ij}]$. Then the mapping

$$\phi : [a_{ij}] \longrightarrow (a_{11}, \cdots, a_{1n}; a_{21}, \cdots, a_{2n}; \cdots; a_{n1}, \cdots, a_{nn})$$

maps $GL(n, \mathbb{R})$ one-to-one and onto an open subset of $\mathbb{R}^{n^2}$. For the image is the complement of $\Delta^{-1}(\{0\})$, where Δ is the (continuous) determinant function from $\mathbb{R}^{n^2}$ into $\mathbb{R}$ and $\{0\}$ is a closed set. Hence ϕ is a homeomorphism and (M, ϕ) is a sub-atlas, (also see the Example II.)

Tangent Vectors and Tangent Spaces

Let M be a C^∞ (or C^r) manifold. Then a real-valued function $f : M \longrightarrow \mathbb{R}$ is smooth (or C^k) if $f \circ \phi^{-1}$ is smooth (or C^k) on $\phi(U)$ for every coordinate chart (U, ϕ). All such functions are denoted $C^\infty(M)$ (or $C^k(M)$).

Similarly $f : M \longrightarrow \mathbb{R}$ is smooth (or C^k) on an open subset $G \subset M$, if $f \circ \phi^{-1}$ is smooth (or C^k) on $\phi(G \cap U)$ for every coordinate chart (U, ϕ).

If, on the other hand N is another smooth (or C^s) manifold, then $f : M \longrightarrow N$ is smooth (or C^k) if it is continuous and for every real-valued function g on N, the composite function $g \circ f : M \longrightarrow \mathbb{R}$ is smooth (or C^k) in the above sense. (Both f and g can be considered as defined on open domains of M and N respectively). Note the independence of positive integers r, k and s.

Remark The last definition also applies to the case where M and N are replaced by topological spaces.

The definition of a tangent vector generalizes the directional derivative in $\mathbb{R}^m$.

If X_q is a vector at a point $q \in \mathbb{R}^m$ and f is a C^∞ function in a neighborhood of q, we define $X_q(f) \doteq X_q.(\nabla f)_q$, where ∇f is the gradient vector field of f (X_q is not necessarily a unit vector like in calculus). Based on the properties of the dot product and the operator ∇ we have

Definition 2 A <u>tangent vector</u> at $q \in M$ is a real-valued function X_q on $C^\infty(q)$ having the following properties :

(1) $X_q(f + q) = X_q(f) + X_q(g)$,
(2) $X_q(\alpha f) = \alpha X_q(f)$, $(\alpha \in \mathbb{R})$,
(3) $X_q(f\,g) = X_q(f)g(q) + f(q)X_q(g)$.

where $C^\infty(q)$ is the set of all real-valued functions that are C^∞ on some neighborhood of q and f, $g \in C^\infty(q)$.

Remark When we identify two functions if they agree on some open set containing q, we have an equivalence relation whose equivalence classes are called the <u>germs</u> of C^∞ functions at q. If we redefine $C^\infty(q)$ as the collection of these germs, it becomes a vector space with the usual vector space operations and also an algebra.

Definition 3 The tangent space to M at q denoted $T_q(M)$ is the set of all mappings $X_q : \mathcal{C}^\infty(q) \longrightarrow \mathbb{R}$ satisfying (1), (2) and (3) of Definition 2.

Notes. (i) $T_q(M)$ is a vector space by the vector space operations defined as
$(X + Y)(f) = X(f) + X(f)$ and $(\alpha X_q)(f) = \alpha X_q(f)$; $(X, Y \in T_q(M))$.
(ii) Let $x_1, \cdots, x_m$ be a local coordinate system about q induced by the chart (U, ϕ) and $q \in U$. For each i, a <u>coordinate vector</u>, denoted $(\frac{\partial}{\partial x_i})_q$ is defined by
$(\frac{\partial}{\partial x_i})_q f = \dfrac{\partial(f \circ \phi^{-1})}{\partial u_i}(\phi(q))$ where differentiation on the right hand side is the usual one in $\mathbb{R}^m$, $(f \in \mathcal{C}^\infty(q)$. The coordinate vectors $e_{iq} \doteq (\frac{\partial}{\partial x_i})_q$, $(i = 1, \cdots, m)$ form a basis for $T_q(M)$, i.e. if $X_q \in T_q(M)$, then

$$X_q = \sum_{i=1}^{m} X_q(x_i)(\frac{\partial}{\partial x_i})_q$$

where $x_i = u_i \circ \phi$ is regarded as an element of $\mathcal{C}^\infty(q)$; thus $T_q(M)$ has dimension m.

(iii) For $q_1 \neq q_2$ T_{q_1} and T_{q_2} are regarded as different finite dimensional vector spaces although of the same dimension ; no isomorphism between them is considered.

Cotangent Vectors, Cotangent Spaces and Differentials

Given a tangent space $T_q(M)$ we can consider its dual, the <u>cotangent space</u> $T^*(M)$, the elements of which are called <u>cotangent vectors</u> (or <u>covectors</u>) . Thus if $\sigma_q \in T_q^*(M)$, it is a linear mapping $\sigma_q : T_q(M) \longrightarrow \mathbb{R}$.

Given the basis $e_{1q}, \cdots, e_{mq}$ of $T_q(M)$, there is a uniquely determined dual basis $w_q^1, \cdots, w_q^m$ satisfying $w_q^i(e_{jq}) = \delta_j^i$ (Kronecker delta, i.e. $\delta_j^i = \begin{cases} 1 & \text{if } i = j \\ 0 & \text{if } i \neq j \end{cases}$).

The components of σ_q relative to this basis are equal to its values on e_{iq}, thus $\sigma_q = \sum_{i=1}^{m} \sigma_q(e_{iq})w_q^i$.

It is customary to denote w_q^i, $(i = 1, \cdots, m)$ by $dx^1, dx^2, \cdots, dx^m$ due to their relations to differentials as we shall see below . Thus the relation defining the dual basis is restated as $dx^i(\frac{\partial}{\partial x_j})_q = \delta_j^i$.

There is a cotangent vector df, associated with each $f \in \mathcal{C}^\infty(q)$, $q \in M$ defined by

$$df(X_q) \doteq X_q(f), \quad X_q \in T_q(M).$$

The linearity of $df : T_q(M) \to \mathbb{R}$ follows from the linearity of the tangent vector X_q.

Denoting $(\frac{\partial}{\partial x_i})_q f = \dfrac{\partial(f \circ \phi^{-1})}{\partial u_i}(\phi(q))$ (see the Note (ii) above) shortly by $(\frac{\partial f}{\partial x_i})_q$, consider the linear combination $\sum_{j=1}^{m}(\frac{\partial f}{\partial x_j})_q dx^j$ which is again a cotangent vector in $T_q^*(M)$. Then we have

$$\langle \sum_{j=1}^{m}(\frac{\partial f}{\partial x_j})_q\, dx^j, (\frac{\partial}{\partial x_i})_q \rangle = \sum_{j=1}^{m}(\frac{\partial f}{\partial x_j})_q < dx^j, (\frac{\partial}{\partial x_i})_q >= \sum_{j=1}^{m}(\frac{\partial f}{\partial x_j})_q\, \delta_j^i = (\frac{\partial f}{\partial x_i})_q$$

$= df(\frac{\partial}{\partial x_i})_q$, showing that :

$$df = \sum_{j=1}^{m} (\frac{\partial f}{\partial x_j})_q dx^j \qquad (*)$$

The expression (*) may be called the <u>differential</u> of f. However it is not the same as the df defined in advanced calculus as a linear transformation. A better name for df in (*) could be the <u>exterior derivative</u> of f.

If $\sigma_q \in T_q^*(M)$, it could be written in terms of the canonical basis as $\sigma_q = \sum_{i=1}^{m} w_q^i dx^i$.

<u>Tangent and Cotangent Bundles</u>

Definition 4 Let M be a $\mathcal{C}^\infty$-manifold of dimension m . The set $T(M) \doteq \bigcup_{q \in M} T_q(M)$ which is partitioned into disjoint subsets $T_q(M)$ indexed by the points of M is called the <u>tangent bundle</u> of M. In fact picking a tangent vector from the tangent bundle $T(M)$ is equivalent to picking a pair (q, X_q) where $q \in M$ and $X_q \in T_q(M)$. The <u>projection mapping</u> $\pi : T(M) \to M$ is defined by $\pi(q, T_q) = q$. If $(q, X_q) \in T(M)$, there exists a coordinate chart (U, ϕ) with $q \in U$ and a coordinate system $(x_1, \cdots, x_m)$. Also $\pi^{-1}(U)$ is the set of all pairs (q, X_q) for which $q \in U$. Let $(q, X_q) \in \pi^{-1}(U)$, then q has coordinates $x_1, \cdots, x_m)$ and X_q is of the form $X_q == \sum_{i=1}^{m} a_i e_{iq} \equiv \sum_{i=1}^{m} a_i (\frac{\partial}{\partial x_i})_q$ in terms of coordinate vectors. The mapping $(q, X_q) \longrightarrow (x_1, \cdots, x_m, a_1, \cdots, a_m)$ is an injection of $\pi^{-1}(U)$ onto an open set of $\mathbb{R}^{2m}$, namely $\phi(U) \times \mathbb{R}^m$. One can take the sets $\pi^{-1}(U)$ as the coordinate charts on $T(M)$. These along with the appropriate one-to-one onto maps $(q, X_q) \to \mathbb{R}^m$ form an atlas of the coordinate charts covering $T(M)$. The $\mathcal{C}^\infty$-compatibility is proved without much difficulty. Thus $T(M)$ is differentiable manifold of dimension $2m$. (Roughly speaking we need m dimensions for the position of q in M and another m coordinates for specifying X_q in $T_q(M)$).

Similarly the union of all cotangent spaces $T_q^*(M)$ is called the <u>cotangent bundle</u> of M and denoted by $T^*(M)$, An element of $T^*(M)$, i.e. (q, X_q^*), can be assigned local coordinates $(x_1, \cdots, x_m)$ which specify the point $q \in M$ at which the covector is taken and $(y_1, \cdots, y_m)$ that are the components of the covector with respect to the dual basis $(dx^1, \cdots, dx^m)$. $T^*(M)$ is thus a differential manifold of dimension $2m$.

<u>Vector Fields on a Manifold</u>

Let M be a manifold of dimension m. At each point q there is a tangent space $T_q(M)$ and its dual space $T_q^*(M)$, both also of dimension m.

Definition 5 A <u>vector field</u> X on a subset A of M is an assignment to each point $q \in A$ of an element $X_q \in T_q(M)$, i.e. $X : M \longrightarrow T(M)$ satisfying $\pi \circ X = i_M$ (the identity on M). (Similar definition for a covector field $X^* : M \longrightarrow T(M)$).

Let $f \in \mathcal{C}^\infty(A)$ and G be an open set in M. Then we say a vector field X is $\mathcal{C}^k$-differentiable on G , if the function Xf defined by $(Xf)(q) \doteq X_q(f)$ is $\mathcal{C}^k$-differentiable on $A \cap G$. If $k = \infty$, we say the vector field X is $\mathcal{C}^\infty$-smooth on G.

X maps the C^∞ function f to the function Xf given above and X is a linear mapping of C^∞ functions also satisfying $X(fg) = (Xf)g + f(Xg)$.

If we have a covector field, i.e. a mapping $q \longrightarrow T_q^*(M)$, then it is called a differential 1-form, the notation would be $\sum_{i=1}^{m} w_q^i dx^i$.

Generalization of tangent and cotangent bundles is the concept of vector bundles.

Vector Bundles (Fiber Bundles)
This is roughly speaking a topological space (often a manifold) with a topological space (often a vector space) attached to each point.

Definition 6 A vector bundle is a triplet (E, B, π) where E and B are topological spaces and the mapping $\pi : E \longrightarrow B$, the projection mapping is onto and continuous.

E is called the total space and B the base space. For each $q \in B$, the set of points $\pi^{-1}(q)$ in E is called the fiber over q.

Frequently we refer to E as the vector bundle over B (or parametrized by B).

EXAMPLES:
(1) The tangent bundle of a manifold M : $B = M$, $E = T(M), \pi : T(M) \to M$, $\pi(q, X_q) = q$.
(2) Circular cylinder C^2 : $B = S^1$, $E = S^1 \times \mathbb{R}$, $\pi S^1 \times \mathbb{R} \to S^1$. Then $\mathbb{R}$ is the fiber over every $q \in S^1$.
(3) Generalization of (2) is the cartesian product bundle $(B \times F, B, \pi)$ where B and F are topological spcaes and $\pi : B \times F \longrightarrow B$ is defined by $\pi(q_{1,2}) = q_1$ for all $q_1 \in B$ and $q_2 \in F$.

A cartesian product bundle is said to be trivial if $B = M$ and $E = M \times V$ where V is a vector space . Then at every point of the manifold a fixed vector space V is attached defining the trivial vector bundle with the projection mapping as given in (3).

Vector bundles are almost every time required to be locally trivial fiber bundles. That means for every point of B there is an open neighborhood U, a natural number k and a homemorphism $\psi : U \times \mathbb{R}^k \longrightarrow \pi^{-1}(U)$ such that for all $q \in U$:

- $(\pi \circ \psi)(q, v) = q$ for all vectors $v \in \mathbb{R}^k$,
- the map $v \longrightarrow \psi(q, v)$ is an isomorphism between the vector spaces $\mathbb{R}^k$ and $\pi^{-1}(\{q\})$.

The local trivialization shows that locally the map π looks like the projection of $U \times \mathbb{R}^k$ on U.

Every fiber $\pi^{-1}(\{q\}$ is a finite dimensional vector space and hence has a dimension k_q. The local trivialization shows that the function $q \to k_q$ is locally constant. If k_q is equal to a constant k on all of B, then k is called the rank of the vector bundle and E is said to be a (trivial) vector bundle of rank k.

Vector bundles of rank 1 are called trivial line bundles and bundles of rank 2 plane bundles.

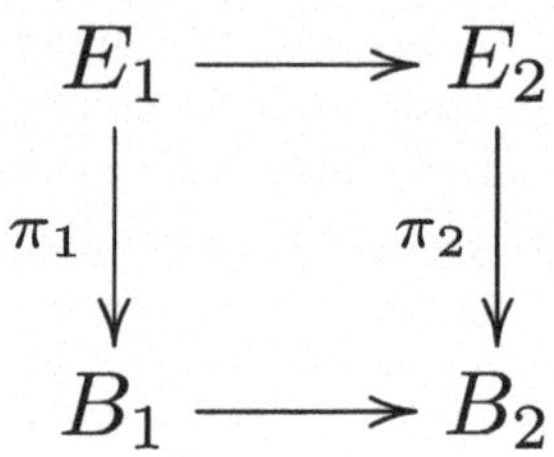

Fig. 2.2 Morphism of two continuously related vector bundles E_1 and E_2

A morphism from the vector bundle E_1 over B_1 to the vector bundle E_2 over B_2 with projection mappings π_1 and π_2 respectively, is given by a pair of continuous maps $f : E_1 \longrightarrow E_2$ and $g : B_1 \longrightarrow B_2$ satisfying $g \circ \pi_1 = \pi_2 \circ F$, i.e. making the above diagram commutative:

For every $q \in B_1$, the map $\pi^{-1}(\{q\}) \longrightarrow \pi_2^{-1}(\{g(q)\})$ induced by f is a linear map between vector spaces. We note that g is determined by f since π_1 is onto and f is then said to cover g (Fig. 2.2).

From the standpoint of global analysis the class of vector bundles together with bundle morphisms form a category.

Sections

Definition 7 A smooth section of the bundle (E, B, π) is a smooth mapping $s : B \longrightarrow E$ for which $\pi \circ s$ is the identity mapping.

The definition means that to each $q \in B$ there is assigned an element of $\pi^{-1}(q)$. A vector field on a manifold can therefore be defined as a section of the tangent bundle $T(M)$.

Note. Similarly C^k-sections of bundles can be defined. For the definition of smooth and C^k mappings $s : B \to E$ see the Remark under the heading "tangent vectors and tangent spaces".

Specializing to manifolds, i.e. when $B = M$ is a smooth manifold, the class of all smooth sections will be denoted by $\mathcal{E}(M; E)$.

Assume that $\mathcal{E}(M; E)$ are smooth sections of a locally trivial fiber bundle $E \to M$. Then $\mathcal{E}(M; E)$ is a vector space under pointwise addition and scalar multiplication. Its neutral element is the zero section which maps every $q \in M$ to the zero element of the vector space $\pi^{-1}(\{q\})$. It is also a module over $C^\infty(M)$, (with respect to pointwise multiplication).

We can also denote functions anf smooth sections of compact support by C_c^∞ and $\mathcal{E}_c(M; E)$.

Global Theory of Generalized Functions

The locally convex spaces $\mathcal{E}(\Omega)$, $\mathcal{D}'(\Omega)$ and $\mathcal{E}'(\Omega)$ can be extended from opens $\Omega \subset \mathbb{R}^m$ to arbitrary manifolds M so as we can talk about distributions or generalized functions on M.

In fact more generally we can do this extension to smooth sections on manifolds rather than smooth functions on them. In this way we obtain distributional sections of E or generalized sections on E. (when E is the trivial bundle rank 1, we retrieve the case of smooth functions on manifolds). For this purpose we fix M to be an m-dimensional manifold and let E be a vector bundle over M of rank k. We can introduce a locally convex topology in $\mathcal{E}(M; E)$ as follows : choose a covering $\mathcal{U} = \{U_\alpha\}$ $(\alpha \in \Gamma)$ of M by opens which are the domains of coordinate charts (U_α, ϕ_α) and also of local trivializations $\psi_\alpha : U_\alpha \times \mathbb{R}^k \longrightarrow \pi^{-1}(U_\alpha)$. Thus smooth sections restricted to U_α can be represented by the cartesian products of k copies of $\mathcal{C}^\infty(\phi_\alpha(U_\alpha))$, i.e. there is an isomorphism $\lambda_\alpha : \mathcal{E}(M; E)|_{U_\alpha} \longrightarrow [\mathcal{C}^\infty(\phi_\alpha(U_\alpha))]^k$. Considering all such restrictions we define an injection

$$\gamma : \mathcal{E}(M; E) \longrightarrow \prod_\alpha [\mathcal{C}^\infty(\phi_\alpha(U_\alpha))]^k \equiv \prod_\alpha [\mathcal{E}(\phi_\alpha(U_\alpha))]^k.$$

Endowing the right-hand side with the product topology , the topology of $\mathcal{E}(M; E)$ is the induced topology under the above injection. This topology can also be described in terms of seminorms : For a smooth function h consider the seminorms $\|h\|_{N,K} = \sup\{(D^\alpha h)(x) : x \in K, \ |\alpha| \leq N\}$ (K a compact subset) which are in fact equivalent to the norms $p_N(h)$.

We note that the covering can be chosen to be countable , (i.e. Γ a countable index set) since manifolds satisfy the second countability axiom. For $K \subset \phi_\alpha(U_\alpha)$ compact and N a non-negative integer and $s \in \mathcal{E}(M; E)$ we restrict it U_α and map by the isomorphism γ_α into $[\mathcal{E}(\phi_\alpha(U_\alpha))]^k$, take its l-th component $(1 \leq l \leq k)$, and apply the seminorm $\|.,.\|_{K,N}$ of $\mathcal{E}(\phi_\alpha(U_\alpha))$ to find $\|s\|_{\alpha,l,K,N} = \|\{\gamma_\alpha(s|U_\alpha)\}_l\|_{N,K}$. Putting together all such seminorms will define the topology on $\mathcal{E}(M; E)$. This topology does not depend on the choices involved and it makes $\mathcal{E}(M; E)$ a Fréchet space. We can introduce bundle charts by the mappings

$$\Phi : \pi^{-1}(U) \longrightarrow \phi(U) \times \mathbb{R}^k$$

$$x \longrightarrow (\phi(\pi(x)), \ \psi^{-1}(x))$$

ψ : local trivializations. Then the local expression s_α of a smooth section $s \in \mathcal{E}(M; E)$ relative to a bundle chart (U_α, Φ_α) given by $s_\alpha \doteq \Phi_\alpha \circ s \circ \phi_\alpha^{-1}$ will be a mapping of the form

$$\phi(U_\alpha) \longrightarrow \phi(U_\alpha) \times \mathbb{R}^k$$

$$x \longrightarrow (x, \ s_{\alpha,1}, \ s_{\alpha,2}, \cdots, s_{\alpha,k})$$

where $s_{\alpha,j}$'s $(j = 1, \cdots, k)$ are the local components of s.

Back to the topology of $\mathcal{E}(M; E)$, a sequence of sections $\{s_n\}_{n \geq 1}$ converges to s in this topology if and only if, for any coordinate chart (U, ϕ) in M and for any compact $K \subset U$, writing $s_n = (s_{\alpha,1}^n, s_{\alpha,2}^n, \cdots, s_{\alpha,k}^n)$ and $s = (s_{\alpha,1}, s_{\alpha,2}, \cdots, s_{\alpha,k})$, all the derivatives $D^\beta(s_{\alpha,j}^n)$ converge uniformly on K to $D^\beta(s_{\alpha,j})$ as $n \to \infty$. (β multi-indices).

Fig. 2.3 Commutativity of imbedding of a compactly supported section into generalized section in two different ways

$$\begin{array}{ccc} \mathcal{D}(M;E) & \longrightarrow & \mathcal{E}(M;E) \\ \downarrow & & \downarrow \\ \mathcal{E}'(M;E) & \longrightarrow & \mathcal{D}'(M;E) \end{array}$$

When E is the trivial line bundle over M, the notation is simplified to $\mathcal{E}(M)$. As in the local case, this is an algebra, also the multiplication of sections by functions makes $\mathcal{E}(M; E)$ into a module over $\mathcal{E}(M)$.

Compactly Supported Smooth Sections

We define $\mathcal{D}(M; E) \doteq \mathcal{E}_c(M; E)$, the space of all compactly supported smooth sections. This space is furnished with a locally convex topology exactly the same way as in the local case, i.e. $\mathcal{D}(M; E) = \bigcup_K \mathcal{E}_K(M; E)$. The union is over all compacts $K \subset M$ and $\mathcal{E}_K(M; E) \subset \mathcal{E}(M; E)$ is the space of smooth sections supported in K endowed with the topology induced from $\mathcal{E}(M; E)$. On $\mathcal{D}(M; E)$ then we can consider the inductive limit topology as in the local case , (c.f. 1.2). Again in the case of trivial line bundle over M, the notation is simplified to $\mathcal{D}(M)$.

Generalized Sections $\mathcal{D}'(M; E)$

This is the space of distributional sections of E, i.e. the space of generalized sections of E. In order to define it we should normally consider the dual of $\mathcal{D}(M; E)$ as in the local case. But in order to secure the canonical imbedding (in parallel to the local case) $\mathcal{E}(M; E) \hookrightarrow \mathcal{D}'(M; E)$ and to make $\mathcal{D}'(M; E)$ a module over $\mathcal{E}(M)$ with continuous multiplication $\mathcal{E}(M) \times \mathcal{D}'(M; E) \longrightarrow \mathcal{D}'(M; E)$ (the product being defined by : $(fu)(s) = u(fs)$, $f \in \mathcal{E}(M)$, $u \in \mathcal{D}'(M; E)$) we rather take the dual of $\mathcal{D}(M; E^v)$, where E^v is defined as $E^v \doteq E^* \otimes D = Hom(E, D)$,(the bundle whose fiber at $x \in M$ is the vector space consisting all linear maps $E_x \to D_x$; D_x's being the section of the (density) bundle D). (For the details see [Ban-Cr]).

Compactly Supported Generalized Sections $\mathcal{E}'(M; E)$

The space of compactly supported distributional sections of E is defined as in the local case (but making the use of again E^v) as $\mathcal{E}'(M; E) \doteq (\mathcal{E}(M; E^v))^*$ by the same pairing as before , one obtains an inclusion

$$\mathcal{D}(M; E) \hookrightarrow \mathcal{E}'(M; E).$$

Hence as in the local case, we obtain the following commutative diagram of inclusions (Fig. 2.3).

EXAMPLE.

(1) When E is the trivial line bundle over M, the notation will be simplified to $\mathcal{D}(M)$, $\mathcal{E}(M)$. Hence as vector spaces $\mathcal{D}(M) = \mathcal{E}_c(M)$, $\mathcal{E}(M) = C^\infty(M)$, then the elements of $\mathcal{D}'(M)$ will be distributions on M.

(2) If E is the trivial bundle over M of rank k , then $\mathcal{D}(M; \mathbb{R}^k) = [\mathcal{D}(M)]^k$, $\mathcal{E}(M; M \times \mathbb{R}^k)$
$= [\mathcal{E}(M)]^k$. On the other hand $\mathcal{D}'(M; \mathbb{R}^k) = [\mathcal{D}'(M)]^k$, $\mathcal{E}'(M; \mathbb{R}^k) = [\mathcal{E}'(M)]^k$.

As in the local theory, distribution $T \in \mathcal{D}'(M; E)$ can be restricted to arbitrary opens $U \subset M$, to yield distributions $T|_U \in \mathcal{D}'(U; E|_U)$. More precisely the restricted map

$$\mathcal{D}'(M; E) \longrightarrow \mathcal{D}'(U; E|_U)$$

is defined as the dual of the map

$$\mathcal{D}'(U; E^v|U) \to \mathcal{D}(M; E^v)$$

which takes a compactly supported section defined on U and extends it by zero outside.

Chapter 3
Non-linear Distributions

3.1 Introduction

The theory of linear generalized functions has as the main priority of indefinite differentiability in analysis. The objects in Schwartz distribution space $\mathcal{D}'$ are infinitely differentiable. In fact in $\mathbb{R}^n$ there is a hierarchy of spaces given by the chain of inclusions

$$\mathcal{A}_n \subset \mathcal{C}^\infty \subset \cdots \mathcal{C}^p \subset \mathcal{C}^0 \subset L^1_{loc} \subset \mathcal{D}' \quad (p \in \mathbb{N}) \quad (*)$$

($\mathcal{A}_n$: the space of analytic functions).

It was stated that (c.f. 1.3. Proposition 1) $\mathcal{D}'$ is -as a matter of fact- the minimal extension of $\mathcal{C}^0$, the space of continuous functions, in the sense that locally every distribution is a weak partial derivative of some continuous function.

An important result which alone would fully justify the use of Schwartz distributions is the Malgrange-Ehrenpreis theorem (2.2, Theorem 1) discussed in Sect. 2.2. According to this result every linear partial differential operator with constant coefficients has a fundamental solution, i.e. if $L = \sum_{|\alpha| \leq p} a_\alpha D^\alpha \equiv P(D)$, (for the notation c.f. 2.2) there exists $E \in \mathcal{D}'(\mathbb{R}^n)$ such that $L\,E = \delta$ and $u = E * g$ is a solution of the partial differential equation $Lu = g$ of order p with constant coefficients under some assumptions on g. Depending on these assumptions the solution is a classical solution or a distributional (weak) solution. More precisely it is a classical (or strict) solution if $u \in \mathcal{C}^p$, otherwise it is distribution solution if $u \in \mathcal{D}' \setminus \mathcal{C}^p$. If the partial differential equation is accompanied by boundary (or initial) conditions and these are singular, e.g. concentrated mass or charge expressed in terms of delta functions, then the solution is also expected to be at least as singular as boundary conditions, i.e. a distribution. Furthermore if g is a given (non-regular) distribution, then $Lu = g$ can be considered as an equation for an unknown distribution u. Then a solution u in distributional sense will satisfy

$$< Lu, \phi > = < g, \phi > \quad \text{for all} \ \ \phi \in \mathcal{D}(\mathbb{R}^n) \quad (\dagger)$$

© The Author(s), under exclusive license to Springer Nature Switzerland AG 2026 97
U. Çapar, *A Guide to Generalized Functions*,
https://doi.org/10.1007/978-3-032-09184-0_3

The formal adjoint L^* of the operator L will be given by $L^*u = \sum_{|\alpha| \leq p}(-1)^{|\alpha|}a_\alpha D^\alpha$ ($\bar{a}_\alpha$ if the coefficients are complex). Then u is a distributional solution if and only if

$$< u, L^*\phi >=< g, \phi > \quad \forall \phi \in \mathcal{D}(\mathbb{R}^n) \qquad (\ddagger)$$

i.e. for each test function ϕ, the action of u on the test function $L^*\phi$ should be the same as the action of g on ϕ.

The definition of a generalized solution is in fact global in character. We may also define the notion of a local distributional solution, i.e. in an open region $\Omega \subset \mathbb{R}^n$. Then (†) or (‡) should hold for every test function $\phi \in \mathcal{D}(\Omega)$.

Encouraged by the success of the Malgrange-Ehrenpreis result, researchers started seeking the conditions to be imposed on g and the coefficients of the differential operator so that a parallel result for the existence of fundamental solution in the case of variable coefficients could be established. However these attempts were disappointed when Lewy produced a very simple example of a linear variable coefficient partial differential equation in $\mathbb{R}^3$ which failed to have a distributional solution in any neighborhood of x in $\mathbb{R}^3$, (see 3.3, A)). Thus the linear theory of distributions was insufficient for tackling even the linear partial differential equations. The fact that it will be inadequate a fortiori for the study of non-linear phenomena, e.g. for the solution of non-linear partial differential equations is illustrated by the following simple example of an initial value shock wave equation:

$$u_t + u_x u = 0 , \quad t > 0, x \in \mathbb{R} \quad (1)$$

$$u(0, x) = h(x) , \quad x \in \mathbb{R} \quad (2)$$

(1) is first order non-linear p.d.e. with a quadratic non-linearity.

$$\text{Suppose } h(x) = \begin{cases} 1 & \text{if } x \leq 0 \\ 1 - x & \text{if } 0 \leq x \leq 1 \\ 0 & \text{if } x \geq 1 \end{cases}.$$

In this case the shock front is

$$\Gamma = \{(t, x) : t \geq 0, \ x \in \mathbb{R}, \ t \geq 1, x = \frac{t + 1}{2}\}.$$

For $0 \leq t \leq 1$ we have the classical solution

$$u(t, x) = \begin{cases} 1 & \text{if } x \leq 0 \\ \frac{x-1}{t-1} & \text{if } 0 \leq x \leq 1 \\ 0 & \text{if } x \geq 1 \end{cases}$$

while for $t \geq 1$ we have a generalized solution

$$u(t, x) = \begin{cases} 1 & \text{if } x < \frac{t+1}{2} \\ 0 & \text{if } x > \frac{t+1}{2} \end{cases} \qquad (3)$$

$u(t, x) \in \Gamma$ is defined at will.

The generalized solution (3) has a jump discontinuity across the shock front Γ. This discontinuity is of the Heaviside function type at $\xi = 0$:

$$H(\xi) = \begin{cases} 0 & \text{if } \xi > 0 \\ 1 & \text{if } \xi < 0 \end{cases}$$

Then the partial derivative u_x in (1) will have across Γ a singularity of the type the Dirac δ_ξ distribution has at $\xi = 0$. Hence the product $u_x\, u$ can not be dealt within the theory of Schwartz distributions since both of the factors have singularities at the same point $\xi = 0$.

As a matter of fact these inadequacies of linear distributions to tackle non-linear phenomena stem from the fact that in $\mathcal{D}'$ there is not a general suitable product rule as discussed in next sections.

A reasonable way to circumvent this difficulty could be to imbed $\mathcal{D}'$ into a differential algebra. However this also has limitations: we should naturally require that the operations in $\mathcal{D}'$ should be preserved in this larger setting which we would prefer to be also an associative algebra. But this may lead to absurd results; e.g. consider the distribution $v.p.\frac{1}{x}$ (c.f. 1.2, Example 4)). We have $(v.p.\frac{1}{x})\,x = 1$. Multiplying both sides by the Dirac delta distribution: $\delta = [(v.p.\frac{1}{x})x]\,\delta = (v.p.\frac{1}{x})\,(x\delta) = (v.p.\frac{1}{x})\,0 = 0$?

In fact the celebrated Schwartz Impossibility Result (see the next section and [Schw-2]) roughly states that under very mild conditions the imbedding of $\mathcal{D}'$ into an associative differential algebra has to give some concessions regarding the natural properties of derivatives and product rules. For instance such an algebra can not extend the pointwise multiplication of continuous functions and it also can not possess an element δ satisfying $x\,\delta = 0$. But $\mathcal{D}'$ does possess such an element δ, namely the Dirac delta function itself.

Nevertheless the consequences of the Schwartz Impossibility Result should not be misinterpreted as such an imbedding is impossible at all. All it states that such an imbedding should give up some properties of certain operations. For instance $x\,\delta$ would no more be zero (δ: Dirac). Among different theories of non-linear distributions, e.g. Colombeau, Egorov, Rosinger etc., Colombeau's differential algebra is regarded warding off the consequences of the Schwartz Impossibility Result in an optimal manner, i.e. many concessions given are minimal in a certain sense.

3.2 Schwartz Impossibility Result

In its original one dimensional form the impossibility result is given as:

Theorem *Suppose $\mathcal{G}$ is an associative algebra with a rule of derivative $D : \mathcal{G} \longrightarrow \mathcal{G}$, i.e. a linear mapping which satisfies the Leibniz rule of the product derivative*

$$D(f \cdot g) = (Df)\, g + f\, (Dg), \quad f, g \in \mathcal{G}$$

Suppose further that

(1) 1, x, $x(\ln |x| - 1)$ and $x^2(\ln |x| - 1)$ are C^0 functions included in $\mathcal{G}$,
(2) Function 1 is the unit element,
(3) The multiplication in $\mathcal{G}$ is such that following commutativity holds:
    $(x\,(\ln |x| - 1))(x) = x^2(\ln |x| - 1)$,
(4) The derivative D applied to 1, x, $x^2(\ln |x| - 1)$ is the usual derivative on $C^1(\mathbb{R})$,

Then there exists no $\delta \in \mathcal{G}$, $\delta \neq 0$ such that $x\delta = 0$.

Note. In (1) last two functions have by definition the value zero at $x = 0$.

Proof Applying (4) to the third function in (1) which is in $\mathcal{G}$ we have

$$D^2(x(\ln |x| - 1)) = \frac{1}{x} \in \mathcal{G}$$

We first show that $\frac{1}{x}$ can be taken as the left inverse x^{-1} for the function x in the algebra $\mathcal{G}$.

By (1) x and $x(\ln |x| - 1) \in \mathcal{G}$. Apply the Leibniz rule to twice differentiation of their product:

$$D^2\left((x(\ln |x| - 1)).(x)\right) = \left(D^2(x(\ln |x| - 1))\right).(x) + (x(\ln |x| - 1)).(D^2 x) + 2(D(x(\ln |x| - 1))).(Dx)$$

using (3)

$$D^2\left(x^2(\ln |x| - 1)\right) = (\frac{1}{x})(x) + 2(D(x(\ln |x| - 1))) \qquad (*)$$

The left hand side is calculated as

$$D[D(x^2(\ln |x| - 1))] = D[2x(\ln |x| - 1) + x] = 2(D(x(\ln |x| - 1))) + 1$$

substituting in (*) we find $(\frac{1}{x})(x) = 1$ showing that $\frac{1}{x}$ is the left inverse of function x denoted by x^{-1} since by (2), 1 is the unit element of the algebra $\mathcal{G}$. Now assuming that there is an element $\delta \in \mathcal{G}$, $\delta \neq 0$ and $x\,\delta = 0$ we obtain the following contradiction:

$$0 = (x^{-1})\,((x).\delta) = ((x^{-1}).(x)).\delta = 1.\delta = \delta. \quad \square$$

The following Corollary is important from the point of non-linear theory of distributions:

Corollary *Suppose in addition to the assumptions of Theorem 1 we have*

(i) $\mathcal{D}' \subset \mathcal{G}$,
(ii) The derivative $D : \mathcal{G} \to \mathcal{G}$ *coincides on* $\mathcal{D}'(\mathbb{R})$ *with the distributional derivative,*

then the multiplication rule in $\mathcal{G}$ *restricted to* $C^0(\mathbb{R})$ *does not coincide with the usual multiplication of continuous functions, i.e. the imbedding in i) can not extend the pointwise multiplication of continuous functions.*

Proof We have in $\mathcal{G}$: $D(x|x|) = |x| + x\,D(|x|)$. Therefore $D^2(x|x|) = 2D(|x|) + x.D^2(|x|)$.

$$\text{If } x_+ = \begin{cases} x & \text{if } x > 0 \\ 0 & \text{if } x \leq 0 \end{cases}$$

then in classical or distributional sense $Dx_+ = H(x)$ (Heaviside function) and by 1.3, Example (i) we have in distributional sense $DH(x) = \delta(x)$. As $|x| = 2x_+ - x$, by the assumption ii) $D(|x|) = 2H(x) - 1$, $D^2(|x|) = 2\delta(x)$. Hence in $\mathcal{G}$

$$D^2(x.|x|) = 2D(|x|) + 2x\delta(x) \quad (\checkmark)$$

Suppose now that the product of continuous functions x and $|x|$ is the same in $\mathcal{G}$, but since $x|x| \in \mathcal{D}'$, the derivative on the left can be carried out in distributional sense

$$D(x.|x|) = |x| + xD(|x|) = |x| + x(2H(x) - 1) = 2|x|;\ \ D^2(x|x|)) = 2D(|x|),$$

substitution in $(\checkmark)$ yields $x\delta(x) = 0$ violates the conclusion of Theorem. $\quad \square$

Note. As it has been emphasized before, the Impossibility Result does not assert that the differential algebra $\mathcal{G}$ can not contain the Dirac δ distribution, nor does it maintain that no differential algebra can be constructed to include $\mathcal{D}'$ as a vector subspace. It says that in any such differential algebra into which $\mathcal{D}'$ is imbedded, the Dirac delta will no longer satisfy $x\,\delta = 0$, or in other words it will have a singularity at $x = 0$ which is of higher order than the function $\frac{1}{x}$.

3.3   Other Limitations and Pathologies of the Linear Distribution Theory

(A) The Lewy Inexistence Result ([Lewy]).
Once the Malgrange-Ehrenpreis statement that every linear partial differential operator with constant coefficients possesses a fundamental solution was established, researchers embarked on the investigation of the conditions under which the Malgrange- Ehrenpreis theorem could be extended to the p.d.e. with variable coefficients. But these hopes ended when Lewy produced his disappointing example which was a very simple first order linear p.d.e. with variable coefficients in $\mathbb{R}^3$, yet failed to have a distributional solution even locally. Lewy's counter-example was the following

$$- i\frac{\partial u}{\partial x_1} + \frac{\partial u}{\partial x_2} - 2(x_1 + ix_2)\frac{\partial u}{\partial x_3} = g(x), \quad x = (x_1, x_2, x_3) \in \mathbb{R}^3. \quad (1)$$

A distribution solution did not exist for a large class of smooth functions $g \in C^\infty(\mathbb{R}^3)$. Furthermore (1) is not an artificial type counter-example, but appears naturally in the theory of complex functions of several variables.

Lewy's example shows that the linear theory of distributions is not adequate even for the study of linear p.d.e., i.e. the solutions should be sought in spaces larger than $\mathcal{D}'$.

(1) is a special case of an m-th order linear p.d.e. with variable coefficients defined in an open domain $\Omega \subset \mathbb{R}^n$. Corresponding differential operator is

$$P(x, D) = \sum_{\alpha \in \mathbb{N}^n} c_\alpha(x)D^\alpha \, ; |\alpha| \leq m, \, x \in \Omega \quad (2)$$

with the coefficients $c_\alpha \in C^\infty(\Omega)$.

The principal part of $P(x, D)$ is by definition $P_m(x, \xi) = \sum_{\alpha \in \mathbb{N}^n, |\alpha|=m} c_\alpha(x)\xi^\alpha$, $|\alpha| = m$, $x \in \Omega$, $\xi \in \mathbb{R}^n$ and its complex conjugate is $\bar{P}(x, \xi) = \sum_{\alpha \in \mathbb{N}^n, |\alpha|=m} \overline{c_\alpha(x)}\xi^\alpha$, $x \in \Omega, \xi \in \mathbb{R}^n$.
Furthermore the commutator of $P(x, D)$ is defined by

$$C_{2m-1}(x, \xi) = i \sum_{1 \leq j \leq n} \left(\frac{\partial P_m(x, \xi)}{\partial \xi_j} \frac{\partial \bar{P}_m(x, \xi)}{\partial x_j} - \frac{\partial P_m(x, \xi)}{\partial x_j} \frac{\partial \bar{P}_m(x, \xi)}{\partial \xi_j} \right), \quad x \in \Omega, \, \xi \in \mathbb{R}^n.$$

A basic necessary condition for solvability is the following

Theorem 1 ([Hörm-2]) *Suppose the linear p.d.e.* $P(x, D)u = g$, $x \in \Omega$ *has a solution* $u \in \mathcal{D}'(\Omega)$ *for every* $\in \mathcal{D}(\Omega)$, *then for every* $x \in \Omega$, $\xi \in \mathbb{R}^n$

$$P_m(x, \xi) = 0 \Longrightarrow C_{2m-1}(x, \xi) = 0.$$

Application to Lewy's example:

$n = 3$, $\Omega = \mathbb{R}^3$, $m = 1$

$P_1(x, \xi) = -i\xi_1 + \xi_2 - 2(x_1 + ix_2)\xi_3$

$\bar{P}_1(x, \xi) = i\xi_1 + \xi_2 - 2(x_1 - ix_2)\xi_3$

$C_1(x, \xi) = i[-i(-2|xi_3) - 2\xi_3 i + 2i\xi_3 - (-2i\xi_3) + 0] = i4i\xi = -4\xi$

But for $x \in \mathbb{R}^3$ and $\xi_1 = -2x_2$, $\xi_2 = 2x_1$ and $\xi_3 = 1$ we have $P_1(x, \xi) = 0$ and $C_1(x, \xi) \neq 0$ violating the necessary condition.

(B) Stability Paradoxes.

There are other disappointments in the linear theory except the Schwartz's Impossibility Result. The so-called "stability paradoxes" point out an anomaly of the linear theory of distributions. The systems of equations (i) and (ii) given by

$$(i) \quad \begin{cases} u = 0 \\ u^2 = 1 \end{cases}$$

$$(ii) \quad \begin{cases} u = 0 \\ u^2 = \delta \end{cases}$$

have weak solutions !

$u_n(x) = \sqrt{2}\cos nx$ converges in distribution to a weak solution $u(x)$ of the first system since

$$\int_{-\infty}^{\infty} \sqrt{2}\cos nx \, \phi(x)\, dx = \frac{\sqrt{2}\sin nx}{n}\phi(x)|_{-\infty}^{\infty} - \frac{\sqrt{2}}{n}\int_{-\infty}^{\infty} \sin nx \, \phi'(x)\, dx \to 0$$

and $\int_{-\infty}^{\infty} 2\cos^2 nx \, \phi(x)\, dx \to \int_{-\infty}^{\infty} \phi(x)\, dx =< 1, \phi >$ as $n \to \infty$.

For a weak solution of the second system consider the sequence $w_n = \sqrt{n}\alpha(nx)$ where $\alpha \in \mathcal{D}(\mathbb{R})$, $\alpha \geq 0$ and $\int_{\mathbb{R}} \alpha^2 \, dx = 1$.

Then for any test function ϕ, $< w_n, \phi >= \sqrt{n}\int_{-\infty}^{\infty} \alpha(nx)\phi(x)\, dx = \frac{1}{\sqrt{n}}\int_{-\infty}^{\infty} \alpha(t)\phi(\frac{t}{n})\, dt \longrightarrow 0$ and $< w_n^2, \phi >= n\int_{-\infty}^{\infty} \alpha^2(nx)\phi(x)\, dx = \int_{-\infty}^{\infty} \alpha^2(t)\phi(\frac{t}{n})\, dt \longrightarrow \phi(0) =< \delta, \phi >$ as $n \to \infty$.

The weak convergence is also convergence in the strong topology in $\mathcal{D}'(\mathbb{R})$ (c.f. 1.2, Theorem 1). Thus the limits $u(x) = \lim u_n(x)$ and $w(x) = \lim w_n(x)$ constitute solutions to the systems (i) and (ii).

3.4   Colombeau's Non-linear Distributions

3.4.1   Sequential Approach

Referring back to the hierarchy of spaces in (*) of 3.1:

Except for L^1_{loc} and $\mathcal{D}'$, the others are suitable for non-linear operations, in particular unrestricted multiplication, they are associative and commutative algebras with unit elements.

From the point of partial differentiability only $\mathcal{A}_n$, $\mathcal{C}^\infty$ and $\mathcal{D}'$ allow indefinite iterations of such operations. Unfortunately however they are not sufficiently large to study non-linear phenomena, in particular non-linear p.d.e.'s. The limitations and inadequacies of the larger space $\mathcal{D}'$ were analyzed in previous sections. We need a space of generalized functions $\mathcal{G}$ larger than the space $\mathcal{D}'$ of the Schwartz distributions with the following properties:

(a) $\mathcal{G}$ must be an associative and commutative differential algebra with a derivative operator $D^\alpha : \mathcal{G} \to \mathcal{G}$.
(b) $D^\alpha|_{\mathcal{D}'}$ is the usual distributional derivative.
(c)   The restriction of the multiplication to $\mathcal{C}^\infty$ must coincide with the usual pointwise product on $\mathcal{C}^\infty(\mathbb{R}^n)$.

Mikusinski related $\mathcal{D}'$ to the uniform limits of smooth regularizations ([Miku-1]) and introduced the concept of regular operations which commute with the limits of the regularizations.

According to Theorem 6, Sect. 1.7 on regularizations: for each test function $\rho \in \mathbb{R}^n$, $\int_{\mathbb{R}^n} \rho \, dx = 1$ and $\epsilon > 0$ $\rho_\epsilon \longrightarrow \delta$ (Dirac measure) where $\rho_\epsilon(x) = \frac{1}{\epsilon^n}\rho(\frac{x}{\epsilon})$. By 1.7, Theorem 7-b) for $T \in \mathcal{D}'$ and each regularized sequence (a mollifying sequence) $\{\rho_\epsilon\}$, we have $f_\epsilon = T * \rho_\epsilon \longrightarrow T$ in $\mathcal{D}'$. Furthermore according to 1.7, Theorem 2-c) $f_\epsilon \in \mathcal{C}^\infty(\mathbb{R}^n)$, (a smooth function). Hence $\{f_\epsilon : 0 < \epsilon < \infty\} \in (\mathcal{C}^\infty(\mathbb{R}^n))^{(0,\infty)}$ (the space of all sequences of smooth functions). For a fixed distribution $T \in \mathcal{D}'(\mathbb{R}^n)$ we can construct different mollifying sequences, say $\{\psi_\epsilon\}$ ($\psi \in \mathcal{D}(\mathbb{R}^n)$). Consequently $T \to \{f_\epsilon\} \in (\mathcal{C}^\infty(\mathbb{R}^n))^{(0,\infty)}$ is a multi-valent association. Because of this multivalent association we may define an improper inclusion $\mathcal{D}'(\mathbb{R}^n)$ $\cdot \subset \cdot$ $(\mathcal{C}^\infty(\mathbb{R}^n)^{(0,\infty)}$.

It should be noted that $(\mathcal{C}^\infty)^{(0,\infty)}$ is a differential algebra with the term-wise operations and comes close to the desired extension of the space of linear distributions $\mathcal{D}'$. All we have to do is to remove the quotation mark $\cdot \subset \cdot$ on the inclusion and change it into a proper inclusion.

This can be done by factorization: Let $\nu(\mathbb{R}^n)$ be the set of all distributionally convergent sequences in $(\mathcal{C}^\infty)^{(0,\infty)}$ and $\nu_0(\mathbb{R}^n)$ be the set of all null sequences. Then clearly $\mathcal{D}' = \nu(\mathbb{R}^n)/\nu_0(\mathbb{R}^n)$. But this is no larger than $\mathcal{D}'$. We observe that $\nu(\mathbb{R}^n)$ is not an algebra, e.g. for $n = 1$ and any real valued sequence converging to the Dirac delta function, the squared sequence will not converge distributionally. As for ν_0, the sequences $\{\sin\frac{x}{\epsilon}\}_{\epsilon>0}$ and $\{\cos\frac{x}{\epsilon}\}_{\epsilon>0}$ converge to zero in distribution as $\epsilon \to 0$, (e.g. $\int_{\mathbb{R}} \sin\frac{x}{\epsilon} \phi(x)\,dx = \epsilon \int_{\mathbb{R}} \sin t \, \phi(\epsilon t)\,dt \to 0$, $\phi \in \mathcal{D}(\mathbb{R})$), but $(\sin\frac{x}{\epsilon})^2 + (\cos\frac{x}{\epsilon})^2 = 1$ shows that $\nu_0(\mathbb{R})$ is not an ideal nor can it be contained in a proper ideal of any algebra. So the strategy should be to enlarge $\nu(\mathbb{R}^n)$ so that it becomes an algebra and to make $\nu_0(\mathbb{R}^n)$ smaller (i.e. identify less) to get an ideal.

In Colombeau's method this is done by a particular choice of an index set $\mathcal{A}$, of a sub-algebra $\mathcal{E}_M(\mathbb{R}^n) \subset (\mathcal{C}^\infty(\mathbb{R}^n))^{\mathcal{A}}$, of an ideal $\mathcal{N}(\mathbb{R}^n)$ whereby the above mentioned multi-valent association becomes univalent and the imbedding $\mathcal{D}' \subset \mathcal{E}_M/\mathcal{N}$ is achieved.

For the index set $\mathcal{A}$:

Definition 1 For $m \in \mathbb{N}_+ = \mathbb{N} \setminus \{0\}$ and $\mathcal{D}(\mathbb{R}^n)$ being the space of $\mathcal{C}^\infty$ complex valued functions with compact support, define the set $\mathcal{A}_m$ as

$$\mathcal{A}_m = \{\psi \in \mathcal{D}(\mathbb{R}^n) : \int_{\mathbb{R}^n} \psi(x)\,dx = 1, \int_{\mathbb{R}^n} x^p \psi(x)\,dx = 0,\ 1 \le |p| \le m\} \quad (1)$$

and let $\mathcal{A} \doteq \mathcal{A}_1$.

It immediately follows that $\mathcal{A} = \mathcal{A}_1 \supset \mathcal{A}_2 \supset \mathcal{A}_3 \cdots$
Some properties of the index set $\mathcal{A}$ is given by

Lemma 1 *(i)* $\mathcal{A}_m \neq \emptyset,\ m \in \mathbb{N}_+,$

(ii) $\bigcap_{m \in \mathbb{N}_+} \mathcal{A}_m = \emptyset,$

(iii) $\psi \in \mathcal{A}_m \Rightarrow \psi_\epsilon = \frac{1}{\epsilon^n} \psi(\frac{x}{\epsilon}) \in \mathcal{A}_m.$

Proof First let us assume $n = 1$. We can choose $\psi \in \mathcal{D}(\mathbb{R})$ with support in $[-1, 1]$ and $\int_{\mathbb{R}} \psi\,dx = 1$. Define $\phi_1 \in \mathcal{D}(\mathbb{R})$ as $\phi_1 = \psi + \lambda_1 D\psi$ where $\lambda_1 \in \mathbb{C}$. We obtain $\phi_1 \in \mathcal{A}_1$ if $\int_{\mathbb{R}} \phi_1(x)\,dx = 1$ and $\int_{\mathbb{R}} x\,\phi_1(x)\,dx = 0$. The first condition is obviously met under the assumptions on ψ. For the second condition we have with integration by parts:

$$0 = \int_{\mathbb{R}} x\,\psi(x)\,dx + \lambda_1 \int_{\mathbb{R}} x D\psi(x)\,dx = \int_{\mathbb{R}} x\,\psi(x)\,dx - \lambda_1$$

yielding $\lambda_1 = \int_{\mathbb{R}} x\,\psi(x)\,dx$.

Similarly by defining $\phi_2 = \phi_1 + \lambda_2 D^2\psi$ we obtain $\phi_2 \in \mathcal{A}_2$ if $\int_{\mathbb{R}} \phi_2(x)\,dx = 1$, $\int_{\mathbb{R}} x\,\phi_2(x)\,dx = \int_{\mathbb{R}} x^2\,\phi_2(x)\,dx = 0$. First two of these equations are verified by the expression of ϕ_1 and integration by parts. For the last, $0 = \int_{\mathbb{R}} x^2\,\phi_2(x)\,dx = \int_{\mathbb{R}} x^2\,\phi_1(x)\,dx + \lambda_2 \int_{\mathbb{R}} x^2\,D^2\psi(x)\,dx = \int_{\mathbb{R}} x^2\phi_1(x)\,dx + 2\lambda_2$ which gives $\lambda_2 = -\frac{1}{2}\int_{\mathbb{R}} x^2\,\phi_1(x)\,dx$.

This procedure can be repeated indefinitely proving (i) for $n = 1$.

If $n \ge 2$, then as we showed the result for $n = 1$, given $m \in \mathbb{N}_+$ there exists $\phi \in \mathcal{D}(\mathbb{R}) \cap \mathcal{A}_m$, i.e. $\phi \in \mathcal{A}_m(\mathbb{R})$. Define $\chi \in \mathcal{D}(\mathbb{R}^n)$ by $\chi(x) = \phi(x_1) \cdots \phi(x_n)$. Then $\chi(x)\,dx = 1$ and for $p = (p_1, \cdots, p_n),\ p_j \in \mathbb{N}_+\ (j = 1, \cdots, n)$ and $1 \le$

$|p| \leq m$ we have $\int_{\mathbb{R}^n} x^p \chi(x)\,dx = \prod_{j=1}^n x_j\,\phi_j(x_j)\,dx_j = 0$ since $\phi \in \mathcal{A}_m(\mathbb{R})$. Thus $\chi \in \mathcal{A}_m(\mathbb{R}^n)$.

For (ii): Assume the contrary and let $\phi \in \bigcap_{m \in \mathbb{N}_+} \mathcal{A}_m$. Consider the Fourier transform of ϕ which is seen to be equal to 1. For $\hat{\phi}(\xi) = \int_{\mathbb{R}^n} \left(\sum_{m=0}^{\infty} \frac{(-2\pi i < \xi, x >)^m}{m!} \right) \phi(x)\,dx$ and in view of the moment condition in Definition 1 $\int_{\mathbb{R}^n} \frac{<\xi,,x>^m}{m!} \phi(x)\,dx = 0$, $(m = 1, 2, \cdots)$. Thus $\hat{\phi}(\xi) = \int_{\mathbb{R}^n} \phi(x)\,dx = 1$. But since $\phi \in \mathcal{D}(\mathbb{R}^n)$, this result is in contradiction with the Paley-Wiener theorem (c.f. [Rud], pp. 181) which implies that $\lim_{|\xi| \to \infty} \hat{\phi}(\xi) = 0$. Thus $\bigcap_{m \in \mathbb{N}_+} \mathcal{A}_m = \emptyset$.

(iii) is obvious by the Definition 1. $\square$

The space of sequences which constitutes the basic framework of Colombeau's theory is

$$\mathcal{E}(\mathbb{R}^n) = (\mathcal{C}^{\infty}(\mathbb{R}^n))^{\mathcal{A}}. \quad (2)$$

$\mathcal{E}(\mathbb{R}^n)$ is clearly an associative, commutative differential algebra with the termwise operations performed on the respective sequences of functions. For every index $\phi \in \mathcal{A}$ there is $f \in \mathcal{C}^{\infty}(\mathbb{R}^n)$, the dependence on both $\phi \in \mathcal{A}$ and $x \in \mathbb{R}^n$ being indicated by writing f in the form $f(\phi, x)$. In fact what matters is the behaviour of f when ϕ converges to the Dirac function along the regularizations $\phi_\epsilon(x) = \epsilon^{-n} \phi(\frac{x}{\epsilon})$.

The sub-algebra $\mathcal{E}_M(\mathbb{R}^n)$ of <u>moderate elements</u> is defined as

Definition 2 $\mathcal{E}_M(\mathbb{R}^n) = \{f \in \mathcal{E}(\mathbb{R}^n) : \forall K \subset\subset \mathbb{R}^n, \forall \alpha \in \mathbb{N}^n, (0 \leq |\alpha| < \infty)\ \exists m \in \mathbb{N}_+$ such that $\forall \phi \in \mathcal{A}_m, \exists \eta > 0, c > 0$ with $|D^\alpha f(\phi_\epsilon, x)| \leq \frac{c}{\epsilon^m}$ if $x \in K$ and $0 < \epsilon < \eta\}$.

<u>Note.</u> Recalling the definition of L^∞ norm, this would be equivalent to $\mathcal{E}_M(\mathbb{R}^n) = \{f \in \mathcal{E}(\mathbb{R}^n) : \forall K \subset\subset \mathbb{R}^n, \forall \alpha \in \mathbb{N}^n, \exists m \in \mathbb{N}$ such that $\forall \phi \in \mathcal{A}_m, \|D^\alpha f(\phi_\epsilon)\|_{L^\infty(K)} = O(\epsilon^{-m})$ as $\epsilon \to 0\}$.

Thus the sub-algebra $\mathcal{E}_M(\mathbb{R}^n)$ consists of families $\{f(\phi)\}_{\phi \in \mathcal{A}}$ which grow at most polynomially in $\frac{1}{\epsilon}$ as $\epsilon \to 0$ when evaluated at ϕ_ϵ, uniformly on compact sets, together with all derivatives.

The ideal $\mathcal{N}(\mathbb{R}^n)$ (or the null-space) in $\mathcal{E}_M(\mathbb{R}^n)$ is defined as

Definition 3 $\mathcal{N}(\mathbb{R}^n) = \{f \in \mathcal{E}_M(\mathbb{R}^n) : \forall K \subset\subset \mathbb{R}^n, \forall \alpha \in \mathbb{N}^n, (0 \leq |\alpha| < \infty)\ \exists l \in \mathbb{N}_+$ such that $\forall \phi \in \mathcal{A}_m\ (m \geq l), \exists \eta > 0, c > 0$ with $|D^\alpha f(\phi_\epsilon, x)| \leq c\,\epsilon^{m-l}$ if $x \in K$ and $0 < \epsilon < \eta\}$.

<u>Notes 1</u>. Again this definition would be equivalent to:
$\mathcal{N}(\mathbb{R}^n) = \{f \in \mathcal{E}_M(\mathbb{R}^n) : \forall K \subset\subset \mathbb{R}^n, \forall \alpha \in \mathbb{N}^n, \exists l \in \mathbb{N}$ such that $\forall m \geq l, \forall \phi \in \mathcal{A}_l\ \|D^\alpha f(\phi_\epsilon)\|_{L^\infty(K)} = O(\epsilon^{m-l})$ holds as $\epsilon \to 0\}$.
2. A slight ameliorization in Definition 3 would be as follows: $|D^\alpha f(\phi_\epsilon, x) \leq c\epsilon^{\beta(m)-l}, \beta \in B$ where $B = \{\beta : \mathbb{N}_+ \to (0, \infty), ; \beta \nearrow, \lim \beta(m) \to \infty\}$, i.e. $O(\epsilon^{m-l})$ is replaced by $O(\epsilon^{\beta(m)-l})$ as $\epsilon \to 0$.
These elements vanish faster than any power of ϵ eventually , $(0 \leq |\alpha| < \infty)$.

To show that $\mathcal{N}(\mathbb{R}^n)$ satisfies the ideal condition assume $f \in \mathcal{E}_M$, $g \in \mathcal{N}$ and use the Leibniz formula $D^\alpha(f\,g) = \sum_{\beta \le \alpha} c_{\alpha\beta}(D^{\alpha-\beta}f)(D^\beta g)$, $(\beta \le \alpha \Rightarrow \beta_j \le \alpha_j$, $j = 1, \cdots, n)$. As there are a finite number of terms it is sufficient to check the ideal condition for a single term $(D^{\alpha-\beta}f)(D^\beta g)$:

(1) $\exists m_1 \in \mathbb{N}_+$ such that $\forall \phi \in \mathcal{A}_{m_1}$, $\|D^{\alpha-\beta}f(\phi_\epsilon)\|_{L^\infty} = O(\epsilon^{-m_1})$;

(2) $\exists m_2 \in \mathbb{N}_+$ such that $\forall q \ge m_2$, $\forall \phi \in \mathcal{A}_q$, $\|D^\beta g(\phi_\epsilon)\|_{L^\infty} = O(\epsilon^{q-m_2})$.

Let $m = m_1 + m_2$. Then $\forall \bar{q} \ge m$, $\forall \phi \in \mathcal{A}_{\bar{q}}$, since $\bar{q} \ge m_2$ and $\mathcal{A}_{\bar{q}} \subset \mathcal{A}_{m_1}$ we have

$$\|(D^{\alpha-\beta}f)(D^\beta g)\|_{L^\infty} \le \|D^{\alpha-\beta}f\|_{L^\infty}\,\|D^\beta g\|_{L^\infty} = O(\epsilon^{-m_1})\,O(\epsilon^{\bar{q}-m_2}) = O(\epsilon^{\bar{q}-m}).$$

If we repeat this argument for each term in the Leibniz formula and take the largest $\bar{q}$ we find that $f.g \in \mathcal{N}$.

Definition 4 Colombeau Algebra of Generalized functions $\mathcal{G}(\mathbb{R}^n) = \mathcal{E}_M(\mathbb{R}^n)/\mathcal{N}(\mathbb{R}^n)$.

This is an associative, commutative algebra. Further by Definitons 1 and 2 it is easily checked that $D^\alpha \mathcal{N} \subset \mathcal{N}$ and $D^\alpha \mathcal{E}_M \subset \mathcal{E}_M$, $\alpha \in \mathbb{N}^n$. Therefore we can define the partial differential operators $D^\alpha : \mathcal{G}(\mathbb{R}^n) \to \mathcal{G}(\mathbb{R}^n)$, $\alpha \in \mathbb{N}^n$ by

$$D^\alpha(f + \mathcal{N}) = D^\alpha f + \mathcal{N}, \ \ f \in \mathcal{E}_M \quad (3)$$

These differential operators on $\mathcal{G}(\mathbb{R}^n)$ satisfy the Leibniz rule of product derivatives.

Frequently we refer to Colombeau generalized functions simply as 'generalized functions' in contrast to Schwartz generalized functions or linear distributions.

Remarks

(1) By the definition of the ideal $\mathcal{N}$, for a function to be a representative of an equivalence class in the quotient space, $\mathcal{E}_M/\mathcal{N}$ what matters is the behavior along the delta sequences ϕ_ϵ, i.e. the values with m large enough and $\epsilon > 0$ small enough.

(2) $\mathcal{E}_M(\mathbb{R}^n) \subset_{\ne} \mathcal{E}(\mathbb{R}^n)$, i.e. class of moderate elements is a proper subclass of the algebra $\mathcal{E}(\mathbb{R}^n)$: Consider $f : \mathcal{A} \times \mathbb{R}^n \longrightarrow \mathbb{C}$ defined by

$$f(\phi, x) = e^{\frac{1}{\Delta(\phi)}}, \phi \in \mathcal{A}, \ x \in \mathbb{R}^n$$

where $\Delta(\phi)$ (the diameter of ϕ) is given with

$$\Delta(\phi) = \sup\{\|x - y\| : x, y \in supp\ \phi\}; \ \ \phi \in \mathcal{A}$$

As $\phi_\epsilon = \frac{1}{\epsilon^n}\phi(\frac{x}{\epsilon})$, $supp\ \phi_\epsilon = \epsilon\ supp\ \phi$ and $\Delta(\phi_\epsilon) = \epsilon\,\Delta(\phi)$, hence $f(\phi_\epsilon, x) = \exp\{\frac{1}{\epsilon\Delta(\phi)}\}$ has an exponential growth (not polynomial) as $\epsilon \to 0$. Thus $f \in \mathcal{E} \setminus \mathcal{E}_M$.

3.4.2 Inclusion Properties

$$\mathcal{C}^\infty \subset \cdots \subset \mathcal{C}^0 \subset \mathcal{D}' \subset \mathcal{G}(\mathbb{R}^n).$$

First we want to prove these imbeddings and study the extensions to $\mathcal{G}(\mathbb{R}^n)$ of algebraic and/or differential operators on $\mathcal{C}^\infty$, $\mathcal{C}^0$ and $\mathcal{D}'$.

(I) The most straightforward imbedding into $\mathcal{G}(\mathbb{R}^n)$ is that of $\mathcal{C}^\infty$. The imbedding is achieved by $\mathcal{C}^\infty \ni f \longrightarrow F \in \mathcal{G}(\mathbb{R}^n)$, where $F = \tilde{f} + \mathcal{N}$ and $\tilde{f} = f(x)$, $\forall \phi \in \mathcal{A}$, $x \in \mathbb{R}^n$, i.e. constant mapping on the index set $\mathcal{A}$. The mapping is well-defined and since $\tilde{f}$ is independent of delta sequences ϕ_ϵ and ϵ, the conditions in 3.4.1, Definition 2 for the moderate growth are trivially satisfied, thus $\tilde{f} \in \mathcal{E}_M$ and $F = \tilde{f} + \mathcal{N} \in \mathcal{G}(\mathbb{R}^n)$.

Furthermore $f \to F$ is one-to-one: For $f_1, f_2 \in \mathcal{C}^\infty$ and $\tilde{f}_1 + \mathcal{N} = \tilde{f}_2 + \mathcal{N}$ implies $\tilde{f}_1 - \tilde{f}_2 \in \mathcal{N}$ or $f_1(x) - f_2(x) \in \mathcal{N}$. Then this is only possible by 3.4.1, Definition 3 if $f_1(x) = f_2(x)$.

We say $\mathcal{C}^\infty$ is a faithful sub-algebra of $\mathcal{G}(\mathbb{R}^n)$ and the mapping $f \to F$ is an algebra homomorphism. $1 \in \mathcal{C}^\infty(\mathbb{R}^n)$ is the unit element of the algebra $\mathcal{G}(\mathbb{R}^n)$.

(II) The imbedding $\mathcal{C}^0 \subset \mathcal{G}(\mathbb{R}^n)$:

For $f \in \mathcal{C}^0$, define $f \longrightarrow F$ by $F = \bar{f} + \mathcal{N} \in \mathcal{G}(\mathbb{R}^n)$ where $\bar{f}(\phi, x) = (f * \check{\phi})(x)$, $(\check{\phi}(x) = \phi(-x)$, c.f. 1.8.1, Proof of Theorem 1a)).

Explicitly $\bar{f}(\phi, x) = \int_{\mathbb{R}^n} f(y)\, \phi(y - x)\, dy = \int_{\mathbb{R}^n} f(x + y)\, \phi(y)\, dy$.

<u>Note.</u> The use of this type of convolution is for the convenience of certain expressions. All properties of convolutions remain the same with obvious slight modifications.

The mapping $'f \to F$ is again well-defined. We can see this by checking that $\bar{f} \in \mathcal{E}_M$. Firstly for each $\phi \in \mathcal{A}$, $\bar{f} \in \mathcal{C}^\infty$ by 1.7, Theorem 1c). On the other hand given $\alpha \in \mathbb{N}^n$

$$D^\alpha \bar{f}(\phi_\epsilon, x) = \frac{(-1)^{|\alpha|}}{\epsilon^{n+|\alpha|}} \int_{\mathbb{R}^n)} f(y)\, D^\alpha \phi(\frac{y - x}{\epsilon})\, dy \quad (*)$$

will indicate that $\bar{f}$ is a moderate function. In fact in view of 3.4.1 Definition 2, supposing that $K \subset\subset \mathbb{R}^n$ is a compact set

$$|D^\alpha \bar{f}(\phi_\epsilon, x)| = \frac{1}{\epsilon^{n+|\alpha|}} \left| \int_{\mathbb{R}^n} f(y) D^\alpha \phi(\frac{y - x}{\epsilon})\, dy \right| \le \frac{C}{\epsilon^{n+|\alpha|}}$$

since the integral is bounded as x varies in K.

Also if $f \in \mathcal{C}^0$ and $\bar{f} \in \mathcal{N}$, then in view of the expression $(*)$, the only way to verify the condition in 3.4.1, Definition 3 to belong to $\mathcal{N}$ is $f = 0$. Hence $f \to F$ is one -to-one.

Remark 1 $\mathcal{C}^0 \subset \mathcal{G}(\mathbb{R}^n)$ is a vector space imbedding, but not imbedding of algebras, i.e. the multiplication in $\mathcal{G}(\mathbb{R}^n)$ when restricted to $\mathcal{C}^0$, does not in general coincide with the usual multiplication of continuous functions. This coincidence would violate the Impossibility Result. However the two rules of multiplications are associated to each other by a cerain equivalence relation included in a particular coupled calculus to be explained in 3.4.8.

(III) $\mathcal{D}' \subset \mathcal{G}(\mathbb{R}^n)$. For $T \in \mathcal{D}'$, $T \longrightarrow F = f_T + \mathcal{N} \in \mathcal{G}(\mathbb{R}^n)$. The imbedding is defined by $f_T(\phi, x) = < T_y, \phi(y - x) >$. This is clearly the extension of the imbedding in II). In fact when T is a regular distribution $f \in C^0$, $f_T \equiv \bar{f}$.

The mapping $T \to F$ is well-defined since firstly $f_T \in \mathcal{E}(\mathbb{R}^n)$ by 1.7, Theorem 1c. Then

$$D^\alpha f_T(\phi_\epsilon, x) = \frac{(-1)^{|\alpha|}}{\epsilon^{n+|\alpha|}} < T_y, D^\alpha \phi(\frac{y - x}{\epsilon}) > \qquad (**)$$

shows that $f_T \in \mathcal{E}_M(\mathbb{R}^n)$. Also if $f_T \in \mathcal{N}$, in view of $(**)$ this is possible only if $f_T = 0$, hence $f \to F$ is injective.

The derivative rule in $\mathcal{G}(\mathbb{R}^n)$ as given by 3.4.1 (3), when restricted to $\mathcal{D}'$ coincides with the distributional derivative in $\mathcal{D}'$. In fact for $T \in \mathcal{D}', T \longrightarrow F = f_T + \mathcal{N} \in \mathcal{G}(\mathbb{R}^n)$, where $f_T(\phi, x) = < T_y, \phi(y - x) >$. Differentiating in $\mathcal{G}$ we find $D^\alpha F = D^\alpha f_T + \mathcal{N} = (-1)^{|\alpha|} < T_y, D^\alpha \phi(y - x) > +\mathcal{N}$.

On the other hand first differentiate T in $\mathcal{D}'$ to find $< D^\alpha T, \phi(x) >= (-1)^{|\alpha|} < T, D^\alpha \phi(x) >$. Now imbed into $\mathcal{G}: D^\varnothing T \to \mathcal{U} = f_{D^\alpha T} + \mathcal{N}$ where $f_{D^\alpha T}(\phi, x) = < D^\alpha T_y, \phi(y - x) >= (-1)^{|\alpha|} < T_y, D^\alpha \phi(y - x) >$.

Hence the following commutative diagram holds (Fig. 3.1):

Remark 2 The imbedding $\mathcal{D}' \hookrightarrow \mathcal{G}(\mathbb{R}^n)$ is an imbedding of vector spaces which extends the distributional partial derivatives.

Examples

(1) Dirac delta function $\delta \in \mathcal{G}(\mathbb{R})$:
Recall x_+ defined by $x_+(x) = 0$ for $x < 0$, $x_+(x) = x$ for $x \geq 0$ and $\delta = D^2 x_+$. According to the Remark above the derivative can be regarded in $\mathcal{G}(\mathbb{R}^n)$. Thus if $\delta = f_\delta + \mathcal{N} \in \mathcal{G}(\mathbb{R})$, imbedding the regular distribution x_+ into $\mathcal{G}$ we have: $f_{x_+}(\phi, x) = \int_0^\infty y\, \phi(y - x)\, dy = \int_{-x}(x + u)\, \phi(u)\, du$; $f_\delta(\phi, x) = D^2(\int_{-x}^0 (x + u)\, \phi(u)\, du) = D \int_{-x}^\infty \phi(u)\, du = \phi(-x)$.

$$\mathcal{D}' \ni T \longrightarrow f_T + \mathcal{N} \in \mathcal{G}(\mathbb{R}^n)$$

$$\mathcal{D}' \ni D^\alpha T \longrightarrow D^\alpha f_T + \mathcal{N} \in \mathcal{G}(\mathbb{R}^n)$$

Fig. 3.1 Commutativity of imbedding the partial derivatives of a Schwartz distributions into the Colombeau algebra in two differet ways

(2) In $\mathcal{D}' :< x\,\delta, \phi >=< \delta, x\,\phi) = 0$, (1.6.1, Operation V) and $\phi \in \mathcal{D}(\mathbb{R})$). However in $\mathcal{G}(\mathbb{R})$ this is not true: $x.\delta = f + \mathcal{N} \in \mathcal{G}(\mathbb{R})$ where $f(\phi, x) = x.\phi(-x)$, (regarding $\mathcal{E}_M$ as an algebra). But $Df(\phi_\epsilon, 0) = \frac{1}{\epsilon}\phi(0)$, $(\phi \in \mathcal{A}, \epsilon > 0)$ violates the condition in 3.4.1, Definition 3 for $x = 0$ and any compact K with $0 \in K$. (In every index set $\mathcal{A}_m$ there exists ϕ with $\phi(0) \neq 0$ which can be proved by some modification of 1.4.1, Lemma 1, c.f. [Ros], Appendix 3, pp. 127). Hence f can not be in the ideal $\mathcal{N}$ and $x\,\delta \neq 0$ in $\mathcal{G}(\mathbb{R})$.

Remark 3 The fact that $x.\delta = 0$ in $\mathcal{D}'(\mathbb{R})$ but $x.\delta \neq 0$ in $\mathcal{G}(\mathbb{R})$ shows that the multiplication of distributions with $\mathcal{C}^\infty$ functions does not extend to $\mathcal{G}(\mathbb{R}^n)$. This shortcoming will be sorted out by the coupled calculus introduced in Sect. 3.4.5.

Remark 4 If $f \in \mathcal{C}^\infty \subset \mathcal{C}^0$, it can be imbedded into $\mathcal{G}(\mathbb{R}^n)$ both as $\tilde{f} + \mathcal{N}$ and $\bar{f} + \mathcal{N}$, (imbedding as a regular distribution would be the same as $\bar{f} + \mathcal{N}$). However it is not two different ways of imbedding since $\bar{f} - \tilde{f} \in \mathcal{N}(\mathbb{R}^n)$ as shown below: (for simplicity take $n = 1$), calling $g = \bar{f} - \tilde{f} \in \mathcal{E}_M$, imbedding II) allows us to write $g(\phi, x) = \int_{\mathbb{R}}(f(x + y) - f(x)\,\phi(y)\,dy$ $(\phi \in \mathcal{A})$.
   Then $D^j g(\phi, x) = \int_{\mathbb{R}}(D^j f(x + y) - D^j f(x))\,\phi(y)\,dy$ $(\checkmark)$
   Given $m \in \mathbb{N}_+$, by the Taylor formula

$$D^j f(x + y) - D^j f(x) = \sum_{1 \leq r \leq m} \frac{1}{r!} y^r D^{j+r} f(x) + \frac{1}{(m+1)!} y^{m+1} D^{j+m+1} f(x + \Theta y)$$

for some $\Theta \in (0, 1)$. Substituting in $(\checkmark)$:

$$D^j g(\phi_\epsilon, x) = \sum_{1 \leq r \leq m} \frac{\epsilon^r}{r!} D^{j+r} f(x) \int_{\mathbb{R}} (\frac{y}{\epsilon})^r \,\phi(\frac{y}{\epsilon}) \,\frac{dy}{\epsilon}$$

$$+ \frac{1}{(m+1)!} \int_{\mathbb{R}} y^{m+1} D^{j+m+1} f(x + \Theta y)\phi(\frac{y}{\epsilon}) \,\frac{dy}{\epsilon}.$$

By 3.4.1, Definition 1 of the index set the integrals under the summation sign will vanish. We find then by a change of variable:

$$D^j g(\phi_\epsilon, x) = \frac{\epsilon^{m+1}}{(m+1)!} \int_{\mathbb{R}} u^{m+1} D^{j+m+1} f(x + \epsilon\Theta u) \,\phi(u) \,du.$$

This shows that $g = \bar{f} - \tilde{f} \in \mathcal{N}$ since by 3.4.1, Definition 1 the integral tends to zero as $\epsilon \to 0$. Thus it is bounded by a constant, hence $|D^j g(\phi_\epsilon, x)| \leq C\epsilon^{\beta(m)-l}$ where $\beta(m) = m + 2$ and $l = 1$.
   The case $n > 1$ is shown similarly using a multi-variable Taylor formula.

(3) In the linear theory $x.\delta = 0$ showing that the singularity of the delta function is lower than that of $\frac{1}{x}$. However since in Colombeau's theory $x.\delta \neq 0$, this is not the case. In fact its singularity is higher than the singularity of $\frac{1}{x^m}$ for every $m \in$

$\mathcal{N}_+$. To see this, in $\mathcal{G}(\mathbb{R})$, $x^m.\delta = g + \mathcal{N}$ where $g(\phi, x) = x^m \phi(-x)$, $\phi \in \mathcal{A}$, $x \in \mathbb{R}$. Thus $D^m g(\phi_\epsilon, 0) = \frac{m!}{\epsilon}\phi(0)$. $\exists \phi \in \mathcal{A}_m$ such that $\phi(0) \neq 0$. This shows $g \notin \mathcal{N}$, hence $x^m.\delta \neq 0$ in $\mathcal{G}(\mathbb{R})$.

(4) To illustrate that the pointwise multiplication in $\mathcal{C}^0$ does not extend to the multiplication in $\mathcal{G}(\mathbb{R})$, consider: $(x_+).(x_-)$ where $x_-(x) = x$ for $x < 0$ and $x_-(x) = 0$ for $x \geq 0$. Obviously in $\mathcal{C}^0(\mathbb{R})$, $(x_+).(x_-) = 0$. Now we show that $(x_+).(x_-) \neq 0$ in $\mathcal{G}(\mathbb{R})$. If $x_+ = f_{x_+} + \mathcal{N}$, $x_- = f_{x_-} + \mathcal{N}$, then

$$(f_{x_-}.f_{x_+})(\phi, x) = \int_{-\infty}^{0} y\,\phi(y - x)\,dy.\int_{0}^{\infty} y\,\phi(y - x)\,dy$$

$$(f_{x_-}.f_{x_+})(\phi_\epsilon, 0) = \int_{-\infty}^{0} \frac{1}{\epsilon}y\phi(\frac{y}{\epsilon})\,dy.\int_{0}^{\infty} \frac{1}{\epsilon}y\,\phi(\frac{y}{\epsilon})\,dy$$

$= \epsilon^2 \int_{\infty}^{0} u\,\phi(u)\,du.\int_{0}^{\infty} u\,\phi(u)\,du$ which implies (see also [Ros], Appendix 3, pp. 127) that $f_{x_-}.f_{x_+} \notin \mathcal{N}$. Thus $x_-.x_+ \neq 0$ in $\mathcal{G}(\mathbb{R}^n)$.

3.4.3 Functional Analytic Approach

Colombeau's differential algebra can also be constructed by the following functional analytic arguments which also reflect its natural character, (c.f. [Ros], Appendix 1, pp. 120–123, [Col-1], pp. 50–66).

A linear distribution $T \in \mathcal{D}'$ is a function of the form $T : \mathcal{D}(\mathbb{R}^n) \longrightarrow \mathbb{C}$ which is linear and also continuous with respect to the topology in $\mathcal{D}(\mathbb{R}^n)$. Then it is natural to attempt to define the product of two distributions $T_1 \in \mathcal{D}'$, $T_2 \in \mathcal{D}'$ as $T = T_1 . T_2 : \mathcal{D}(\mathbb{R}^n) \longrightarrow \mathbb{C}$ by the usual product of complex valued functions, i.e. the monomial

$$T(\phi) = T_1(\phi).T_2(\phi), \quad \phi \in \mathcal{D}(\mathbb{R}^n) \quad (1)$$

T defined like this will not of course be linear on $\mathcal{D}(\mathbb{R}^n)$, but we may be inclined to visualize it as a smooth function on $\mathcal{D}(\mathbb{R}^n)$, i.e. an element of the algebra $\mathcal{C}^\infty(\mathcal{D}(\mathbb{R}^n))$ whereby the imbedding $\mathcal{D}'(\mathbb{R}^n) \subset \mathcal{C}^\infty(\mathcal{D}(\mathbb{R}^n))$ would have been possibly achieved. However this imbedding is involved with two problems:

Firstly we have to define a suitable rule of differentiation in $\mathcal{C}^\infty(\mathcal{D}(\mathbb{R}^n))$ which should yield the usual distributional derivative when restricted to $\mathcal{D}'(\mathbb{R}^n)$. Colombeau defines this rule as follows: for $R \in \mathcal{C}^\infty(\mathcal{D}(\mathbb{R}^n))$

$$(\frac{\partial R}{\partial x_i})(\phi) = - < R'(\phi), \frac{\partial \phi}{\partial x_i} >, \ \phi \in \mathcal{D}(\mathbb{R}^n) \quad (2)$$

Here $R'(\phi)$ is a derivative in the Fréchet sense, e.g. $R'(\phi) \in L(\mathcal{D}(\mathbb{R}^n), \mathbb{C})$ such that $w(\psi) \doteq R(\phi + \psi) - R(\phi) - <R'(\phi), \psi>$ is tangent to zero, i.e. for any neighborhhod containing 0, there exists $V \subset \mathcal{D}(\mathbb{R}^n))$, a neighborhood containing the origin, such that $w(tV) = o(t)$. If $R \in \mathcal{D}'(\phi)$, then $R'(\phi) = R$, thus by (2) we retrieve the distributional derivative $(D_{x_i} R)(\phi) = - <R, D_{x_i}\phi>$, $(D_{x_i} \equiv \frac{\partial}{\partial x_i})$.

<u>Notes</u>. (1) If $R \in \mathcal{C}^\infty(\mathcal{E}'(\mathbb{R}^n))$ we define $\frac{\partial R}{\partial x_i} \in \mathcal{C}^\infty(\mathcal{E}'(\mathbb{R}^n))$ by the same formula (2) when ϕ ranges in $\mathcal{E}'(\mathbb{R}^n)$ and then $\frac{\partial \phi}{\partial x_i} \in \mathcal{E}'(\mathbb{R}^n)$ on the right-hand side should be understood in the distributional sense, $(\mathcal{E}'(\mathbb{R}^n)$: Distributions of compact support, c.f. 1.5).
(2) If $D^\alpha = \frac{\partial^{\alpha_1 + \cdots + \partial_n}}{\partial x_1^{\alpha_1} \cdots \partial x_n^{\alpha_n}}$, $\alpha = (\alpha_1, \cdots, \alpha_n)$, $D^\alpha R \in \mathcal{C}^\infty(\mathcal{D}(\mathbb{R}^n))$ (or $\in \mathcal{C}^\infty(\mathcal{E}'(\mathbb{R}^n))$ can be calculated by the above.

Remark 1 If $R_1, R_2 \in \mathcal{C}^\infty(\mathcal{D}(\mathbb{R}^n))$,	one has by (2): $\frac{\partial}{\partial x_i}(R_1 R_2)(\phi) = - <(R_1 R_2)'(\phi), \partial \frac{\partial \phi}{\partial x_i}> = -R_1(\phi) <R_2'(\phi), \frac{\partial \phi}{\partial x_i}> -R_2(\phi) <R_1'(\phi), \frac{\partial(\phi)}{\partial x_i}> = (R_1 \frac{\partial R_2}{\partial x_i} + R_2 \frac{\partial R_1}{\partial x_i})(\phi)$; thus Leibniz's formula for the derivation of a product holds.

A second problem of the imbedding $\mathcal{D}'(\mathbb{R}^n) \subset \mathcal{C}^\infty(\mathcal{D}(\mathbb{R}^n))$ is that the multiplication rule of (1) does not even extend the product of smooth functions. That means if two smooth functions f_1, $f_2 \in \mathcal{C}^\infty(\mathbb{R}^n)$ are regarded as regular distributions then (1) would be false since

$$(f_1.f_2)(\phi) = \int_{\mathbb{R}^n} f_1(x) f_2(x) \phi(x)\,dx \neq \int_{\mathbb{R}^n} f_1(x) \phi(x)\,dx \int_{\mathbb{R}^n} f_2(x) \phi(x)\,dx, \quad (3)$$

To get around this difficulty we try to introduce a suitable quotient structure in $\mathcal{C}^\infty(\mathcal{D}(\mathbb{R}^n))$ which will factor out the difference between the right and left sides of (3).

Firstly $\mathcal{D}(\mathbb{R}^n)$ is dense in $\mathcal{E}'(\mathbb{R}^n)$,   (c.f. 1.2, Theorem 4; 1.5 Proposition 1. Note 4). Therefore by restricting an element $f \in \mathcal{C}^\infty(\mathcal{E}'(\mathbb{R}^n))$ to $\mathcal{D}(\mathbb{R}^n)$ we obtain an element $f|_{\mathcal{D}}$ of $\mathcal{C}^\infty(\mathcal{D}(\mathbb{R}^n))$. Furthermore the imbedding $\mathcal{C}^\infty(\mathcal{E}'(\mathbb{R}^n)) \subset \mathcal{C}^\infty(\mathcal{D}(\mathbb{R}^n))$ is injective ([Col-2]).

Then the strategy is to try to correct the lack of identity in (3) in the smaller space $\mathcal{C}^\infty(\mathcal{E}'(\mathbb{R}^n))$. For that purpose we recall the reflexivity of $\mathcal{C}^\infty(\mathbb{R}^n)$. Hence every linear and continuous functional $F \in \mathcal{E}''(\mathbb{R}^n)$ can be represented by some $f \in \mathcal{C}^\infty(\mathbb{R}^n)$. Thus for every $T \in \mathcal{E}'$ we have $<F, T> = <T, f>$. To single out the representative f, take $T = \delta_x$. Then

$$<F, \delta_x> = <\delta_x, f> = f(x) \quad (4)$$

Due to (4), elements of $\mathcal{E}''$ can be multiplied like ordinary functions: For $F_1, F_2 \in \mathcal{E}''$, define $F = F_1 . F_2$ by

$$<F, \delta_x> = <F_1, \delta_x> . <F_2, \delta_x>$$

which is nothing but the usual multiplication of the corresponding functions in $C^\infty(\mathbb{R}^n)$. In view of partial derivation on $\mathcal{E}'(\mathbb{R}^n)$ defined by (2), we have

$$\mathcal{E}''(\mathbb{R}^n) \subset C^\infty(\mathcal{E}'(\mathbb{R}^n)).$$

(4) suggests the following equivalence relation in $C^\infty(\mathbb{R}^n)$:
$F_1 \equiv F_2 \iff F_1(\delta_x) = F_2(\delta_x) \ \forall x \in \mathbb{R}^n.$

To see how could this equivalence relation be utilized, consider the linear mapping $A : C^\infty(\mathcal{E}'(\mathbb{R}^n)) \to C^\infty(\mathbb{R}^n)$:

$$(A(F))(x) = F(\delta_x), \ F \in C^\infty(\mathcal{E}'(\mathbb{R}^n)), \ x \in \mathbb{R}^n \quad (5)$$

which is in fact the extension of (4) from $\mathcal{E}''(\mathbb{R}^n) = L(\mathcal{E}'(\mathbb{R}^n), \mathbb{C})$ to $C^\infty(\mathcal{E}'(\mathbb{R}^n))$.

This is well-defined:
$\mathbb{R}^n \ni x \longrightarrow \delta_x \in \mathcal{E}'(\mathbb{R}^n)$ is C^∞-smooth since

$$\frac{1}{\xi} < \delta_{x_1,\cdots,x_i+\xi,\cdots,x_n} - \delta_{x_1,\cdots,x_n}, \phi >= \frac{1}{\xi}[\phi(x_1,\cdots,x_i,\cdots,x_n) - \phi(x_1,\cdots,x_n)]$$

$$\longrightarrow \frac{\partial}{\partial x_i}\phi(x_1,\cdots,x_n) = - < \frac{\partial}{\partial x_i}\delta_x, \phi > \quad \text{as } \xi \to 0 \quad (6)$$

(derivative in the distributional sens) and when the derivative is iterated we see that $x \to \delta_x$ can be differentiated of any order.

This implies that $A(F) \in C^\infty(\mathbb{R}^n)$.

Remark 2 The new definition of derivative given by (2) is also justified by the mapping A:

If $F \in C^\infty(\mathcal{E}'(\mathbb{R}^n))$, then $A(\frac{\partial F}{\partial x_i}) = \frac{\partial}{\partial x_i}(AF)$, i.e. the new concept of derivative corresponds to the usual derivative in $C^\infty(\mathbb{R}^n)$. This is shown as follows:

By (5) and (2) $(A(\frac{\partial F}{\partial x_i}))(x) = (\frac{\partial F}{\partial x_i})(\delta_x) = - < F'(\delta_x), \frac{\partial \delta_x}{\partial x_i} >\equiv -F'(\delta_x).\frac{\partial \delta_x}{\partial x_i}$. On the other hand by the property of usual derivative $\frac{\partial}{\partial x_i}(AF)(x) = F'(\delta_x).\frac{\partial}{\partial x_i}(x \longrightarrow \delta_x)$ and by (6) this is also equal to $-F'(\delta_x).\frac{\partial \delta_x}{\partial x_i}$.

We have the following commutative diagram (Fig. 3.2):

If $KerA = \{F \in C^\infty(\mathcal{E}'(\mathbb{R}^n)) : F \equiv 0\}$
then it follows that

$$C^\infty(\mathcal{E}'(\mathbb{R}^n))/KerA \cong C^\infty(\mathbb{R}^n)$$

yields isomorphic algebras, (A is surjective). In this way we have found a suitable factorization on $C^\infty(\mathcal{E}'(\mathbb{R}^n))$ by the $KerA$ which sorts out the difficulty inherent in (3). $KerA$ is an ideal since if $T \in C^\infty(\mathcal{E}'(\mathbb{R}^n))$, $F \in KerA$, then $(TF)(\delta_x) = T(\delta_x)F(\delta_x) = 0$.

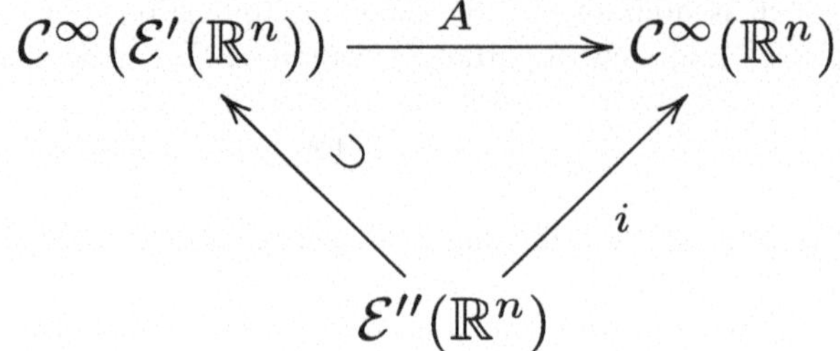

Fig. 3.2 The imbedding of the derivative of a compactly supported distribution into the space of smooth functions on n dimensional Eucledean space in two different ways

Now in order to correct the multiplication problem in the larger space $\mathcal{C}^\infty(\mathbb{R}^n)$ or in a sub-algebra thereof, we should find an ideal $\mathcal{N}$ in the larger space such that its intersection with $\mathcal{C}^\infty(\mathcal{E}'(\mathbb{R}^n))$ should be $Ker\,A$.

Colombeau gave the following characterization of the ideal $Ker\,A$:

Proposition 1 *Let $F \in \mathcal{C}^\infty(\mathcal{E}'(\mathbb{R}^n))$. Then $F \in Ker\,A$ (i.e. $F \equiv 0$) if and only if for every $q \in \mathbb{N}_+$ and $\phi \in \mathcal{A}_q$, (c.f. 3.4.1, Definition 1) and compact $K \subset \mathbb{R}^n$ there exist $c > 0$, $\eta > 0$ such that $|F(\phi_{\epsilon,x})| \le c\epsilon^{q+1}$ if $0 < \epsilon < \eta$ and $x \in K$, $(\phi_{\epsilon,x}(y) = \frac{1}{\epsilon^n}\phi(\frac{y-x}{\epsilon}))$.*

Sketchy Proof. $\phi_{\epsilon,x} \in \mathcal{D}(\mathbb{R}^n) \subset \mathcal{E}'(\mathbb{R}^n)$ for all $x \in K$. As $\{\phi_{\epsilon,x}\}$ is a delta sequence (c.f. 1.2, Example 7) the following mean value type result can be stated, ([Col-2], 1.3.2):

$$F(\phi_{\epsilon,x}) - F(\delta_x) \in \overline{\{< F'(\delta_x + t(\phi_{\epsilon,x} - \delta_x)), (\phi_{\epsilon,x} - \delta_x) >\}_{0 \le t \le 1}}$$

For $\eta > 0$ small enough the set $\{\delta_x + t(\phi_{\epsilon,x} - \delta_x) : x \in K, 0 \le t \le 1, 0 \le \epsilon \le \eta\}$ is bounded in $\mathcal{E}'(\mathbb{R}^n)$, therefore contained in a compact set of $\mathcal{E}'(\mathbb{R}^n)$. Thus it suffices to prove that there exists a bounded subset $B \subset \mathcal{E}'(\mathbb{R}^n)$ such that $\phi_{\epsilon,x} - \delta_x \in \epsilon^{q+1} B$.

If $f \in \mathcal{C}^\infty(\mathbb{R}^n)$ and $\phi \in \mathcal{A}_q$, using $\int \phi_{\epsilon,x}(y)dy = 1$ we find $< \phi_{\epsilon,x} - \delta_x, f >= \int_{\mathbb{R}^n}(f(y) - f(x))\,\phi_{\epsilon,x}(y)\,dy$. By the change of variable $\mu = \frac{y-x}{\epsilon}$ gives $< \phi_{\epsilon,x} - \delta_x, f >= \int_{\mathbb{R}^n}(f(x + \epsilon\mu) - f(x))\,\phi(\mu)\,d\mu$. Then Taylor's formula applied to f yields:

$$< \phi_{\epsilon,x} - \delta_x, f >= \sum_{|k|}^{q} \frac{\epsilon^{|k|}}{k!} \int_{\mathbb{R}^n} (f^{(k)}(x)\mu^k)\,\phi(\mu)\,d\mu + \int_{\mathbb{R}^n} r_x(\epsilon\mu)\phi(\mu)\,d\mu$$

where $|r_x(\epsilon\mu)| \le \frac{1}{(q+1)!}\epsilon^{q+1} \sup_{t \in K'} \|f^{(q+1)}(t)\| a^{q+1}$. As ϕ has compact support μ ranges in a compact set K' if $\epsilon > 0$ is small enough and if x ranges in K. Then $|\mu|_{\mathbb{R}^n} \le a$ for some $a > 0$. Since $\phi \in \mathcal{A}_q$, the moments are zero; thus

$$| < \phi_{\epsilon,x} - \delta_x, f > | = |\int_{\mathbb{R}^n} r_x(\epsilon\mu)\,\phi(\mu)\,d\mu| \le \frac{a^{q+1}}{(q+1)!} \sup_{t \in K'} \|f^{(q+1)}(t)\| \epsilon^{q+1} \int_{\mathbb{R}^n} |\phi(\mu)|\,d\mu.$$

If we set $V = \{f \in \mathcal{C}^\infty(\mathbb{R}^n) : \sup_{t \in K'} \|f^{(q+1)}(t)\| \le 1\}$, then V is a 0-neighborhood in $\mathcal{C}^\infty(\mathbb{R}^n)$ and if $B = V^\circ$ denotes the polar set of V, (i.e. $V^\circ = \{F \in \mathcal{E}'(\mathbb{R}^n) : | < F, f > | \le 1, \ f \in V\}$), this means $\exists c > 0$ such that $\phi_{\epsilon,x} - \delta_x \in cB$

The converse of the proposition, namely the inequality $|F(\phi_{\epsilon,x})| \le c\epsilon^{q+1}$ implies $F(\delta_x) = 0$ obviously holds: Take $q = 0$, then as $\epsilon \to 0$, $\phi_{\epsilon,x} \to \delta_x$ $\qquad\square$

We search for an ideal of $\mathcal{C}^\infty(\mathcal{D}(\mathbb{R}^n))$ such that its intersection with $\mathcal{C}^\infty(\mathcal{E}'(\mathbb{R}^n))$ would be $Ker A$. Characterization of $Ker A$ by Proposition 1 would play a leading role in this direction. However for a given $R \in \mathcal{C}^\infty(\mathcal{D}(\mathbb{R}^n))$ we should concentrate on the behavior of $R(\phi_{\epsilon,x})$ in this characterization. Given $\phi \in \mathcal{A}_q$, $x \in \mathbb{R}^n$ the product of an element of $Ker A \subset \mathcal{C}^\infty(\mathcal{E}'(\mathbb{R}^n)) \subset \mathcal{C}^\infty(\mathcal{D}(\mathbb{R}^n))$ and an element of $\mathcal{C}^\infty(\mathcal{D}(\mathbb{R}^n))$ may exhibit a very fast growth in $\frac{1}{\epsilon}$ when $\epsilon \to 0$, thus violating the growth condition in Proposition 1. Therefore we are led to consider the elements of $\mathcal{C}^\infty(\mathcal{D}(\mathbb{R}^n))$ that have a moderate growth in $\frac{1}{\epsilon}$. In this way we recover 3.4.1, Definition 2, i.e. the sub-algebra of moderate functions. The form of the sub-algebra we arrive at would be now:

$$\mathcal{E}(\mathbb{R}^n) = \{R \in \mathcal{C}^\infty(\mathcal{D}(\mathbb{R}^n)) : \ \forall K \subset\subset \mathbb{R}^n, \ \forall \alpha \in \mathbb{N}^n, \ (0 \le |\alpha| < \infty), \ \exists m \in \mathbb{N},$$

such that $\forall \phi \in \mathcal{A}_m$, $\exists \eta > 0, c > 0 \Longrightarrow \forall x \in K, \ \forall \epsilon \in (0, \eta), \ |D^\alpha R(\phi_{\epsilon,x})| \le \frac{c}{\epsilon^m}\}$.

Remark 3 (1) This form of the moderate sub-algebra is equivalent to 3.4.1, Definition 2, if we identify $f(\phi, x) \in [\mathcal{C}^\infty(\mathbb{R}^n)]^{\mathcal{A}_m}$ with $R(\phi_x) \in \mathcal{C}^\infty(\mathcal{A}_m)$ (where $\phi_x(\lambda) = \phi(\lambda - x)$) and $f(\phi_\epsilon, x)$ with $R(\phi_{\epsilon,x})$.

(2) Still equivalently, for every K and D^α there are $m_1, m_2 \in \mathbb{N}_+$ such that $\forall \phi \in \mathcal{A}_{m_1}$, $\exists \eta > 0, c > 0$ with $|D^\alpha R(\phi_{\epsilon,x})| \le c\frac{1}{\epsilon^{m_2}}$, $0 < \epsilon < \eta$, (recover m above by setting $m = \max\{m_1, m_2\}$).

Theorem 1 *(a)* $\quad R \in \mathcal{E}_M \Longrightarrow D^\beta R \in \mathcal{E}_M$ *for every multi-index β.*

(b) The product of two moderate elements in $\mathcal{C}(\mathcal{D}(\mathbb{R}^n))$ is moderate.

(c) Any element $F \in \mathcal{C}^\infty(\mathcal{E}'(\mathbb{R}^n))$ is moderate.

(d) Any linear distribution is moderate.

Proof (a) Given K, $\forall \alpha \in \mathbb{N}^n$ by the moderateness of R, there exists $m \in \mathbb{N}_+$ such that $\forall \phi \in \mathcal{A}_m$, $\exists \eta > 0, c > 0$ with $|D^{\beta+\alpha} R(\phi_{\epsilon,x})| \le \frac{c}{\epsilon^m}$ if $x \in K$ and $0 < \epsilon < \eta$.
(b) Let R and S be moderate functions. Use the Leibniz's formula

$$D^\alpha(R.S) = \sum_{\beta \le \alpha} c_{\alpha\beta}(D^{\alpha-\beta}R)(D^\beta S).$$

For each $\alpha, \beta (0 \le \beta \le \alpha)$, find $m_i (i = 1, 2)$ in accordance with the definition of moderateness. Then each term in the Leibniz formula can be estimated by $\text{Constant} \times \frac{1}{\epsilon^{\max m_i}}$.

(c) Let $F \in \mathcal{C}^\infty(\mathcal{E}'(\mathbb{R}^n))$. As $\phi_{\epsilon,x}$ is a delta sequence with singularity at x, we have $\phi_{\epsilon,x} \to \delta$ as $x \to 0$. Since $\delta \in \mathcal{E}'(\mathbb{R}^n)$ and $x \to \delta_x$ is $\mathcal{C}^\infty$-smooth by the note following (5), $D^\alpha F(\phi_{\epsilon,x})$ is bounded.

(d) It is sufficient to prove that any derivative in the sense of distribution of a continuous function is moderate. For we know that any distribution (linear) is locally such a function, (c.f. 1.3, Proposition 1). Thus given $T \in \mathcal{D}'$ and $T = D^\alpha g$ for some $g \in \mathcal{C}^\infty(\mathbb{R}^n)$, $<D^\alpha g, \phi_{\epsilon,x}> = (-1)^{|\alpha|} <g, D^\alpha \phi_{\epsilon,x}> = (-1)^{|\alpha|} \int_{\mathbb{R}^n} g(\lambda) \frac{1}{\epsilon} D^\alpha \phi(\frac{\lambda - x}{\epsilon}) \, d\lambda$. For $\mu = \frac{\lambda - x}{\epsilon}$, by the change of variable in the multiple integral we find

$$< D^\alpha g, \phi_{\epsilon,x} > = (-1)^{|\alpha|} \int_{\mathbb{R}^n} g(x + \epsilon\mu) \frac{1}{\epsilon^{|\alpha|}} D^\alpha \phi(\mu) \, d\mu \ \text{ and}$$

$$| < D^\alpha g, \phi_{\epsilon,x} > | \leq \frac{1}{\epsilon^{|\alpha|}} \int_{\mathbb{R}^n} |g(x + \epsilon\mu)| \, |D^\alpha \phi(\mu)| \, d\mu \leq \frac{c}{\epsilon^{|\alpha|}}$$

for some $c > 0$ depending on ϕ and g, $(m = |\alpha|)$, $\square$

To finalize the functional analytic approach culminating at 3.4.1, Definition 4) of the Colombeau's algebra we return back to finding an ideal $\mathcal{N}$ of $\mathcal{E}_M(\mathcal{D}(\mathbb{R}^n))$ such that $\mathcal{N} \bigcap \mathcal{C}^\infty(\mathcal{E}'(\mathbb{R}^n)) = Ker A$. For this purpose we might consider the ideal of $\mathcal{E}_M(\mathcal{D}(\mathbb{R}^n))$ spanned by $Ker A$, but $\mathcal{N}$ of 3.4.1, Definition 3 is larger and more appropriate. It was shown to be an ideal. Also it is easily shown that $R \in \mathcal{N} \Rightarrow D^\alpha R \in \mathcal{N}$. If $R \in \mathcal{C}^\infty(\mathcal{E}'(\mathbb{R}^n)) \bigcap \mathcal{N}$, then $R(\phi_{\epsilon,x}) \longrightarrow R(\delta_x)$ in $\mathcal{C}^\infty(\mathcal{E}'(\mathbb{R}^n))$. But R belongs to $\mathcal{N}$ and Remark 3 also applies to 3.4.1, Definition 3; thus for $|\alpha| = 0$, $|R(\phi_{\epsilon,x}| \leq c\epsilon^{m-l}$. Taking $\epsilon \to 0$ we find $|R(\delta_x)| = 0$. Hence $R \in Ker A$.

Conversely if $R \in Ker A$, then $D^\alpha R \in Ker A$. This is seen by Remark 2 above since it gives $A(\frac{\partial}{\partial x_i} R)(x) = \frac{\partial}{\partial x_i}(AR(x)) = 0$ and we can continue differentiating. Then we may use the characterization of $Ker A$ by Proposition 1 to conclude that $R \in \mathcal{N}$, (take $l = 0$). Thus $Ker A = \mathcal{N} \cap \mathcal{C}^\infty(\mathcal{E}'(\mathbb{R}^n))$.

3.4.4   *Generalized Complex Numbers*

Unlike linear generalized functions (Schwartz distributions), the point values of Colombeau generalized functions can be defined. Accordingly if $x \in \mathbb{R}^n$, $F \in \mathcal{G}(\mathbb{R}^n)$ and K (compact), the objects like $F(x)$, $\int_K F(x) \, dx$ can be defined and will belong to an algebra, namely the algebra of generalized complex numbers.

If $F = f + \mathcal{N} \in \mathcal{G}(\mathbb{R}^n)$, it would be natural to define $F(x)$ for a fixed $x \in \mathbb{R}^n$ through the mapping

$$\mathcal{A} \ni \phi \longrightarrow h(\phi) = f(\phi, x) \in \mathbb{C} \qquad (*)$$

$(\mathcal{A} \equiv \mathcal{A}_\infty$ is the functional index set (Sect. 3.4)). However $h \to F(x)$ is not injective. We factor out the non-injectivity by a suitable quotient structure similar to the one in the definition of generalized functions.

Let us denote $\mathcal{E}_0 = \mathbb{C}^\mathcal{A}$ which is an associative and commutative algebra.

Definition 1 $\mathcal{E}_{M^\circ} = \{h \in \mathcal{E}_0 : \exists m \in \mathbb{N}_+ \text{ such that } \forall \phi \in \mathcal{A}_m(\mathbb{R}^n), \exists \eta > 0, c > 0 \text{ with } |h(\phi_\epsilon)| \leq \frac{c}{\epsilon^m} \text{ if } 0 < \epsilon < \eta\}$.
$\mathcal{E}_{M^\circ}$ is a sub-algebra in $\mathcal{E}_0$.
We also define

Definition 2 $\mathcal{N}_0 = \{h \in \mathcal{E}_{M^\circ} : \exists l \in \mathbb{N}_+ \text{ such that } \forall \phi \in \mathcal{A}_m(\mathbb{R}^n)\ (m \geq l)\ \exists \eta > 0, c > 0 \text{ with } |h(\phi_\epsilon)| \leq c\, \epsilon^{\beta(m)-l} \text{ if } 0 < \epsilon < \eta\}$.
(For β see 3.4.1, the Note to Definition 3).
Again it follows that $\mathcal{N}_0$ is an ideal in $\mathcal{E}_{M^\circ}$ and $\mathcal{E}_{M^\circ} \subset_{\neq} \mathcal{E}_0$. Also $\mathcal{N}_0$ is not an ideal in $\mathcal{E}_0$.

Definition 3 The associative and commutative algebra $\overline{\mathbb{C}}$ of generalized complex numbers is defined as $\overline{\mathbb{C}} = \mathcal{E}_{M^\circ}/\mathcal{N}_0$.

<u>Notes</u>. (1) We notice that compared to 3.4.1, Definitions 2 and 3 there are correspondences $\mathcal{E}_0 \leftrightarrow \mathcal{E}$, $\mathcal{E}_{M^\circ} \leftrightarrow \mathcal{E}_M$ and $\mathcal{N}_0 \leftrightarrow \mathcal{N}$, so that the generalized complex numbers can be regarded as (Colombeau) generalized functions of order zero.
(2) Considering the definition of the index sets $\mathcal{A}_m$, $\overline{\mathbb{C}}$ depends on the dimension n of the underlying space $\mathbb{R}^n$. However this situation will not cause any problem in the sequel.

<u>Inclusion of Complex Numbers in $\overline{\mathbb{C}}$</u>.

For $z \in \mathbb{C}$, this inclusion is achieved by $z \longrightarrow \tilde{z} + \mathcal{N}_0 \in \overline{\mathbb{C}}$, where $\tilde{z}(\phi) = z$, $\forall \phi \in \mathcal{A}$. the inclusion is obviously injective, so that $\mathbb{C} \subset \overline{\mathbb{C}}$ is an imbedding of algebras.
An important relation is the following:

Definition 4   (Association with complex numbers). A complex number z is said to be associated with the generalized complex number $\tilde{z} \in \overline{\mathbb{C}}$ denoted

$$\tilde{z} \vdash z$$

if there is a representation of $\tilde{z}$ as $\tilde{z} = h + \mathcal{N}_0 \in \overline{\mathbb{C}}$ such that $\exists m \in \mathbb{N}_+$ satisfying $\lim_{\epsilon \downarrow 0} h(\phi_\epsilon) = z$ for every $\phi \in \mathcal{A}_m$.

<u>Note</u>. Not all $\tilde{z} \in \overline{\mathbb{C}}$ has an associated $z \in \mathbb{C}$. Take for instance Example 1 of 3.4.2, where δ was the delta function of one dimension, where as a generalized function it was represented by $\delta = f_\delta + \mathcal{N}$ with $f_\delta(\phi, x) = \phi(-x)$. Thus as indicated by (*) above (see also Definition 6 below) the point value of δ at $x = 0$ will be the generalized complex number $\tilde{z} = h + \mathcal{N}_0$, $h(\phi) = \phi(0)$. However then $h(\phi_\epsilon) = \frac{1}{\epsilon}$ does not have any limit as $\epsilon \to 0$.

<u>Remarks</u>: (1) If $\tilde{z} = z \in \mathbb{C}$, then $\tilde{z}(\phi) = z\ \forall \phi \in \mathcal{A}$, thus $\tilde{z} \vdash z$.
(2) The association $\tilde{z} \vdash$ is univalent, i.e. $\tilde{z} \vdash z_1$ and $\tilde{z} \vdash z_2$, then $z_1 = z_2$, $(z_1, z_2 \in \mathbb{C})$. This is because if $\tilde{z} = h_1 + \mathcal{N}_0 = h_2 + \mathcal{N}_0$, then we have $h_1 - h_2 \in \mathcal{N}_0$, thus by Definition 3, $\exists l \in \mathbb{N}_+$, for $m \geq l$ $|h_1(\phi_\epsilon) - h_2(\phi_\epsilon)| = |(h_1 - h_2)(\phi_\epsilon)| \leq c\, \epsilon^{\beta(m)-l}$

for sufficiently small $\epsilon > 0$ showing that $\lim_{\epsilon\downarrow 0} h_1(\phi_\epsilon) = \lim_{\epsilon\downarrow 0} h_2(\phi_\epsilon)$, hence $z_1 = z_2$, that means $\bar{z} \vdash z$ is well defined. However it is not injective:

Examples Let $\quad \bar{z} = h + \mathcal{N}_0 \in \overline{\mathbb{C}} = \mathcal{E}_{M^\circ}/\mathcal{N}_0 \quad$ be $\quad$ given $\quad$ by $\quad h(\phi) = \int_{\mathbb{R}} |x| \phi(x)\, dx$, $\phi \in \mathcal{A}_0$. We find $h(\phi_\epsilon) = \int_{\mathbb{R}} |x| \frac{1}{\epsilon} \phi(\frac{x}{\epsilon})\, dx = \int_{\mathbb{R}} |u| \phi(u)\, du = \epsilon\, h(\phi)$. This shows both $h \in \mathcal{E}_{M^\circ}$ and at the same time $\bar{z} \vdash z$, $z = 0 \in \mathbb{C}$. Also $h(\phi_\epsilon) = \epsilon\, h(\phi)$ does not satisfy the ideal condition of Definition 2, (unless $h(\phi) = 0$, for all $\phi \in \mathcal{A}$ which is not the case according to [Ros], Appendix 3, which states that for any $m \in \mathbb{N}_+$, there exists $\phi \in \mathcal{A}_m$ such that $h(\phi) = \int_{\mathbb{R}} |x| \phi(x)\, dx \neq 0$). Thus $h \notin \mathcal{N}_0$, accordingly $\bar{z} \neq 0 \in \overline{\mathbb{C}}$ showing the non-injectivity.

Definition 5 $\quad \overline{\mathbb{C}}_0$ is the set of all generalized complex numbers having associated complex numbers, i.e.

$$\overline{\mathbb{C}}_0 = \{\bar{z} \in \overline{\mathbb{C}} : \exists z \in \mathbb{C}, \ \bar{z} \vdash z\}.$$

<u>Note</u>. We have $\mathbb{C} \subset_{\neq} \overline{\mathbb{C}}_0 \subset_{\neq} \overline{\mathbb{C}}$ are imbeddings of algebras and if $\bar{z} \in \overline{\mathbb{C}}_0$, then $\bar{z} \vdash z$ is an algebra homomorphism.

Firstly it is easily verified that $\bar{z}_1 \vdash z_1$, $\bar{z}_2 \vdash z_2 \implies \bar{z}_1 + \bar{z}_2 \vdash z_1 + z_2$ and $\bar{z}_1.\bar{z}_2 \vdash z_1.z_2$, so that $\overline{\mathbb{C}}_0$, $\overline{\mathbb{C}}$ are algebras. By the Note to Definition 4, $\overline{\mathbb{C}}_0$ is a proper subalgebra of $\overline{\mathbb{C}}$ and the Example of Remark 2 above shows that $\mathbb{C}$ is a proper sub-algebra of $\overline{\mathbb{C}}_0$.

Definition 6 (Point values of generalized functions). If $f \in \mathcal{E}(\mathbb{R}^n)$ and $x \in \mathbb{R}^n$ we let $f_x(\phi) = f(\phi, x)$, $\phi \in \mathcal{A}$. Given a generalized function $F = f + \mathcal{N} \in \mathcal{G}(\mathbb{R}^n)$, $f \in \mathcal{E}_M$, the value $F(x)$ of F is by definition the generalized complex number $F(x) = f_x + \mathcal{N}_0 \in \overline{\mathbb{C}} = \mathcal{E}_{M^\circ}/\mathcal{N}_0$.

By 3.4.1, Definitions 2 and 3 we can see that $f \in \mathcal{E}_M \Rightarrow f_x \in \mathcal{E}_{M^\circ}$ and $f \in \mathcal{N} \Rightarrow f_x \in \mathcal{N}_0$, so that the point value $F(x)$ is well defined.

Examples (1) $\delta \in \mathcal{D}'$, $x = 0$, $\delta = f_\delta + \mathcal{N}$, $f_\delta(x) = \phi(-x)$ giving $\delta(0) = h + \mathcal{N}_0 \in \overline{\mathbb{C}}$, where $h(\phi) = \phi(0)$, $\phi \in \mathcal{A}$ and by the Note to Definition 4, $\delta(0) \notin \overline{\mathbb{C}}_0$.
(2) Reconsider the generalized function $F = x^m \delta \in \mathcal{G}(\mathbb{R})$ of 3.4.2, Example 3 yielding for the point value $F(x) = h + \mathcal{N}_0 \in \overline{\mathbb{C}}$ where $h(\phi) = x^m \phi(-x)$, $\phi \in \mathcal{A}$. As $h(\phi_\epsilon) = \frac{x^m}{\epsilon} \phi(-\frac{x}{\epsilon}) \longrightarrow 0$ as $\epsilon \to 0$ we have $h \in \mathcal{N}_0$ and $F(x) = 0 \in \overline{\mathbb{C}}$ although $x^m.\delta \neq 0 \in \mathcal{G}(\mathbb{R})$, (c.f. 3.4.2). Hence $x^m.\delta$, $m \in \mathbb{N}_+$ does not vanish in $\mathcal{G}(\mathbb{R})$ in spite of all of its point values do vanish in $\overline{\mathbb{C}}$.

<u>Conclusion</u>. The knowledge of all point values of a generalized function will not suffice to determine it.
(3) If $F = f \in \mathcal{C}^\infty(\mathbb{R}^n) \subset \mathcal{G}(\mathbb{R}^n)$, i.e. the generalized function is a $\mathcal{C}^\infty$-smooth function, then its point values at any $x \in \mathbb{R}^n$ coincides with the usual value $f(x)$. In fact by 3.4.2, $F = \tilde{f} + \mathcal{N} \in \mathcal{G}(\mathbb{R}^n)$, $\tilde{f}(\phi, x) = f(x)$, $\phi \in \mathcal{A}$, $x \in \mathbb{R}^n$. Thus $F(x) = f(x) \in \overline{\mathbb{C}}$.

(4) Suppose that the generalized function corresponds to a continuous function, i.e.
$F = f \in \mathcal{C}^\circ(\mathbb{R}^n) \subset \mathcal{G}(\mathbb{R}^n)$. Then $F(x) \in \overline{\mathbb{C}}_0$ and $F(x) \vdash f(x)$.
In fact by the inclusion result of 3.4.2, II):
$F = \bar{f} + \mathcal{N} \in \mathcal{G}(\mathbb{R}^n)$, $\bar{f} \in \mathcal{E}_M$ with $\bar{f}(x, y) = \int_{\mathbb{R}^n} f(x + y)\,\phi(y)\,dy$, $\phi \in$
$\mathcal{A}$, $x \in \mathbb{R}^n$ and the point value $F(x)$ at a fixed point $x \in \mathbb{R}^n$ will be given by
$F(x) = h + \mathcal{N}_0 \in \overline{\mathbb{C}}$ where $h(\phi) = \int_{\mathbb{R}^n} f(x + y)\,\phi(y)\,dy$, thus $h(\phi_\epsilon) =$
$\int_{\mathbb{R}^n} f(x + \epsilon u)\,\phi(u)\,du$, $(u = \frac{y}{\epsilon})$ and as $\phi \in \mathcal{A}$, we have $\lim_{\epsilon \downarrow 0} h(\phi_\epsilon) = f(x)$,
i.e. $F(x) \vdash f(x)$.

We also need the concept of the integral of generalized functions in the sequel.

Definition 7 (Integrals of generalized functions). Suppose given a generalized function $F = f + \mathcal{N} \in \mathcal{G}(\mathbb{R}^n)$ and $K \subset\subset \mathbb{R}^n$ compact, then the integral of F on K is defined as the generalized complex number

$$\int_K F(x)\,dx = h + \mathcal{N}_0 \in \overline{\mathbb{C}} = \mathcal{E}_{M^\circ}/\mathcal{N}_0$$

where $h(\phi) = \int_K f(\phi, x)\,dx$, $\phi \in \mathcal{A}$. The integral is well-defined, indeed $f(\phi, .) \in$
$\mathcal{C}^\infty(\mathbb{R}^n)$ since $f \in \mathcal{E}_M$. Furthermore $h(\phi_\epsilon) = \int_K f(\phi_\epsilon, x)\,dx$, $\phi \in \mathcal{A}$, $\epsilon > 0$ and in
view of 3.4.1, Definitions 1 and 2 above we have $h \in \mathcal{E}_{M^\circ}$.
Also the definition of the integral does not depend on f since $f \in \mathcal{N} \Rightarrow h \in \mathcal{N}_0$.

Notes. (Special Cases). (1) If F is like in the Example 3 above, i.e. $F = f \in$
$\mathcal{C}^\infty(\mathbb{R}^n) \subset \mathcal{G}(\mathbb{R}^n)$, then the integral of F coincides with the notion of the ordinary
integral. This is because, in this case $F = \tilde{f} + \mathcal{N}$, $\tilde{f}(\phi, x) = f(x)$, $\phi \in \mathcal{A}$, thus
$\int_K F(x)\,dx = \tilde{z} + \mathcal{N}_0 \in \overline{\mathbb{C}}$, $\tilde{z}(\phi) = \int_K f(x)\,dx$ and by the imbedding of $\mathbb{C} \subset \overline{\mathbb{C}}$
we have $\int_K F(x)\,dx = \int_K f(x)\,dx$.
(2) If on the other hand $F = f \in \mathcal{C}^\circ(\mathbb{R}^n) \subset \mathcal{G}(\mathbb{R}^n)$ like in the Example 4 above, then
for every compact $K \subset\subset \mathbb{R}^n$ we have

$$\int_K F(x)\,dx \in \overline{\mathbb{C}}_0 \quad \text{and} \quad \int_K F(x)\,dx \vdash \int_K f(x)\,dx,$$

i.e. the following commutative diagram holds (Fig. 3.3):
By the inclusion 3.4.2, II of a continuous function into $\mathcal{G}(\mathbb{R}^n)$:
$F = \bar{f} + \mathcal{N}$, $\bar{f} \in \mathcal{E}_M$ and $\bar{f}(\phi, x) = \int_{\mathbb{R}^n} f(x + y)\,\phi(y)\,dy$; $(\phi \in \mathcal{A}, x \in \mathbb{R}^n)$. By

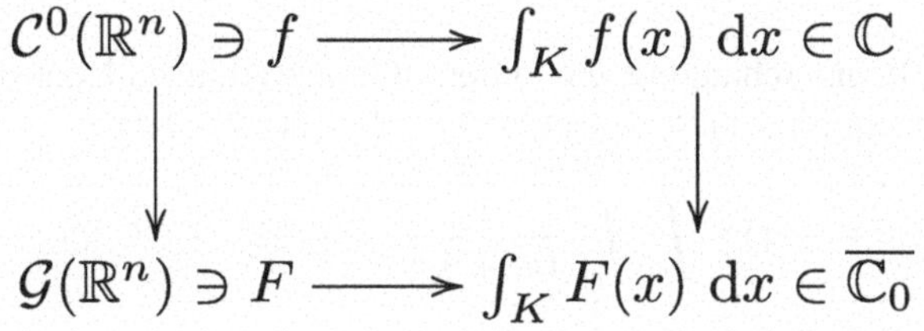

Fig. 3.3 Finding the integral
of a continuous function as a
generalized complex number
in Colombeau algebras in
two different ways

$$\begin{array}{ccc} \mathcal{C}^0(\mathbb{R}^n) \ni f & \longrightarrow & \int_K f(x)\,dx \in \mathbb{C} \\ \downarrow & & \downarrow \\ \mathcal{G}(\mathbb{R}^n) \ni F & \longrightarrow & \int_K F(x)\,dx \in \overline{\mathbb{C}}_0 \end{array}$$

Definition 7:
$\int_K F(x)\,dx = h + \mathcal{N}_0 = \int_K \left(\int_{\mathbb{R}^n} f(x+y)\,\phi(y)\,dy \right) dx + \mathcal{N}_0;$	let	$\epsilon > 0,$

then	$h(\phi_\epsilon) = \int_K \left(\int_{\mathbb{R}^n} f(x+\epsilon y)\,\phi(y)\,dy \right) dx \xrightarrow{\epsilon \to 0} \int_K f(x)\,\left(\int_{\mathbb{R}^n} \phi(y)\,dy \right) dx = \int_K f(x)\,dx,$

(By Definition 1) and by Definition 4 this means $\int_K F(x)\,dx \vdash \int_K f(x)\,dx$ $\qquad \square$

(3) Let $F \in \mathcal{G}(\mathbb{R}^n)$, $\xi \in \mathcal{D}(\mathbb{R}^n)$ and $K \subset\subset \mathbb{R}^n$ be given such that $supp\,\xi \subset K$. Suppose the product $\xi.F$ be defined in $\mathcal{G}(\mathbb{R}^n)$, (i.e. if $F = f + \mathcal{N}$, then $\xi.F = \xi.f + \mathcal{N}$). The integral of $\xi.F$ on K or on any compact set containing $supp\,\xi$ would be the same, thus we define $\int_{\mathbb{R}^n}(\xi.F)(x)\,dx \doteq \int_K (\xi.F)(x)\,dx.$

Theorem *Suppose $\xi \in \mathcal{D}(\mathbb{R}^n)$ and $T \in \mathcal{D}'(\mathbb{R}^n)$. Then $\int_{\mathbb{R}^n}(\xi.T)(x)\,dx = < T, \xi > \in \mathbb{C} \subset \overline{\mathbb{C}}$, (the product $\xi.T$ is considered in $\mathcal{G}(\mathbb{R}^n)$).*

Proof According to 1.3, Proposition.1 every linear distribution $T \in \mathcal{D}'(\mathbb{R}^n)$ is locally $D^\alpha f$ of some continuous function f for some multi-index α. We can select a suitable compact set K, such that $supp\xi \subset K^\circ$ (K° : the interior of K). Then $\int_{\mathbb{R}^n}(\xi, T)(x)\,dx = \int_{\mathbb{R}^n}(\xi.D^\alpha f)(x)\,dx.$

(D^α is the partial derivative operator in $\mathcal{G}(\mathbb{R}^n)$. Considering $\xi.D^\alpha f$ as a member of $\mathcal{G}(\mathbb{R}^n)$, $\xi.D^\alpha f = g + \mathcal{N}$ where g will be in view 3.4.2, II):

$g(\phi, x) = (-1)^{|\alpha|}\xi(x) \int_{\mathbb{R}^n} f(y)\,D^\alpha\phi(y-x)\,dy$, $\phi \in \mathcal{E}_M, x \in \mathbb{R}^n$. Thus by the Note 3) preceding the Theorem:

$\int_{\mathbb{R}^n}(\xi.T)(x)\,dx = h + \mathcal{N}_0 \in \overline{\mathbb{C}}$ with $h(\phi) = (-1)^{|\alpha|} \int_{\mathbb{R}^n}\int_{\mathbb{R}^n} \xi(x)\,f(y)\,D^\alpha\phi(y-x)\,dx\,dy$, $\phi \in \mathcal{A}.$

Integration by parts yields:

$h(\phi) = (-1)^{|\alpha|} \int_{\mathbb{R}^n}\int_{\mathbb{R}^n} D^\alpha\xi(x)\,f(y)\,\phi(y-x)\,dx\,dy$

and by a suitable change of variable we have

$h(\phi) = (-1)|\alpha| \int_{\mathbb{R}^n}\int_{\mathbb{R}^n} D^\alpha\xi(x-y)\,f(x)\,\phi(y)\,dx\,dy$ so that

$h(\phi_\epsilon) = (-1)^{|\alpha|} \int_{\mathbb{R}^n}\int_{\mathbb{R}^n} D^\alpha\xi(x-\epsilon y)\,f(x)\,\phi(y)\,dx\,dy$ and as $\int \phi(y)\,dy = 1$

(3.4.1, Definition 1)): $\lim_{\epsilon\downarrow 0} h(\phi_\epsilon) = (-1)^{|\alpha|} \int_{\mathbb{R}^n} D^\alpha\xi(x)\,f(x)\,dx.$

But then observe that in the sense of distributional derivative

$\lim_{\epsilon\downarrow 0} h(\phi_\epsilon) = < D^\alpha f, \xi > = < T, \xi > \in \mathbb{C},$ $\qquad (*)$

On the other hand by the Taylor formula

$$h(\phi_\epsilon) = (-1)^{|\alpha|} \int_{\mathbb{R}^n} \int_{\mathbb{R}^n} \{ D^\alpha\xi(x) + \sum_{1 \leq q \leq m} \frac{1}{q!}(-\epsilon y)^q\, D^{\alpha+q}\xi(x)$$

$$+ \frac{1}{(m+1)!}(-\epsilon y)^{m+1}\, D^{\alpha+q+1}\xi(x - \Theta\epsilon y) \} \times f(x)\,\phi(y)\,dx\,dy, \quad (0 < \theta < 1).$$

By interchanging the order of integration and considering $\int y^q\,dy = 0$, (3.4.1, Definition 1): $h(\phi_\epsilon = (1)^{|\alpha|} \int_{\mathbb{R}^n} D^\alpha\xi(x)\,f(x)\,dx$

$$+ (-1)^{|\alpha|} \int_{\mathbb{R}^n} \int_{\mathbb{R}^n} \frac{1}{(m+1)!}(-\epsilon y)^{m+1}\, D^{\alpha+q+1}\xi(x - \Theta\,\epsilon\,y)\, f(x)\,\phi(y)\,dx\,dy,$$

using (*):

$$|h(\phi_\epsilon) - <T,\xi>| = \frac{\epsilon^{m+1}}{(m+1)!} | \int_{\mathbb{R}^n} \int_{\mathbb{R}^n} y^{m+1} \, D^{\alpha+q+1}\xi(x - \Theta\epsilon y)\, f(x)\, \phi(y)\, dx\, dy|.$$

By an analysis similar to that of 3.4.2, Remark 4 we have $h(\phi) - <T,\xi> \in \mathcal{N}_0$; thus $\int_{\mathbb{R}^n}(\xi.T)(x)\, dx$ is represented by the complex number $<T,\xi>$. $\qquad\square$

Application. If $T = \delta$ in one dimension, then for every $x \in \mathcal{D}(\mathbb{R})$
we have $\int_{\mathbb{R}}(\xi.\delta)(x)\, dx = <\delta,\xi> = \xi(0)$.

Note. If $T \in \mathcal{D}'(\mathbb{R})$ is a regular distribution, i.e. a locally integrable function, then $\int_{\mathbb{R}^n}(\xi.T)(x)\, dx = <T,\xi> = \int_{\mathbb{R}^n} f_T(x)\,\xi(x)\, dx$ becomes an ordinary integral.

3.4.5 *Coupled Calculus*

As indicated in 3.4.2, II) Remark 1, $\mathcal{C}^\circ(\mathbb{R}^n) \subset \mathcal{G}(\mathbb{R}^n)$ is a vector space imbedding but not an imbedding of algebras. That means the usual multiplication of continuous functions in $\mathcal{C}^\circ$ does not extend to $\mathcal{G}(\mathbb{R}^n)$. This was one of the consequences of the Schwartz's Impossibility Result. Among the other theories of nonlinear generalized functions, the one of Colombeau's is regarded to tackle such deficiencies and pathologies in an optimal manner. In fact this is achieved by a coupled calculus peculiar to Colombeau distributions.

The starting point is the introduction of certain equivalence relations. We first notice that whenever $T \in \mathcal{D}'(\mathbb{R}^n)$, the complex numbers $\int_{\mathbb{R}^n}(\xi.T)(x)\, dx \in \mathbb{C}$ yield a local characterization of T. Conversely by 1.4, Theorem 1, given such a local characterization, it determines a unique distribution.

By the Theorem of 3.4.4 zero distribution in $\mathcal{D}'(\mathbb{R}^n)$ can be characterized by

$$T = 0 \in \mathcal{D}'(\mathbb{R}^n) \Longleftrightarrow \int_{\mathbb{R}^n}(\xi.T)(x)\, dx = 0, \;\; \forall\xi \in \mathcal{D}(\mathbb{R}^n).$$

However this does not generalize to nonlinear distributions in $\mathcal{G}(\mathbb{R}^n)$. We rather have

Definition 1 $F \in \mathcal{G}(\mathbb{R}^n)$ is called <u>test null</u>, denoted $F \sim 0$ if for every $\xi \in \mathcal{D}(\mathbb{R}^n)$ we have $\int_{\mathbb{R}^n}(\xi.F)(x)\, dx = 0 \in \overline{\mathbb{C}}$, (the product is computed in $\mathcal{G}(\mathbb{R}^n)$.)

$F_1, F_2 \in \mathcal{G}(\mathbb{R}^n)$ are called <u>test equal</u> if $F_1 - F_2$ is test null. Then this is denoted by $F_1 \sim F_2$. Being test equal is clearly an equivalence relation which is a bit too strong. A weaker relation is introduced after Definition 2.

Definition 2 $T \in \mathcal{D}'(\mathbb{R}^n)$ is said to be <u>associated</u> with a generalized function $F \in \mathcal{G}(\mathbb{R}^n)$, denoted by $F \models T$ if for every $\xi \in \mathcal{D}(\mathbb{R}^n)$

$$\int_{\mathbb{R}^n} (\xi.F)(x)\,dx \vdash \int_{\mathbb{R}^n} (\xi.T)(x)\,dx.$$

(For the relation $\vdash$, association of complex numbers see 3.4.5, Definition 4 and the theorem).

<u>Note</u>. By 3.4.4, Definition 7 this association would mean

$$\lim_{\epsilon \downarrow 0} \int_{\mathbb{R}^n} f_F(\phi_\epsilon, x)\,\psi(x)\,dx = <T, \psi> \quad (F = f_F + \mathcal{N})$$

for all $\phi \in \mathcal{A}(\mathbb{R}^n)$. This can be interpreted as follows: If we reduce the information contained in F to the level of linear distribution theory, i.e. consider only its limiting action on test functions ϕ then F behaves like T.

Definition 3 Two generalized functions F_1, $F_2 \in \mathcal{G}(\mathbb{R}^n)$ are said to be associated if $F_1 - F_2 \models 0 \in \mathcal{D}'(\mathbb{R}^n)$, i.e. $F_1 - F_2$ has 0 as the associated linear distribution. This relation is denoted by $F_1 \approx F_2$ and is clearly an equivalence relation.

Theorem *(Properties of the relations $\sim$ and $\approx$)*

(a) If $F \in \mathcal{G}(\mathbb{R}^n)$ and $T \in \mathcal{D}'(\mathbb{R}^n)$, then

    (1) $F \models T \Longleftrightarrow F \approx T$
    (2) $T \sim 0 \Longrightarrow T = 0, \quad T \approx 0 \Longrightarrow T = 0$.

(b) Let f_1, f_2 be continuous functions in $C^\circ(\mathbb{R}^n)$. $f_1.f_2$ defines a regular distribution $T_{f_1.f_2} \in \mathcal{D}'(\mathbb{R})$. Then $T_{f_1}.T_{f_2} \in \mathcal{G}(\mathbb{R}^n)$ is related to $T_{f_1.f_2}$ by $T_{f_1}.T_{f_2} \models T_{f_1.f_2}$, i.e. $T_{f_1.f_2}$ is the linear distribution associated to the generalized functionb $T_{f_1}.T_{f_2}$.

(c) Let $h \in C^\infty(\mathbb{R}^n)$ and $T \in \mathcal{D}'(\mathbb{R}^n)$. The product $h.T$ can be regarded both in $\mathcal{D}'(\mathbb{R}^n)$ and also in $\mathcal{G}(\mathbb{R}^n)$, i.e. $S = h.T \in \mathcal{D}'(\mathbb{R}^n)$ and $F = h.T \in \mathcal{G}(\mathbb{R}^n)$. Then $S \sim F$.

Proof (a) Follows directly from the Definitions 1, 2 and 3 and also by the fact that $T \in \mathcal{D}'(\mathbb{R}^n)$, $\int_{\mathbb{R}^n}(\xi.T)(x)\,dx = 0$ for all $\xi \in \mathcal{D}(\mathbb{R}^n)$ characterizes zero linear distribution.
Proof of (b) and (c) also utilizes these definitions and the inclusion properties as covered in 3.4.2. (For the details see [Ros]. Theorem 7, 89–92).

<u>Note</u>. Part (b) of the Theorem implies that if $F_i = T_{f_i} \in \mathcal{G}(\mathbb{R}^n)$, $(i = 1, 2)$. then the product rules in $C^\circ(\mathbb{R}^n)$ and $\mathcal{G}(\mathbb{R}^n)$ are linked by the equivalence relation $F_1.F_2 \approx f_1.f_2$. Thus in Colombeau's theory one of the main difficulties posed by the Schwartz's Impossibility Result and faced by any theory of nonlinear generalized

functions are overcome with the least concession through this additional structure. However we should note that in general the relation $F_1.F_2 \sim f_1.f_2$ is wrong.

<u>Remark</u>. a-2) is wrong for $T \in \mathcal{G}(\mathbb{R}^n)$, that means being test-null or being associated with zero linear distribution does not mean that T is zero generalized function, i.e. $F \sim 0 \not\Rightarrow F = 0$, $F \approx 0 \not\Rightarrow F = 0$, $(F \in \mathcal{G}(\mathbb{R}^n))$.

<u>COUNTER-EXAMPLE</u>. Let $F = x.\delta \in \mathcal{F}(\mathbb{R})$,(the multiplication is in $\mathcal{G}(\mathbb{R})$.) But $x.\delta$ as a linear distribution is zero, $(< x.\delta, \ \phi >=< \delta, x.\phi >= (x\phi)(0) = 0, \ \phi \in \mathcal{D}(\mathbb{R}))$. Thus by the 3.4.4, Theorem. $\int_{\mathbb{R}} (\psi.F)(x)\,dx = 0 \in \mathbb{C}$. However by 3.4.2, Example 3) $x.\delta \neq 0$ in $\mathcal{G}(\mathbb{R})$, showing $F \sim 0 \not\Rightarrow F = 0$ also the second implication is wrong since $F \sim 0 \Rightarrow F \approx 0$.

Examples (1) We consider H^r, $r \in \mathbb{N}_+$; (H : Heaviside function and the product in $\mathcal{G}(\mathbb{R})$). Then $H^r \approx H$.

Proof In fact $H = D\,x_+$, (c.f. 3.2, Corollary). Considering the imbedding of the regular distribution T_{x_+} into $\mathcal{G}(\mathbb{R})$ we have $T_{x_+} = f_{x_+} + \mathcal{N}$ where $f_{x_+}(\phi, x) = \int_{\mathbb{R}} x_+(y + x)\phi(y)\,dy = \int_{-x}^{\infty} (y + x)\,\phi(y)\,dy$. Thus $H = h + \mathcal{N}$ and considering $\phi(+\infty) = 0$ we have $h(\phi, x) = \frac{d}{dx} f_{x_+}(\phi, x) = \int_{-x}^{\infty} \phi(y)\,dy$ and $H^r = h^r + \mathcal{N} \in \mathcal{G}(\mathbb{R})$.

For a given $\phi \in \mathcal{A}$ let $\eta \in C^\infty(\mathbb{R})$ be defined by $\eta(x) = \int_{-x}^{\infty} \phi(y)\,dy$, $x \in \mathbb{R}$.

$$\text{then } \eta(-\infty) = 0, \ \eta(+\infty) = 1, \ \eta'(x) = \phi(x). \qquad (*)$$

Recalling that $\phi_\epsilon(x) = \frac{1}{\epsilon}\phi(\frac{x}{\epsilon})$:
$h(\phi_\epsilon, x) = \int_{-x}^{\infty} \frac{1}{\epsilon}\phi(\frac{y}{\epsilon})\,dy = \int_{-\frac{x}{\epsilon}}^{\infty} \phi(u)\,du = 1 - \eta(-\frac{x}{\epsilon})$.
For $\psi \in \mathcal{D}(\mathbb{R})$:

$$\int_{\mathbb{R}} \psi(H^r - H)(x)dx = g + \mathcal{N}_0 \in \overline{\mathbb{C}}, \qquad (**)$$

with 3.4.4, Definition 7:
$g(\phi_\epsilon) = \int_{\mathbb{R}} \psi(x)[(1 - \eta(-\frac{x}{\epsilon}))^r - (1 - \eta(-\frac{x}{\epsilon}))]\,dx, \ (\epsilon > 0)$.

If we call the inside the bracket of the integrand by k_ϵ and use $(*)$:

$$\lim_{\epsilon \downarrow 0} k_\epsilon(x) = \begin{cases} 0 & \text{if } x \neq 0 \\ (1 - \eta(0))^r - (1 - \eta(0)) & \text{if } x = 0 \end{cases}.$$

If $M = \int_{\mathbb{R}} |\phi(y)|\,dy < \infty$, then clearly $|k_\epsilon(x)| \leq (1 + M)^r + (1 + M)$.

Thus by the Lebesgue bounded convergence theorem we can enter the limit inside the integral to find $\lim_{\epsilon \downarrow 0} g(\phi_\epsilon) = 0$.

Then by $(**)$ we have $\int_{\mathbb{R}} \psi.(H^r - H))(x)\,dx \vdash 0$, $\psi \in \mathcal{D}(\mathbb{R})$ showing $H^r \approx H$. $\qquad \square$

<u>Note</u>. By differentiating both sides of $H^r \approx H$ and using the fact that $F_1 \approx F_2 \Rightarrow D^\alpha F_1 \approx D^\alpha F_2$ we find $r\,H^{r-1}.\delta \approx \delta$ and when $r = 2$, $H.\delta \approx \frac{1}{2}\delta$.

Corollary *The equivalence relations $\sim$ and $\approx$ are not compatible with the multiplication in $\mathcal{G}(\mathbb{R}^n)$.*

Proof We show this by two counter-examples:

<u>COUNTER EXAMPLE 1.</u> Let $F_1 = F_2 = x.\delta \in \mathcal{G}(\mathbb{R})$ where the product is considered in $\mathcal{G}(\mathbb{R})$. By Theorem -c) we have $F_1 \sim 0$, $F_2 \sim 0$. On the other hand by 3.4.2, Example 1 $F_1.F_2 = f + \mathcal{N} \in \mathcal{G}(\mathbb{R})$ where $f(\phi, x) = x^2\phi^2(-x)$, $\phi \in \mathcal{A}$, $x \in \mathbb{R}$. Hence for $\xi \in \mathcal{D}(\mathbb{R})$:
$\int_{\mathbb{R}}(\xi.F_1.F_2)(x)\,dx = h + \mathcal{N}_0 \in \overline{\mathbb{C}}$ with $h(\phi) = \int_{\mathbb{R}}\xi(x)\,x^2\phi^2(-x)\,dx$ and
$h(\phi_\epsilon) = \int_{\mathbb{R}}\xi(x)\,x^2\frac{1}{\epsilon^2}\,\phi^2(-\frac{x}{\epsilon})\,dx = \epsilon\int_{\mathbb{R}}\xi(-\epsilon u)\,u^2\,\phi^2(u)\,du$.
As this does not vanish faster than any power of ϵ (c.f. 3.4.1, Definition 3)) we have $h \notin \mathcal{N}_0$, hence $F_1.F_2 \sim 0$ does not hold.

<u>COUNTER EXAMPLE 2.</u> Let $l \in \mathbb{N}_+$, $r \in \mathbb{N}_+$, by Example 1 and its Note:
$H^l \approx H$, $(r+1)H^r.\delta \approx \delta$ $(\sqrt{})$
The product of the left-hand sides of the above relations will be $(r+1)\,H^{r+l}.\delta$; applying the second relation of $(\sqrt{})$ once more for $r \longleftrightarrow r+l$:

$$(r+1)H^{r+l}.\delta \approx \frac{r+1}{r+l+1}\,\delta,$$

on the other hand the product of the right-hand sides of $(\sqrt{})$ yields $H.\delta \approx \frac{1}{2}\delta$ (Note of Example 1).
As $\frac{r+1}{r+l+1} \neq \frac{1}{2}$ unless $l = r+1$, we have that $H^l.(r+1)H^r.\delta \approx H.\delta$ does not hold. Hence the product of generalized functions in $(\sqrt{})$ does not preserve the relation $\approx$ $\square$

   In particular for $l = r = 1$: $h^2.\delta \approx \frac{1}{2}H.\delta$ does not hold.

3.4.6 *Generalized Nonlinear Operations on $\mathcal{G}(\mathbb{R}^n)$*

Due to the algebraic structure of $\mathcal{G}(\mathbb{R}^n)$, one can perform polynomial operations and obtain new generalized functions. Moreover it is possible to perform non-linear operations more general than the polynomial ones as long as the growth of the representatives are controlled.

Theorem *Let $h \in \mathbb{R}^{2r} \longrightarrow \mathbb{C}$ be a slowly increasing function $\mathcal{O}(\mathbb{R}^{2r}) \subset \mathcal{C}^\infty$, (c.f. 1.5, Definition 2). Let $F : \mathcal{A}(\mathbb{R}^{2n}) \times \mathbb{R}^n \longrightarrow \mathbb{C}$ be defined by $F(\phi, x) = h[f_1(\phi, x), \cdots, f_r(\phi, x)].(\mathbb{R}^{2r}$ is identified with $\mathbb{C}^r)$, where each $f_i \in \mathcal{E}_M(\mathbb{R}^n)$ is a representative of $T_i \in \mathcal{G}(\mathbb{R}^n)$, $(1 \leq i \leq r)$. Then $F \in \mathcal{E}_M(\mathbb{R}^n)$;(its class in $\mathcal{G}(\mathbb{R}^n)$ will be denoted by $h(T_1, \cdots, T_r))$.*

Proof By 1.5 Definition 2' there are $C' > 0$ and $N \in \mathbb{N}$ such that for all $x \in \mathbb{R}^{2r}$

$$|h(x)| \leq C'(1 + \|x\|)^N, \quad (\|.\| \text{ in } \in \mathbb{R}^{2r}) \quad (\dagger)$$

Also given $K \subset\subset \mathbb{R}^n$, utilizing 3.4.1, Definition 2, there is $m \in \mathbb{N}$ such that for all $\phi \in \mathcal{A}_m(\mathbb{R}^n)$ there are $c > 0$ and $\eta > 0$ satisfying

$$|f_i(\phi_\epsilon, x)| \le c\epsilon^{-m} \qquad (\ddagger)$$

for all $x \in K$, $0 < \epsilon < \eta$ and $1 \le i \le r$. Then for each $\phi \in \mathcal{A}_m(\mathbb{R}^n)$ we have by $(\dagger)$ and $(\ddagger)$

$$|F(\phi_\epsilon, x)| \le |h(f_1(\phi_\epsilon, x), \cdots, f_r(\phi_\epsilon, x)| \le C'(1 + \|f_1(\phi_\epsilon, x), \cdots, f_r(\phi_\epsilon, x)\|)^N$$

$$\le C'(1 + rc\epsilon^{-m})^N \le C'(1 + rc)^N \epsilon^{-mN}$$

for all $x \in K$ and $0 < \epsilon < \eta$.

For the first order derivatives of F we have, by the chain rule:

$$\frac{\partial}{\partial x_i} F(\phi, x) = \sum_{j=1}^{r} \partial_j h(f_1(\phi, x), \cdots, f_r(\phi, x)) \frac{\partial f_j}{\partial x_i}(\phi, x)$$

where $\partial_j h(z_1, \cdots, z_r)$ denotes the partial derivatives of h with respect to the complex variable z_j. Again using bounds for $\partial_j h$ and $\frac{\partial f_j}{\partial x_i}$ we get a bound like $k\epsilon^{-N}$ for $|\frac{\partial h(\phi_\epsilon, x)}{\partial x_i}|$ uniformly for $x \in K$. Similar reasonings apply to any partial x-derivative of h implying that $h \in \mathcal{E}(\mathbb{R}^n)$. $\square$

<u>Remark</u>. The class $h(T_1, \cdots, T_r)$ does not depend on the choice of $(f_1, \cdots, f_r)$. To show this assume that another set of representatives $\bar{f}_i \in \mathcal{E}_M$ has been selected. If $\overline{F}(\phi, x) = h(\bar{f}_1(\phi, x), \cdots, \bar{f}_r(\phi, x))$ we prove without much difficulty that $F - \overline{F} \in \mathcal{N}(\mathbb{R}^n)$. This is achieved by first applying the mean value theorem to $F - \overline{F}$ and then estimating $(F - \overline{F})(\phi_\epsilon, x)$ and its partial derivatives. For the details see [Bia].

3.4.7 *Generalized Functions in Arbitrary Domains and the Sheaf Property*

Let Ω be an open subset of $\mathbb{R}^n$. There are mainly two approaches for defining algebras of generalized functions on Ω.

(A) Follow a construction parallel to the above subsections with some technical modifications.

(B) Use the sheaf property of Colombeau generalized functions.

(A) Let $\mathcal{A}(\Omega)$ be defined as follows

$$\mathcal{A}(\Omega) \doteq \{(\phi, x) \in \mathcal{A} \times \Omega : \phi(., -x) \in \mathcal{D}(\Omega)\}$$

i.e. those test functions which will be included in $\mathcal{D}(\Omega)$ after a shift of x units.

Also define

$$\Omega(\phi) \doteq \{x \in \Omega : (\phi, x) \in \mathcal{A}(\Omega)\}$$

Now the role of $\mathcal{E}(\mathbb{R}^n)$ of the Sect. 3.4.1 will be taken up by

$$\mathcal{E}(\Omega) = \{f : \mathcal{A}(\Omega) \to \mathbb{C} : \forall \phi \in \mathcal{A}, \ f((\phi, \cdot)) \in C^\infty(\Omega(\phi))\}.$$

Clearly $\mathcal{E}(\Omega)$ is a differential algebra and $C^\infty(\Omega) \subset \mathcal{E}(\Omega)$, through the imbedding $C^\infty(\Omega) \ni f \to \tilde{f} \in \mathcal{E}(\Omega)$ where $\tilde{f}(\phi, x) = f(x)$, $(\phi, x) \in \mathcal{A}(\Omega)$.
We reconstruct the moderate members of $\mathcal{E}(\Omega)$ in parallel to 3.4.1.

Definition 1 The subalgebra $\mathcal{E}_M(\Omega)$ of moderate members in $\mathcal{E}(\Omega)$
consists of all $f \in \mathcal{E}(\Omega)$ such that for all $K \subset\subset \Omega$, $p \in \mathbb{N}^n$ there is $m \in \mathbb{N}_+$ satisfying the following: $\forall \phi \in \mathcal{A}_m \ \exists \eta, c > 0$ such that $\forall x \in K$, $\epsilon \in (0, \eta)$ $(\phi_\epsilon, x) \in \mathcal{A}(\Omega)$ and $|D_x^p f(\phi_\epsilon, x)| \leq \frac{c}{\epsilon^m}$.

<u>Note.</u> As the distance between K is positive, obviously for ϵ small enough and $x \in K$, $\phi(\cdot, -x) \in \mathcal{D}(\Omega)$; i.e. , $(\phi_\epsilon, x) \in \mathcal{A}(\Omega)$.

Definition 2 The ideal $\mathcal{N}(\Omega)$ in $\mathcal{E}(\Omega)$ consists of all $f \in \mathcal{E}_M(\Omega)$ such that $\forall K \subset\subset \Omega$, $p \in \mathbb{R}^n \ \exists l \in \mathbb{N}_+$, $\beta \in B$ satisfying $\forall m \in \mathbb{N}_+$, $m \geq l$, $\phi \in \mathcal{A}_m \ \exists \eta, c > 0$, $\forall x \in K$, $\epsilon \in (0, \eta)$ with $(\phi_\epsilon, x) \in \mathcal{A}(\Omega)$ and $|D_x^p f(\phi_\epsilon, x)| \leq c \, \epsilon^{\beta(m)-l}$

<u>Note.</u> For B c.f. 3.4.1 Note 2 following Definition 3, i.e. $B = \{\beta : \mathbb{N}_+ \to (0, \infty) : \beta \nearrow, \lim \beta(m) = \infty\}$.
Then $\mathcal{G}(\Omega) = \mathcal{E}_M(\Omega)/\mathcal{N}(\Omega)$.

It is easily shown that $\mathcal{G}(\Omega)$ has similar properties with those of $\mathcal{G}(\mathbb{R}^n)$ as discussed in Sects. 3.4.1 and 3.4.2. One of the basic results was the imbedding $C^0(\mathbb{R}^n) \subset \mathcal{G}(\mathbb{R}^n)$. The counterpart will be $C^0(\Omega) \subset \mathcal{G}(\Omega)$ defined by $C^0(\Omega) \ni f \to \bar{f} + \mathcal{N}(\Omega) \in \mathcal{G}(\Omega)$ where $\bar{f}(\phi, x) = \int_\Omega f(y)\, \phi(y - x)\, dy$, $(\phi, x) \in \mathcal{A}(\Omega)$. The integral is well-defined since by the definition of $\mathcal{A}(\Omega)$, $\phi(\cdot - x) \in \mathcal{D}(\Omega)$. Normally $C^0(\Omega)$ is not integrable. We only have $C^0(\Omega) \in L^1_{loc}(\Omega)$.

<u>Remark.</u> In the case of $\mathcal{G}(\mathbb{R}^n)$, the index set $\mathcal{A}$ as well as $\mathbb{R}^n$ had unrestricted ranges for the independent variables ϕ and x respectively. In the case of $\mathcal{G}(\Omega)$ this is not true. In fact (ϕ_ϵ, x) should vary in $\mathcal{A}$, i.e. $\phi_\epsilon(\cdot\cdot - x) \in \mathcal{D}(\Omega)$.
  (B) Let $\Omega(\text{open}) \subset \mathbb{R}^n$. Let us define the functional index set $\mathcal{A}_m$, $m \geq 1$, $\mathcal{A}_1 \equiv \mathcal{A}$ as in 3.4.1, Definition 1. Consider instead of $\mathcal{E}(\mathbb{R}^n)$, $\mathcal{E}(\Omega) = (C^\infty(\Omega))^{\mathcal{A}}$. Using component-wise operations, it is again a differential algebra. Regularizations and the factorization are carried out as in 3.4.1. In other words the subalgebra of moderate members $\mathcal{E}_M(\Omega)$ and the ideal $\mathcal{N}(\Omega) \in \mathcal{E}_M(\Omega)$ are defined as in the Sect. 3.4.1. with obvious modifications. Thus the factor algebra $\mathcal{G}(\Omega) = \mathcal{E}_M(\Omega)/\mathcal{N}(\Omega)$ is defined.
  Then, however we have a difficulty in defining the inclusion of $\mathcal{D}'(\Omega)$ into $\mathcal{G}(\Omega)$. ($\mathcal{D}'(\Omega)$ is the set of continuous linear forms on $\mathcal{D}(\Omega)$ of compactly supported

functions). The imbedding $\mathcal{D}' \ni T \to (T * \phi \,|\, \Omega)_{\phi \in \mathcal{A}}$ will not be well-defined as $supp\, \phi \cap \Omega$ may not be a compact set.

In order to circumvent this difficulty we observe that the collection $\mathfrak{S} = \{\mathcal{G}(\Omega)\}_{\Omega(\text{open}) \subset \mathbb{R}^n}$ has a sheaf structure.

Definition 3 Let X be a topological space and let $\mathfrak{S}$ be a set of spaces S. Suppose given a mapping $X \supset U(open) \overset{\sigma}{\to} \sigma(U) = S \in \mathfrak{S}$, we call $S = \sigma(U)$ <u>a section over U</u>. We have restriction mappings: for $U \subset V$

$$\rho_{U,V} : \sigma(V) \to \sigma(U) \text{ denoted by } \sigma(V)|_U .$$

Then $(\sigma, \rho_{U,V})$ is called a <u>sheaf of sections</u> over X if the following four conditions ara satisfied:

(1) $\rho_{U,U} = id_{\sigma(U)} \to \sigma(U)$,

(2) For every open sets $U, V, W \subset X$ such that $U \subset V \subset W$ we have $\rho_{U,V} \circ \rho_{V,W} = \rho_{U,W}$,

(3) (Locality) Let $U = \bigcup_{i \in I} U_i$ be the union of open sets $U_i \subset X$. Let $s, t \in \sigma(\cup_{i \in I} U_i)$. Then $\rho_{U_i,U}\, s = \rho_{U_i,U}\, t$ for all $i \in I$ implies $s = t$,

(4) (Glueing) Under the conditions of (3) and $s_i \in \sigma(U_i)$, if $\forall i, j \in I$, $U_i \cap U_j \neq \emptyset \implies \rho_{U_i \cap U_j, U_i} s_i = \rho_{U_i \cap U_j, U_j} s_j$, then $\exists s \in \sigma(\cup_{i \in I} U_i)$, $\forall i \in I$ $\rho_{U_i.U} s = s_i$.

If $(\sigma, \rho_{U,V})$ satisfies 1) and 2) it is called a <u>presheaf</u>.

Colombeau's algebras of generalized functions possess a natural sheaf of sections structure if we take with the above notations $X = \mathbb{R}^n$, for open $\Omega \subset \mathbb{R}^n$ $\sigma(\Omega) = \mathcal{G}(\Omega)$ is a section over Ω.

For open sets $\Omega \subset \psi \subset \mathbb{R}^n$ we define $\rho_{\Omega,\psi} : \mathcal{G}(\psi) \to \mathcal{G}(\Omega)$ by $\rho_{\Omega,\psi} F = F|_\Omega$, $F \in \mathcal{G}(\psi)$.

Now if $T \in (\mathcal{C}^\infty(\Omega))'$, i.e. a distribution with compact support, then

$$T \longrightarrow (T * \phi) \,|\Omega)_{\phi \in \mathcal{A}} + \mathcal{N}(\Omega)$$

is an injection of $(\mathcal{C}^\infty(\Omega))'$. Using the sheaf properties of the Colombeau algebras it can be extended to an injection of $\mathcal{D}'(\Omega)$ into $\mathcal{G}(\Omega)$ and this injection renders $\mathcal{C}^\infty(\Omega)$ a faithful algebra, (c.f. [Col-1]).

3.4.8 Other Index Systems and Simplified Colombeau Algebras

Colombeau generalized functions were defined in $\mathbb{R}^n$ or on an open subset of $\mathbb{R}^n$ with a fixed orthonormal basis, i.e. the classical basis. The set $\bar{A}_q(E_n)$ introduced by Biagioni ([Bia]) and defined below is independent of an orthonormal basis in E_n. In the following E_n denotes a n dimensional real vector space with norm $|\,.\,|_{E_n}$ deduced from an inner product $<,>$.

Definition 1 Set $\bar{A}_0(\mathbb{R}) = \{\phi_1 \in \mathcal{D}(\mathbb{R}) : \phi_1 \text{ is even, constant in a 0-neighborhood and } \int_0^\infty \phi_1(s) ds = \frac{1}{2}\}$.

$(\mathcal{D}(\mathbb{R})$: all complex-valued infinitely differentiable functions with a compact support in $\mathbb{R}$).

Definition 2 For $q = 1, 2, \cdots$
$$\bar{A}_q(\mathbb{R}) = \{\phi_1 \in \bar{A}_0(\mathbb{R}) : \int_0^\infty s^{j/m}\phi_1(s)ds = 0, \ 1 \leq j \leq q, \ 1 \leq m \leq q\}$$

Remarks Notice that $\bar{A}_0 \supset \bar{A}_1 \supset \cdots \cdots \bar{A}_{q-1} \supset \bar{A}_q \cdots$
The n-dimensional spherical coordinates mapping $T_n : \mathbb{R}^n \to \mathbb{R}^n$ is given by
$$x_1 = \rho \cos \alpha_1$$
$$x_2 = \rho \sin \alpha_1 \cos \alpha_2$$
$$x_3 = \rho \sin \alpha_1 \sin \alpha_2 \cos \alpha_3$$
$$\cdots \cdots$$
$$\cdots \cdots$$
$$x_{n-2} = \rho \sin \alpha_1 \sin \alpha_2 \cdots \sin \alpha_{n-3} \cos \alpha_{n-2}$$
$$x_{n-1} = \rho \sin \alpha_1 \sin \alpha_2 \cdots \cdots \sin \alpha_{n-2} \cos \theta$$
$$x_n = \rho \sin \alpha_1 \sin \alpha_2 \cdots \cdots \sin \alpha_{n-2} \sin \theta$$
$\rho \in [0, 1]$, $\alpha_i \in [0, \pi] \, (i = 1, n - 2)$, $\theta \in [0, 2\pi]$
T_n maps $Q = (\rho, \alpha_1, \cdots, \alpha_{n-2}, \theta) \in \mathbb{R}^n$ onto the closed unit ball in $\mathbb{R}^n$
The Jacobian of T_n is of the form $\rho^{n-1}\Gamma(\alpha_1, \cdots, \alpha_{n-2}, \theta)$ for some function Γ. Also

$$x_1^2 + x_2^2 + \cdots + x_n^2 = \rho^2 \qquad (\dagger)$$

Note that for $n = 3$
$$x_1 = \rho \cos \alpha_1$$
$$x_2 = \rho \sin \alpha_1 \cos \theta$$
$$x_3 = \rho \sin \alpha_1 \sin \theta$$
is the usual spherical coordinates in three dimensions. To show $(\dagger)$ first we find $x_{n-1}^2 + x_n^2 = \rho^2 \sin \alpha_1^2 + \cdots + \sin \alpha_{n-2}^2$ and continue by calculating x_{n-2}^2, x_{n-3}^2 etc. We deduce that the constant $\Omega_n = \int_0^\pi \cdots \int_0^\pi \Gamma(\alpha_1 \cdots, \alpha_{n-2}, \theta)d\alpha_1 \cdots d\alpha_{n-2} \, d\theta$ is the $(n-1)-$dimensional area of the sphere S_{n-1} of radius 1 centered at the origin in $\mathbb{R}^n$.

Definition 3 If $q = 0, 1, 2, \cdots$, we set

$$\bar{A}_q(E_n) = \{\phi \in \mathcal{D}(E_n) \ \text{such that} \ \phi(x) = \frac{2n}{\Omega_n}\phi_1(|x|^n) \ \text{for some} \ \phi_1 \in \bar{A}_q(\mathbb{R})\}.$$

Notice that ϕ is C^∞ since ϕ_1 is constant in a 0-neighborhood

Lemma 1 If $\phi \in \bar{A}_q(E_n) i \, q = 0, 1, 2, \cdots$, then $\int_{E_n} \phi(x)dx = 1$ and if $q \geq n$, $P \in \mathcal{P}(^i E_n)$, $1 \leq i \leq q$ $\int_{E_n} P(x)\phi(x)dx = 0$.
$(^i\mathcal{P}(E_n)$ is the real vector space of the i-homogeneous complex-valued polynomials in E_n.)

Proof $\int_{E_n} \phi(x)dx = \frac{2n}{\Omega_n} \int_{E_n} \phi_1(|x|^n)dx = \frac{2n}{\Omega_n} \int_0^\infty \int_0^\pi \cdots \int_0^\pi \int_0^{2\pi} \rho^{n-1}\phi_1(\rho^n)\Gamma$
$(\alpha_1, \cdots, \alpha_{n-2}, \theta d\rho d\alpha_1 \cdots, \alpha_{n-2}d\theta = 2n \int_0^\infty \rho^{n-1}\phi_1(\rho^n)$
$d\rho = 2 \int_0^\infty \phi_1(\sigma)d\sigma = 1$.

If $P \in \mathcal{P}(^i E_n)$ with i odd, then clearly $\int_{E_n} P(x)\phi(x)dx = 0$ if $\phi \in \bar{\mathcal{A}}_q(E_n)$, since
ϕ is an even function.

Let $j \in \mathbb{N}$ be such that $2 \leq 2j \leq q$. Then if $\phi \in \bar{\mathcal{A}}_q(E_n)$, $q \geq n \int_{E_n} P(x)\phi(x)dx =$
$\frac{2n}{\Omega_n} \int_{E_n} P(x)\phi_1(|x|^n)dx = \frac{2n}{\Omega_n} \int_0^\infty \rho^{n-1+2j} phi_1(\rho^n)d\rho.c_n$
where c_n is a constant obtained by integration on $\alpha_1, \cdots, \alpha_{n-2}, \theta$
thus $\int_{E_n} P(x)\phi(x)dx = \frac{2c_n}{\Omega_n} \int_0^\infty \sigma^{2j/n}\phi_1(\sigma)d\sigma = 0$ by the definition of $\bar{\mathcal{A}}_q(\mathbb{R})$ $\square$
(Notation: Recall that for $\epsilon > 0$ and $\phi \in \mathcal{D}(E_m)$, by the notation in 3.4.1,
$\phi_\epsilon(x) = \frac{1}{\epsilon^m}\phi(\frac{x}{\epsilon}9$.

Proposition 1 *For $m, n = 1, 2, \cdots$ there is a bijection $I_n^m : \bar{\mathcal{A}}_0(E_m) \to \bar{\mathcal{A}}_0$ with
the following properties:*

(a) $I_n^m(\bar{\mathcal{A}}_q(E_m)) = \bar{\mathcal{A}}_q(E_n)$; $q = 1, 2 \cdots$,
(b) I_m^m is the identity of $\bar{\mathcal{A}}_0(E_m)$,
(c) $I_n^m(\phi)_\epsilon = (I_n^m\phi)_{\epsilon^{m/n}}$ for all $\phi \in \bar{\mathcal{A}}_0(E_m)$ and $\epsilon > 0$,
(d) $I_n^p \circ I_p^m = I_n^m$ for all $p = 1, 2, \cdots$.

Proof By the definition of $\bar{\mathcal{A}}_q(E_n)$ there is a bijection $I_n : \bar{\mathcal{A}}_0(\mathbb{R}) \to \bar{\mathcal{A}}_0 \to \bar{\mathcal{A}}_0(E_n)$
defined by $(I_n\phi_1)(x) = \frac{2n}{\Omega_n}\phi_1(|x|)^n$. $\forall \phi_1 \in \bar{\mathcal{A}}_0(\mathbb{R})$, $x \in E_n$.
For $m, n = 1, 2 \cdots$ define $I_n^m : \bar{\mathcal{A}}_0(E_m) \to \bar{\mathcal{A}}_0(E_n)$ by $I_n^m = I_n \circ (I_m)^{-1}$.
This I_n^m clearly satisfies a), b) and d). For c) consider the following equality for
$\phi, \in cal A_0(\mathbb{R})$ and $\epsilon > 0$: $I_m[(\phi_1)_\epsilon] = I_m(\phi_1)_{\epsilon^{1/\epsilon}}$. To see this the right-hand side
is $\frac{2m}{\Omega_m}(\frac{1}{\epsilon^{1/\epsilon}})^m\phi_1(|\frac{x}{\epsilon^{1/m}}|^m) = \frac{2m}{\Omega_m}\frac{1}{\epsilon}\phi_1(\frac{|x|^m}{\epsilon}) = I_m[(\phi_1)_\epsilon]$.
Then if $\phi \in \bar{\mathcal{A}}_0(E_m)$ $[(I_m)^{-1}(\phi)]_\epsilon = (I_m^{-1}(\phi_{\epsilon^{1/m}})$. Thus $I_n^m(\phi_\epsilon) = (I_n \circ (I_m^{-1}(\phi_\epsilon) =$
$I_n[I_m^{-1}(\phi_\epsilon)] = I_n([I_m^{-1}(\phi)]_{\epsilon^m} = [I_n \circ (I_m)^{-1}(\phi)]_{\epsilon^{m/n}} = (I_n^m\phi)_{\epsilon^{m/n}}$ $\square$

APPLICATIONS.

(1) If Ω is an open subset of E_n, we denote by $\mathcal{E}(\Omega)$, the set $[\mathcal{C}^\infty]^{\bar{\mathcal{A}}_0(E_n)}$, i.e.the set
of all functions in the form $f(\phi, x)$ which is $\mathcal{C}^\infty$ for fixed $\phi \in \bar{\mathcal{A}}_0(E_n)$.
$\mathcal{E}(\Omega)$ is again an algebra of which $\mathcal{C}^\infty(\Omega)$ is a subalgebra with the inclusion
$f \in \mathcal{C}^\infty(\Omega) \to [(\phi, x) \to f(x)] \in \mathcal{E}(\Omega)$. Now the subalgebra $\mathcal{E}_M(\Omega)$ of moderate
members and the ideal $\mathcal{N}(\Omega)$ is constructed following the lines in 3.4.1. Then the
Colombeau generalized functions is the quotient $\mathcal{G}(\Omega) = \mathcal{E}_M(\Omega)/\mathcal{N}(\Omega)$ (This is
also related to the previous Subsection Colombeau algebras on arbitrary domains).
It is verified that $\mathcal{E}_M(\Omega)$ and $\mathcal{N}(\Omega)$ do not depend on the chosen orthonormal basis.
Let $\mathcal{B} = \{b_1, \cdots b_n\}$ and $\mathcal{C} = \{c_1, \cdots, c_n\}$ be two orthonormal bases of E_n. Then
we have by the change of variables formula
$D_\mathcal{C}^\alpha = \sum_{|\beta|=|\alpha|} D_\mathcal{B}^\beta$ where $\{k_\beta\}_{\beta \in \mathbb{N}^n |\beta|=|\alpha|}$ is a family of real numbers. Using this
relation we conclude that a function belongs to $\mathcal{E}_{M,\mathcal{B}}(\Omega)$ ($\mathcal{N}_\mathcal{B}(\Omega)$) if and only if it
belongs to $\mathcal{E}_{M,\mathcal{C}}(\Omega)$ ($\mathcal{N}_\mathcal{C}(\Omega)$)
(2) The inclusion of $\mathcal{C}^0(\Omega)$, therefore that of $\mathcal{D}'(\Omega)$ is done in parallel to
Sect. 3.4.2, i.e. if $f \in \mathcal{C}^0(\Omega)$, define $f \to T = \bar{f} + \mathcal{N}(\Omega) \in \mathcal{G}(\Omega)$ where

$\bar{f}(\phi, x) = \int_{E_n} f(y)\phi(y - x)dy = \int_{E_n} f(x + y)\phi(y)dy, \quad \phi \in \bar{A}_0(E_n), x \in E_n.$

(3) The generalized complex numbers defined in 3.4.4 (Definition 3) depended on the dimension n of the underlying space $\mathbb{R}^n$. In the Biagioni index system they can be defined independent of the space dimension. In fact instead of starting by the algebra $\mathcal{E}_0 = \mathbb{C}^{\mathcal{A}}$, we start by the algebra $\mathbb{C}^{\bar{A}_0(\mathbb{R})}$.

Then $\qquad \mathcal{E}_{M^\circ} = \{h \in \mathcal{E}_0 : \exists n \in \mathbb{N}_+ \text{ such that } \forall \phi \in \bar{A}_n(\mathbb{R}), \exists \eta > 0, \epsilon > 0 \text{ with}$ $|h(\phi_\epsilon)| \leq \frac{c}{\epsilon^n} \text{ if } 0 < \epsilon < \eta\}$ is a moderate subalgebra of $\mathcal{E}_0$.

We define an ideal of $\mathcal{M}^\circ$ by

$\mathcal{N}_0 = \{h \in \mathcal{M}^\circ : \exists n \in \mathbb{N}_+, \beta \in B \text{ such that } \forall \phi \in \bar{A}_q(\mathbb{R}), q \geq n, \exists \eta > 0, c > 0$ with $|h(\phi_\epsilon)| \leq c^{\beta(q)-n} \text{ if } 0 < \epsilon < \eta\}$.

For B see 3.4.1, Note 2 to Definition 3).

Then $\bar{\mathbb{C}} = \mathcal{M}_\circ/\mathcal{N}_0$ is the associative and commutative algebra of generalized complex numbers.

Accordingly if Ω (open)$\subset E_n$ $x \in \Omega$ and $F \in \mathcal{G}(\Omega)$ is a generalized function with a representative $f \in \mathcal{E}_M(\Omega)$, the point-value of F at x is given by $F(x) = f(I_n^1\phi, x(+\mathcal{N}_0 \text{ for all } \phi \in \bar{A}_0(\mathbb{R}).$

In a similar fashion the integral of F on $K \subset\subset \Omega$ will be defined by $\int_K F(x)dx = \int_K f(I_n^1\phi, x)dx$

Simplified Colombeau Algebra

The simplified Colombeau algebra $\mathcal{G}_s(\Omega)$ results from an attempt to free $\mathcal{G}(\Omega)$ from the functional index set $\bar{A}_0(E_n)$(or $\mathcal{A}(\mathbb{R}^n)$).

We still have a canonical inclusion of $\mathcal{C}^\infty$ into $\mathcal{G}_s(\Omega)$, but we do not have a privileged natural inclusion of $\mathcal{C}^0(\Omega)$ and therefore that of $\mathcal{D}'(\Omega)$ into $\mathcal{G}_s(\Omega)$. We follow the index system of Biagioni. A more general and abstract approach takes place in 3.4.10 and 3.5.1.

Definition 4 Define $\bar{\mathcal{A}}_{0,1}$ as

$$\bar{\mathcal{A}}_{0,1} = \{\phi \in \bar{A}_0(E_n) \text{ such that } diam(supp\,\phi) = 1.$$

Remark There is a bijection $T : \bar{\mathcal{A}}_{0,1}(E_n) \to \bar{A}_0$ defined by

$$T : \bar{\mathcal{A}}_{0,1} \times (0, \infty) \to \bar{A}_0(E_n)$$

$$(\phi, \epsilon) \to \phi_\epsilon$$

It is obviously injective. In the other direction given $\phi \in \bar{A}_0(E_n)$, it suffices to take $\epsilon = diam(supp\,\phi)$. Then for $\psi = \phi_{1/\epsilon}$ we have $diam(supp\,\psi = 1$, and $T(\psi, \epsilon) = \phi$.

Denote by $\mathcal{E}_s$ the set of all functions

$$R : (0, \infty) \times \Omega \to \bar{\mathbb{C}}$$

$$(\epsilon, x) \to R(\epsilon, x).$$

which are C^∞ in the variable x for each $\epsilon > 0$. As the behaviour of R when $\epsilon \to 0$, matters, we could as well replace ∞ by 1.

An element R of $\mathcal{E}_s$ may be considered as an element of $\mathcal{E}(\Omega)$ which is independent of $\phi \in \bar{\mathcal{A}}_{0,1}(E_n)$. In view of the Remark above this is shown by defining a mapping M as

$$M : \mathcal{E}_s(\Omega) \to \mathcal{E}(\Omega)$$

$$R \to [(\phi_\epsilon, x) \to R(\epsilon, x)]. \quad (1)$$

Then denote by $\mathcal{E}_{M,s}$ the subspace of $\mathcal{E}_s(\Omega)$

$$\mathcal{E}_{M,s}(\Omega) = M^{-1}(\mathcal{E}_M(\Omega)), \quad (2)$$

that is an element R of $\mathcal{E}_s$ is in $\mathcal{E}_s \iff$ given $K \subset\subset \Omega$ and $n \in \mathbb{N}^n$ there are $\alpha \in \mathbb{N}$, $c > 0$ and $\eta > 0$ such that $|(D_\mathcal{B}^\alpha R(\epsilon, x)| \leq c\epsilon^{-n}$ for all $x \in K$ and $0 < \epsilon < \eta$. ($\mathcal{B}$ is an ortonormal basis in E_n).

We also have the ideal $\mathcal{N}_s$ which is a subspace of $\mathcal{E}_s(\Omega)$ defined by

$$\mathcal{N}_s(\Omega) = M^{-1}(\mathcal{N}(\Omega)) \quad (3)$$

that is $R \in \mathcal{E}_s(\Omega)$ is in $\mathcal{N}(\Omega) \iff$ given $K \subset\subset \Omega)$ and α, there are $n \in \mathbb{N}$ and $\gamma \in B$ such that $\forall q \in \mathbb{N}$ with $q \geq n$ there are $c > 0$ and $\eta > 0$ such that $|D_\mathcal{B}^\alpha R(\epsilon, x)| \leq c\,\epsilon^{\gamma(q)-n}$ for all $x \in K$, $0 < \epsilon < \eta$.

The simplified Colombeau algebra $\mathcal{G}_s(\Omega)$ is the quotient space

$$\mathcal{G}_s(\Omega) = \mathcal{E}_{M,s}(\Omega)/\mathcal{N}_s(\Omega). \quad (4)$$

<u>Conclusions</u>
(a) The algebra $\mathcal{G}_s(\Omega)$ is a subalgebra of $\mathcal{G}(\Omega)$ via the map M, that is if $\bar{M}$ is defined by

$$\bar{M} : \mathcal{G}_s(\Omega) \to \mathcal{G}(\Omega)$$

$$G \to class \text{ of } M(R)$$

where $R \in \mathcal{E}_s(\Omega)$ is a representative of G, then $\bar{M}$ is injective.
(b) There is a natural injection from $C^\infty(\Omega)$ into $\mathcal{E}_{M,s}(\Omega)$ by $f \in C^\infty(\Omega) \to [(\epsilon, x) \to f(x)] \in \mathcal{E}_{M,s}(\Omega)$.
(c) To each continuous function on Ω (and thus to each distribution in $\mathcal{D}'(\Omega)$) there corresponds an infinite number of generalized functions in $\mathcal{G}_s(\Omega)$.

Let $\{\chi_\epsilon\}_{\epsilon>0}$ be a family of functions such that $\chi_\epsilon \in \mathcal{D}(\Omega)$ and $\chi_\epsilon(x) = 1$ if $x \in \Omega$, $|x| \leq \frac{1}{\epsilon}$ and $d(x, \delta\Omega \geq \epsilon$. If there is not such $x \in \Omega$, then set $\chi_\epsilon = 0$.

Let also $i : \bar{\mathcal{A}}_0(E_n) \to \mathbb{R}$ associating to each $\phi \in \bar{\mathcal{A}}_0(E_n)$, the number $i = diam(supp\phi)$.

Now consider $f \in \mathcal{C}^0(\Omega)$, to each $\phi \in \mathcal{D}(\Omega)$ with $\int \phi(y)dy = 1$ there corresponds a generalized function in $\mathcal{G}_s(\Omega)$, class of

$$(\epsilon, x) \to \int f(x + \epsilon y)\, \chi_{i(\phi_\epsilon)}(x + \epsilon y)\phi(y)dy$$

(c.f. [Bia], 1.8.3). As $\epsilon \to 0$, $i(\phi_\epsilon) \to 0$, $|x| \le \frac{1}{\epsilon}$, hence $\chi_{i(\phi_\epsilon)} \to 1$ and the integral approaches to $\int f(x + \epsilon y)\phi(y)dy$ (compare to 3.4.2 II)

However these infinite classes in $\mathcal{G}_s(\Omega)$ are associated (c.f. 3.5, Definition 7).
<u>Note.</u> The subalgebra $\bar{\mathbb{C}}_s$ of $\mathbb{C}$ is constructed with the same simplifications as done above for $\mathcal{G}_s(\Omega)$.

3.4.9 *Topological Structures on Generalized Complex Numbers $\bar{\mathbb{C}}$ and $\mathcal{G}(\Omega)$*

Biagioni ([Bia], 1980) and Egorov [Ego], 1990) state that so far the topologies in $\bar{\mathbb{C}}$ and $\mathcal{G}(\Omega)$ have not played a role in applications. However the study of measure-theoretical properties or probability distributions on generalized functions (investigated in Chap. 4) are intimately related to the topological structures. In $\bar{\mathbb{C}}$ a base of neighborhoods of the origin is introduced by the following sets (for simplicity $n = 1$ is taken):

$$Q_\mu = \{\bar{x} \in \bar{\mathbb{C}} : \exists x \in \mathcal{E}_{M^\circ}, \text{ a representative of } \bar{x} \text{ with } |x(\phi)| < \mu \; \forall \phi \in \mathcal{A}(\mathbb{R})\}. \quad (1)$$

(Recall that $\mathcal{E}_{M^\circ}$ is the subalgebra of members of moderate growth in $\mathcal{E}_0 = \mathbb{C}^{\mathcal{A}}(\mathbb{R})$, (c.f. 3.4.4 Definition 1.)

As, for the quotient algebra only the values $x(\phi_\epsilon)$ with N large enough and ϵ small enough are significant, an equivalent definition of Q_μ would be as follows:

$$Q_\mu = \{\bar{x} \in \bar{\mathbb{C}} : \exists x^* \in \mathcal{E}_{M^\circ} \text{ a representative of } \bar{x} \text{ and } N \in \mathbb{N} \text{ such that } \forall \phi \in \mathcal{A}_N(\mathbb{R}) \; \exists \eta > 0 \quad (2)$$

$$\text{with } |x^*(\phi_\epsilon| < \mu \text{ for } 0 < \epsilon < \eta\}.$$

<u>Note.</u> Clearly we have $Q_\mu = \mu Q_1$.

Definition 1 Any subset $\theta \subset \bar{\mathbb{C}}$ is said to be <u>open</u> if for all $\bar{x} \in \theta$ there exists $\mu > 0$ such that $\bar{x} + Q_\mu \subset \theta$.

These open sets define a topology on $\bar{\mathbb{C}}$. For this topology the family $\{\bar{x} + Q_\mu, \mu > 0\}$ is a base of neighborhoods of $\bar{x}$.
If $z \in \mathbb{C} \subset \bar{\mathbb{C}}$, then as a generalized complex number it has a representation $z = \bar{z} + \mathcal{N}_0$, $\bar{z}(\phi) = z \; \forall \phi \in \mathcal{A}$. Then Q_μ will be the μ-neighborhood of the origin in

complex plane.
Consider

$$\|\bar{x}\|^{-} \;\dot{=}\; \begin{cases} \inf\{\mu > 0 : \bar{x} \in Q_\mu\} & \text{if } \bar{x} \in Q_\mu \text{ for some } \mu \\ +\infty & \text{if } \bar{x} \notin Q_\mu \text{for all } \mu > 0 \end{cases} \qquad (3)$$

Examples

1. $\forall z \in \mathbb{C}$, $\|z\|^{-} = |z|$ since in this case Q_μ is a neighborhood of the origin. Hence the topology on $\bar{\mathbb{C}}$ induces the usual one on $\mathbb{C}$.

2. $\|\delta(0)\|^{-} = +\infty$.   (As $\delta(0) = h + \mathcal{N}_0$, $h(\phi) = \phi(0)$, $\forall \phi \in \mathcal{A}$,   for   no $\mu > 0$ $\delta(0) \in Q_\mu$.) This also tells that $\mathbb{C}$ is not dense in $\bar{\mathbb{C}}$.

3. $\|\exp(i\delta(0))\| = 1$ if $\mathcal{A}(\mathbb{R})$ has only real-valued functions. ($|\exp(i\delta(0))| = 1$ yields value 1 for the infimum in the definition of $\| . \|^{-}$.

Lemma 1 *The function $\| . \|^{-}$ is an $\bar{\mathbb{R}}$-valued seminorm.*

Proof (i) By definition we have $0 \leq \|\bar{z}\|^{-} \leq \infty$.

(ii) Denote $R_{\bar{z}} = \{\mu : \bar{z} \in Q_\mu\} = \{\mu : \exists z = z(\mu)$ a representative of $\bar{z}$, $|z(\phi)| < \mu \forall \phi \in \mathbb{R}$. (So that if $R_{\bar{z}} \neq \emptyset$, $\|\bar{z}\| = \inf R_{\bar{z}}$, otherwise it is $+\infty$). Then $R_{c\bar{z}} = |c| R_{\bar{z}}$. Taking infimums on both sides we have $\|c\bar{z}\|^{-} = |c| \|\bar{z}\|^{-}$.

(iii) We have by the ordinary triangular inequality $R_{\bar{z}_1} + R_{\bar{z}_2} \subset R_{\bar{z}_1 + \bar{z}_2}$, so that $\inf R_{\bar{z}_1 + \bar{z}_2} \leq \inf(R_{\bar{z}_1} + R_{\bar{z}_2}) = \inf R_{\bar{z}_1} + \inf R_{\bar{z}_2}$, yields the triangular inequality.

In the case that either of $\|\bar{z}_1\|^{-}$ or $\|\bar{z}_2\|^{-}$ is $+\infty$, then $\|\bar{z}_1 + \bar{z}_2\|^{-} \leq \|\bar{z}_1\|^{-} + \|\bar{z}_2\|^{-}$ is always true. If on the other hand $R_{\bar{z}_1 + \bar{z}_2} = \emptyset$, then the triangular inequality will take the form $+\infty = +\infty$.

Remarks

1. $\{\| . \|^{-}\}$ create the same topology on $\bar{\mathbb{C}}$.

2. Q_μ is convex and balanced, but the family $\{Q_\mu, \mu > 0\}$ is a non-absorbing base of the origin, thus the defined topology of $\bar{\mathbb{C}}$ can not be a topological vector space topology, since such a topology has as a base of the origin consisting of balanced and absorbing sets. (e.g. $\|\delta(0)\|^{-} = +\infty$, thus it can not be absorbed), (c.f. GLOSSARY).

3. The function $\|\bar{z}_1 - \bar{z}_2\|^{-}$ is a $\bar{\mathbb{R}}$-valued pseudo-metric on $\bar{\mathbb{C}}$.

4. $|.\|^-$ is a continuous mapping and $\|\bar{z}_1 - \bar{z}_2\|^-$ is a $\bar{\mathbb{R}}$-valued pseudo-metric on $\bar{\mathbb{C}}$.

5. The topology defined above can also be derived from the uniform structure induced by the filter $\mathcal{U}$ spanned by the sets $U_\mu = \{(\bar{x}, \bar{y} \in \bar{\mathbb{C}} \times \bar{\mathbb{C}} : \|\bar{x} - \bar{y}\|^- < \mu\}$, (c.f. [Bour], Chap. II, Sect. 1, Proposition 1).

for 'filter' see the GLOSSARY or the next subsection).

Lemma 2 $\bar{\mathbb{C}}$ *is a disconnected space in its topology.*

Proof The sets $\bar{\mathbb{C}}_\infty = \{\bar{x} \in \bar{\mathbb{C}} : \|\bar{x}\|^- = +\infty\}$ and $\bar{\mathbb{C}}_f = \{\bar{x} \in \bar{\mathbb{C}} \|\bar{x}|^- < \infty\}$ are two disconnections. If $\bar{z}_0 \in \bar{\mathbb{C}}_\infty$ and $\bar{w} \in Q_\mu$ for some $\mu > 0$, then $\|\bar{z}_0 + \bar{w}\|^- = +\infty$, Thus $\bar{z}_0$ is an interior point of $\bar{\mathbb{C}}_\infty$.

<u>Remark.</u>The restriction of $\|.\|^-$ to $\bar{\mathbb{C}}_f$ is a seminorm and induces the topology of $\bar{\mathbb{C}}_f$ as inherited from $\bar{\mathbb{C}}$. Thus $\bar{\mathbb{C}}_f$ becomes a non-Hausdorff (see Lemma 4 below) locally convex topological vector space with a single semi-norm.

Lemma 3 *We have the following implications between three zero type behaviors of the generalized complex numbers:*

$$\bar{z} \in \mathcal{N}_0 \implies \bar{z} \vdash 0 \implies \|\bar{z}\|^- = 0$$

Proof Only the second implication is non-trivial. By Definition 3 of Sect. 3.4.4 and (3), for any $\delta > 0$ $\bar{z}$ has a representation $\bar{z} = z^* + \mathcal{N}_0$ such that $\exists N \in \mathbb{N}$, for all $\phi \in \mathcal{A}_N$ there is $\eta > 0$, $|x^*(\phi_\epsilon)| < \delta$ for $0 < \epsilon < \eta$. Thus $\bar{z} \in Q_\delta$ for all $\delta > 0$ showing that $\|\bar{z}\|^- = 0$.

Lemma 4 $\bar{\mathbb{C}}_f$ *is not a Hausdorff space.*

Proof Consider the Example following the Definition 4 of 3.4.4: $\bar{z} = h + \mathcal{N}_0$ $h(\phi) = \int_{\mathbb{R}} |x| \phi(x) dx$, $\phi \in \mathcal{A}(\mathbb{R})$. Then $h(\phi_\epsilon) = \epsilon h(\phi)$, showing $\bar{z} \vdash 0$, so by Lemma 3 $\|\bar{z}\|^- = 0$, but as shown in 3.4.4 $\bar{z} \notin \mathcal{N}_0$. Then by [Rob-R], Chap. I, nb. 4, Proposition 8, $\bar{\mathbb{C}}_f$ can not be Hausdorff since this Proposition states that a topological vector pace is Hausdorff if and only if given any nozero element there exists a seminorm which takes non-zero value on this element, but $\|.\|$ is the only seminorm on $\bar{\mathbb{C}}$.

Theorem 1 *In* $\bar{\mathbb{C}}$ *every Cauchy sequence converges to a point in* $\bar{\mathbb{C}}$.

Proof A sequence $\{\bar{z}\} \subset \bar{\mathbb{C}}$ if given $\epsilon > 0$ there is $N \in \mathbb{N}$, such that for $m, n \geq N$ we have $\bar{z}_m - \bar{z}_n \in Q_\epsilon$. This means by (1) $\exists (z_m - z_n)$, a representative of $\bar{z}_m - \bar{z}_n$ such that $|z_m(\phi) - z_n(\phi)| < \epsilon$ for all $\phi \in \mathcal{A}(\mathbb{R}^n)$. This shows $\{z_n(\phi)\}$ is a Cauchy sequence of complex numbers. Therefore $\lim_{n \to \infty} z_n(\phi) = z(\phi)$. Hence $\bar{z}_n = z_n(\phi) + \mathcal{N}_0$, $\lim_{n \to \infty} = z(\phi) + \mathcal{N}_0 \in \bar{\mathbb{C}}$ $\square$

<u>Note.</u> This means that $\bar{\mathbb{C}}_f$ is complete since a pseudo-metrizable uniform space is complete if and only if every Cauchy sequence in the space converges to a point ([Kel], Theorem 24, p. 193).

Remarks The following facts can be proved without much difficulty:

(1) $\bar{\mathbb{C}}_f$ and $\bar{\mathbb{C}}_\infty$ are not finite dimensional sets.
(2) A first countable space $\bar{\mathbb{C}}_f$ is barrelled and bornological (c.f. GLOSSARY).

Topology on $\mathcal{G}(\Omega)$

For Ω (Open) $\subset \mathbb{R}^n$ (or E_n) a topology on $\mathcal{G}(\Omega)$ is defined in a way similar to $\bar{\mathbb{C}}$.
If $K \subset\subset \Omega$ is a compact subset of Ω and $k \in \mathbb{N}$, $\mu > 0$ we set

$$Q_{K,k,\mu} = \{G \in \mathcal{G}(\Omega) : \exists R \in \mathcal{E}_M(\Omega), \quad \text{a representative of } G \text{ such that } \sup |(D^\alpha R)(\phi, x)| \leq \mu,$$

$$\alpha \in \mathbb{N}^n, \ |\alpha| \leq k, \ \phi \in \mathcal{A}(\mathbb{R}^n), \ x \in K\} \quad (3)$$

Definition 2 Any subset X of $\mathcal{G}(\Omega)$ is said to be open if for each $G \in X$ there are
$K \subset\subset \Omega$, $k \in \mathbb{N}$ and $\mu > 0$ such that $G + Q_{K,k,\mu} \in X$.
These open sets define a topology on $\mathcal{G}(\Omega)$ (say standard topology) where the family

$$\{G + Q_{K,k,\mu}, \ K \subset\subset \Omega, k \in \mathbb{N}, \ \mu > 0\}$$

is a basis of neighborhoods of G. The function

$$\|G\|_{K,k} = \begin{cases} \inf \mu > 0 : G \in Q_{K,k,\mu} \text{ if } G \in Q_{K,k,\mu} \text{ for some } \mu > 0 \\ +\infty \qquad\qquad\qquad \text{if } G \notin Q_{K,k,\mu} \text{ for all } \mu > 0 \end{cases} \quad (4)$$

is a pseudo-metric on $\mathcal{G}(\Omega)$. This topology on $\mathcal{G}(\Omega)$ induces on $\bar{\mathbb{C}}$ its own topology,
therefore it is not Hausdorff.

The standard topology can also be defined from the uniform structure induced by
the filter $\mathcal{U}$ spanned by the sets

$$U_{K,k,\mu} = \{(G, H) \in \mathcal{G}(\Omega) \times \mathcal{G}(\Omega) : \|G - H\|^-_{K,k} < \mu\}$$

with $K \subset\subset \Omega$, $k \in \mathbb{N}$ and $\mu > 0$.
   $\mathcal{G}(\Omega)$ is not connected in the standard topology. In fact

$$\mathcal{G}_f(\Omega) = \{G \in \mathcal{G}(\Omega) : \exists K \subset\subset \Omega, k > 0, \ \|G\|^-_{K,k} < \infty\}$$

$$\mathcal{G}_\infty = \{R \in \mathcal{G}(\Omega) : \ \|R\|^-_{K,k} = +\infty\}$$

are two disconnections.
   The topology on $\mathcal{G}(\Omega)$ induces on $C^\infty(\Omega)$ its own Fréchet topology since $\| \cdot \|^-_{K,k}$
coincides on $C^\infty(\Omega)$ with its seminorms

$$p_{K,k}(f) = \sup_{|\alpha| \leq k} \sup_{x \in K} |D^\alpha F(x)|. \quad (*)$$

These seminorms create a locally convex topology on $C^\infty(\Omega)$ which is complete and metrizable, i.e. a Fréchet topology.

Recall that to show this we fix compact sets $K_N \subset\subset \Omega$, $N \in \mathbb{N}$ in such a way that K_N lies in the interior of K_{N+1}. Consider the above seminorms of the form $\{p_{K_n,N}\}$. Then this is a countable separating family of seminorms. The metric $d(f,g) = \sum_{N=1}^\infty \frac{2^{-N} p_{K_n,N(f-g)}}{(1+p_{K_n,N})(f-g)}$ creates the same topology on $C^\infty(\Omega)$ as (*).

We have $p_{K_N,N}(f) = \sup_{|\alpha|\le N} \sup_{x\in K} |D^\alpha f(x)|$ and a local base is given by the sets $V_N(f) = \{f \in C^\infty(\Omega): p_{K_N,N}(f) < 1/N\}$. If $\{f_i\}$ is a Cauchy sequence in $C^\infty(\Omega)$ and if N is fixed, then $|D^\alpha f_i(x) - D^\alpha f_j(x)| < 1/N$ on K_N for i and j sufficiently large. For a fixed x_0 the Cauchy sequence $\{D^\alpha f_i(x_0)\}$ converges to a complex number, say $g_\alpha(x_0)$. Thus $D^\alpha f_i(x)$ converges uniformly on compact sets to $g_\alpha(x)$. It is now evident that $g_\alpha \in C^\infty(\Omega)$. In particular $f_i(x) \to g_0(x)$ and $g_\alpha = D^\alpha g_0$. Hence $C^\infty(\Omega)$ is a Fréchet space. The same is true of such its closed subspaces $\mathcal{D}_K$ which is the set of smooth functions whose support lies in K.

<u>Remarks.</u> (1) If we consider the elements of $\bar{\mathbb{C}}$ as constant generalized functions, then the topology on $\mathcal{G}(\Omega)$ induces on $\bar{\mathbb{C}}$ its own topology.

(2) The topology on $\mathcal{G}(\Omega)$ is not a topological vector space topology since it has non-absorbing zero neighborhoods, e.g. an element $G \in \mathcal{G}(\Omega)$ with $\|G\|_{K,k}^- = +\infty$ can not be absorbed by the neighborhoods $Q_{K,k,\mu}$ of the origin.

(3) Like $\bar{\mathbb{C}}$, the topology on $\mathcal{G}(\Omega)$ arises from a uniform structure: in $\mathcal{G}(\Omega) \times \mathcal{G}(\Omega)$. Consider the filter $\mathcal{U}$ spanned by the sets $U_{K,k,\mu} = \{(G,H) \in \mathcal{G}(\Omega) \times \mathcal{G}(\Omega): \|G-H\|_{K,k} < \mu\}$ with $K \subset\subset \Omega$, $k \in \mathbb{N}$ and $\mu > 0$. The filter $\mathcal{U}$ defines a uniform structure on $\mathcal{G}(\Omega)$ and the topology on $\mathcal{G}(\Omega)$ arises from this uniform structure.

(4) This topology on $\mathcal{G}(\Omega)$ is the natural generalization via the inclusion $C^\infty(\Omega) \subset \mathcal{G}(\Omega)$ of the Fréchet space topology of the space $C^\infty(\Omega)$.

The counter part of the weak convergence of distributions would be

$$\int G_n(x)\psi(x)dx \xrightarrow{\bar{\mathbb{C}}} \int G(x)\psi(x)dx \quad \text{for all } \psi \in \mathcal{D}(\Omega)$$

3.4.10 Topological $\bar{\mathbb{C}}$-Modules, Garetto's Work, Duality

The ring $\bar{\mathbb{C}}$ of generalized complex numbers is redefined in the spirit of simplified Colombeau algebras (i.e. freed of functional index set $\mathcal{A}$) factorizing the algebra

$$\mathcal{E}_M = \{(u_\epsilon)_\epsilon \in \mathbb{C}^{[0,1]}: \exists N \in \mathbb{N}, |u_\epsilon| = O(\epsilon^{-N}) \text{ as } \epsilon \to 0\}$$

with respect to the ideal

$$\mathcal{N} = \{(u_\epsilon)_\epsilon \in \mathbb{C}^{[0,1]}: \forall q \in \mathbb{N}, |u_\epsilon| = O(\epsilon^q) \text{ as } \epsilon \to 0\}$$

$\bar{\mathbb{C}}$ is trivially a module over itself. It can be endowed with a structure of a topological ring via a valuation. To achieve this end introduce the following function on $\mathcal{E}_M$:

$$v : \mathcal{E}_M \longrightarrow (-\infty, +\infty] : (u_\epsilon)_\epsilon \longrightarrow \sup\{b \in \mathbb{R} : |u_\epsilon| = O(\epsilon^b) \text{ as } \epsilon \to 0\} \quad (1)$$

It satisfies the following properties:

(i) $v((u_\epsilon)_\epsilon) = +\infty \Longleftrightarrow (u_\epsilon)_\epsilon \in \mathcal{N}$,

(ii) $v((u_\epsilon)_\epsilon(w_\epsilon)_\epsilon) \geq v((u_\epsilon)_\epsilon) + (w_\epsilon)_\epsilon)$

(iii) $v((u_\epsilon)_\epsilon + (w_\epsilon)_\epsilon) \geq \min\{v((u_\epsilon), v((w_\epsilon)_\epsilon)\}$

Clearly (ii) and (iii) become equalities if at least one or both terms are of the form $(c\epsilon^b, c \in \mathbb{C}, b \in \mathbb{R})$ respectively.

Also if $(u_\epsilon - u'_\epsilon \in \mathcal{N}$, combining (i) with (iii) we find that $v((u_\epsilon)_\epsilon) = v((u'_\epsilon)_\epsilon)$. Thus independently of the representative the valuation of a generalized complex number u can be defined by (1):

$$v_{\bar{\mathbb{C}}}(u) \doteq v((u_\epsilon)_\epsilon) \quad (2)$$

Let us now define a function $|\,.\,|_e$ on $\bar{\mathbb{C}}$ as

$$|\,.\,|_e : \bar{\mathbb{C}} \longrightarrow [0, +\infty) : u \longrightarrow |u|_e = e^{-v_{\bar{\mathbb{C}}}}, \; u \in \bar{\mathbb{C}} \quad (3)$$

($|\,.\,|_e$ is an ultra pseudo-norm (c.f. GLOSSARY)).

The coarsest topology on $\bar{\mathbb{C}}$ such that the map $|\,.\,|$ is continuous is also compatible with the ring structure. This topology is named the <u>sharp topology</u> on $\bar{\mathbb{C}}$. The compatibility means that the operations of addition and the product in $\bar{\mathbb{C}}$ are continuous with respect to this topology.

<u>Note</u>. As $\bar{\mathbb{C}}$ is also trivially a module over itself, we may say that the module operators on $\bar{\mathbb{C}}$ are continuous. For $u, w \in \bar{\mathbb{C}}$ the ultrametric pseudo-distance between u and w is defined as

$$\delta_e(u, w) = e^{-v_{\bar{\mathbb{C}}}(u-w)} \quad (4)$$

The ultrametric uniform structure constructed in this manner is called the <u>sharp uniform structure on $\bar{\mathbb{C}}$</u>

Theorem 1 $\bar{\mathbb{C}}$ *in the topology given by the pseudo-norm* $|\,.\,|_e$ *is complete.*

Proof In general one should show that every Cauchy filter on $\bar{\mathbb{C}}$ converges to a point of $\bar{\mathbb{C}}$. But as $\bar{\mathbb{C}}$ is a pseudo metrizable uniform space, it is sufficient to prove that every Cauchy sequence $(u_n)_n$ converges. (See the Note to Theorem 1, 3.4.9 or [Kel], Theorem 24, p. 193). Considering such a Cauchy sequence, given $\eta > 0$ there exists $N \in \mathbb{N}$ such that for all $m, p \geq N$, $|u_m - u_n|_e \leq \eta$. In terms of the valuation

$v_{\bar{\mathbb{C}}}$, that would mean we can extract a subsequence $(u_{n_k})_k$ such that $v_{\bar{\mathbb{C}}}(((u_{n_{k+1},\epsilon} - u_{n_k,\epsilon})_\epsilon) > k$ for all $k \in \mathbb{N}$. This implies that we can find $\epsilon \downarrow 0$, $\epsilon \le 1/2^k$ such that $|u_{n_{k+1}\epsilon} - u_{n_k,k}| \le \epsilon^k$ on $(0, \epsilon_k)$. Let

$$h_{k,\epsilon} = \begin{cases} u_{n_{k+1},\epsilon} - u_{n_k,\epsilon} & \epsilon \in (0, \epsilon_k) \\ 0 & \epsilon \in [\epsilon_k, 1] \end{cases}$$

$(h_{k,\epsilon}) \in \mathcal{E}_M$ since $|h_{k,\epsilon}| \le \epsilon^k$ in the interval $(0, 1]$, moreover the sum $u_\epsilon \doteq u_{n_0,\epsilon} + \sum_{k=0}^{\infty} h_{k,\epsilon}$ is locally finite and by

$$|u_\epsilon| \le |u_{n_0,\epsilon}| + \sum_{k=0}^{\infty} |h_{k,\epsilon}| \le |u_{n_0,\epsilon}| + \sum_{k=0}^{\infty} \frac{1}{2^k}$$

it defines the representative of a generalized complex number $u = [(u_\epsilon)_\epsilon]$. The sequence u_{n_k} tends to u in $\bar{\mathbb{C}}$. In fact for all $\bar{k} \ge 1$, the estimate

$$|u_{n_k,\epsilon} - u_\epsilon| = \left| -\sum_{k=\bar{k}}^{\infty} h_{k,\epsilon} \right| \le \sum_{k=\bar{k}}^{\infty} \epsilon^{k-1} \epsilon_k \le \epsilon^{k-1} \sum_{k=\bar{k}}^{\infty} \frac{1}{2^k},$$

valid on the interval $(0, \epsilon_{k-1})$, yields $v((u_{n_k,\epsilon} - u_\epsilon)_\epsilon) \longrightarrow +\infty$. Thus $(u_n)_n$ is a Cauchy sequence with a convergent subsequence and it converges to the same point $u \in \bar{\mathbb{C}}$. ([Gar-1]). $\square$

Remark Compared with the pseudo-norm $\| . \|^-$ of Sect. 3.4.9 δ_e takes always a finite value. But if $\|\bar{z}\|^- = +\infty$, $\|\bar{w}\|^- = +\infty$, the pseudo-distance $\|\bar{z} - \bar{w}\|^-$ will be undefined.

$\bar{\mathbb{C}}$ is also an algebra, with $\mathbb{C}$ as the scalar field. But Garetto ([Gar-1]) stresses its trivial $\bar{\mathbb{C}}$-module feature in order to pass on to the study of its generalizations as topological $\bar{\mathbb{C}}$-modules.

A topology τ on a $\bar{\mathbb{C}}$-module $\mathcal{G}$ is said to be $\bar{\mathbb{C}}$-linear if the addition $\mathcal{G} \times \mathcal{G} \longrightarrow \mathcal{G} : (u, v) \longrightarrow u + v$ and the product $\bar{\mathbb{C}} \times \mathcal{G} \longrightarrow \mathcal{G} : (\lambda, u) \longrightarrow \lambda u$ are continuous operations. A topological $\bar{\mathbb{C}}$-module $\mathcal{G}$ is a $\bar{\mathbb{C}}$- module with a $\bar{\mathbb{C}}$-linear topology.

Following terminology is adopted from the theory of topological vector spaces:

Definition 1 Let A be a subset of a topological $\bar{\mathbb{C}}$-module $\mathcal{G}$. Then A is
(i) $\bar{\mathbb{C}}$-absorbent if for all $u \in \mathcal{G}$ there exists $a \in \mathbb{R}$ such that $u \in [(\epsilon^b)_\epsilon]A$ for all $b \le a$,
(ii) $\bar{\mathbb{C}}$-balanced if $\lambda A \subseteq A$ for all $\lambda \in \bar{\mathbb{C}}$ with $|\lambda|_e \le 1$,
(iii) $\bar{\mathbb{C}}$-convex if $A + A \subseteq A$ and $[(\epsilon^b)_\epsilon]A \subseteq A$ for all $b \ge 0$
(iv) absolutely convex if it is both $\bar{\mathbb{C}}$-balanced and $\bar{\mathbb{C}}$-convex.
(v) bounded if it is absorbed by all neighborhoods of the origin. In the case A is $\bar{\mathbb{C}}$-balanced, the convexity is equivalent to the following: For all $\lambda, \mu \in \bar{\mathbb{C}}$ and $|\lambda|_e \le q1, |\mu|_e \le 1$, then $\lambda A + \mu A \subseteq A$.

The $\bar{\mathbb{C}}$-convexity can not be considered as a generalization of the corresponding concept in the vector spaces.

If A contains O, then it is convex if and only if $[(\epsilon^{b_1})_\epsilon]A + [(\epsilon^{b_2})_\epsilon]A \subseteq A$ for all $b_1, b_2 \geq 0$.

Garetto then settles down to investigate the topological aspects of a $\bar{\mathbb{C}}$-module modeling on the classical approach to topological vector spaces and locally convex spaces by the celebrated texts like Robertson & Robertson ([Rob-R]) and Horvath ([Hor]).

The following result is in parallel to Proposition 3, Chap. I of [Rob-R]:

Let $\mathcal{G}$ be a topological $\bar{\mathbb{C}}$-module and $\mathcal{U}$ be a base of neighborhoods of the origin. Then for each $U \in \mathcal{U}$

  (i) U is absorbent,
 (ii)   there exists $V \in \mathcal{U}$ with $V + V \subset U$,
(iii)   there exists a balanced neighborhood W of the origin such that $W \subseteq U$.

Definition 2 A locally convex topological $\bar{\mathbb{C}}$-module is a topological $\bar{\mathbb{C}}$-module which has a base of $\bar{\mathbb{C}}$- convex neighborhoods of the origin.

Like in the standard theory of locally convex topological vector spaces, a locally convex topological $\bar{\mathbb{C}}$-module $\mathcal{G}$ has a base of absolutely convex and absorbent neighborhoods of the origin, (c.f. Proposition 1.7 [Gar-1], also Theorem 2, Chap. I [Rob-R]).

In classical locally convex topological spaces seminorms play a central role in defining different topologies like weak and strong topologies, topology of uniform convergence on bounded sets. The counterpart of a seminorm in topological $\bar{\mathbb{C}}$-module is an ultra-pseudo seminorm which requires the use of valuations.

Definition 3 Let $\mathcal{G}$ be a $\bar{\mathbb{C}}$-module. A valuation on $\mathcal{G}$ is a function $v : \mathcal{G} \longrightarrow (-\infty, +\infty]$ such that

  (i) $v(0) = +\infty$,

 (ii) $v(\lambda u) \geq v_{\bar{\mathbb{C}}}(\lambda) + v(u)$ for all $\lambda \in \bar{\mathbb{C}}$, $u \in \mathcal{G}$,

(ii)' $v(\lambda u) = v_{\bar{\mathbb{C}}}(\lambda) + v(u)$ for all $\lambda = [(c\epsilon^a)_\epsilon]$, $c \in \mathbb{C}$, $a \in \mathbb{R}$, $u \in \mathcal{G}$,

(iii) $v(u + w) \geq \min\{v(u), v(w)\}$.

Example Let A be an absolutely convex and absorbent subset of a $\bar{\mathbb{C}}$-module $\mathcal{G}$. Then $v_A(u) = \sup\{b \in \mathbb{R} : u \in [(\epsilon^b)_\epsilon A\}$.

An ultra pseudo-seminorm on $\mathcal{G}$ is a function $\mathcal{P} : \mathcal{G} \longrightarrow [0, +\infty)$ such that

(i) $\mathcal{P}(0) = 0$,

(ii) $\mathcal{P}(\lambda u) \leq |\lambda|_e \mathcal{P}(u)$ for all $\lambda \in \bar{\mathbb{C}}$, $u \in \mathcal{G}$,

(ii)' $\mathcal{P}(\lambda u) = |\lambda|_e \mathcal{P}(u)$ for all $\lambda = [(c\epsilon^a \epsilon], c \in \mathbb{C}, a \in \mathbb{R}, u \in \mathcal{G}$,

(iii) $\mathcal{P}(u + v) \leq \max\{\mathcal{P}(u), \mathcal{P}(v)\}$.

$\mathcal{P}(u) = e^{-v(u)}$ is a typical example of an ultra pseudo-seminorm obtained by means of a valuation on $\mathcal{G}$. An ultra pseudo-norm is an ultra pseudo-seminorm $\mathcal{P}$ such that $\mathcal{P}(u) = 0$ implies $u = 0$. $|\,.\,|_e$ introduced before is an ultra pseudo-norm on $\bar{\mathbb{C}}$.

Let $\{\mathcal{P}_i\}_{i \in I}$ be a family of ultra pseudo-seminorms on a $\bar{\mathbb{C}}$-module $\mathcal{G}$. The topology induced by $\{\mathcal{P}_i\}$ on $\mathcal{G}$, i.e. the coarsest topology such that each ultra pseudo-norm is continuous, induces the structure of locally convex topology on $\bar{\mathbb{C}}$-module on $\mathcal{G}$.

Duality for Topological $\bar{\mathbb{C}}$-modules, Weak Topologies

$L(\mathcal{G}, \bar{\mathbb{C}})$ and $\mathcal{L}(\mathcal{G}, \bar{\mathbb{C}})$ will denote algebraic and topological duals of a topological $\bar{\mathbb{C}}$-module respectively, i.e. in the second case their elements consist of $\bar{\mathbb{C}}$-linear and continuous $\bar{\mathbb{C}}$-linear maps $\mathcal{G} \longrightarrow \mathbb{C}$ respectively.

Definition 4 The $\bar{\mathbb{C}}$-modules $\mathcal{G}$ and $\mathcal{H}$ form a dual pair if there exists a $\bar{\mathbb{C}}$-bilinear form $b : \mathcal{G} \times \mathcal{H} \longrightarrow \mathbb{C}, (u, v) \longrightarrow b(u, v)$
The pair separates points of $\mathcal{G}$ if for all $u \neq 0$ in $\mathcal{G}$, there exists $v \in \mathcal{H}$ such that $b(u, v) \neq 0$. An analogous definition for the separation of points of $\mathcal{H}$. The pairing is separated if it separates points of both $\mathcal{G}$ and $\mathcal{H}$.

A pairing $(\mathcal{G}, \mathcal{H}, b)$ defines a topology on the first two components via the $\bar{\mathbb{C}}$-bilinear form b. The <u>weak topology</u> on $\mathcal{G}$ is the coarsest topology $\sigma(\mathcal{G}, \mathcal{H})$ in $\mathcal{G}$ such that each map $b(\,.\,, v) : \mathcal{G} \longrightarrow \bar{\mathbb{C}} : u \longrightarrow b(u, v)$, for v varying in $\mathcal{H}$ is continuous. Every $b(\,.\,, v)$ is $\bar{\mathbb{C}}$-linear and continuous iff the ultra pseudseminorm $\mathcal{P}_v : \mathcal{G} \longrightarrow |b(u, v)|_e$ is continuous. Hence $\sigma(\mathcal{G}, \mathcal{H})$ is the topology induced by the family of ultra pseudo-seminorms $\mathcal{P}_{v \in \mathcal{H}}$ and it provides the structure of a locally-convex topological $\bar{\mathbb{C}}$-module on $\mathcal{G}$. (Similar statements by interchanging $\mathcal{G}$ and $\mathcal{H}$).

Both $(\mathcal{G}, L(\mathcal{G}, \bar{\mathbb{C}}))$ and $(\mathcal{G}, \mathcal{L}(\mathcal{G}, \bar{\mathbb{C}}))$ are dual pairs via the bilinear form $(u, T) = T(u)$, T being an element of the dual.

For proving the next result on the completeness of the weak topologies we can not use Theorem 24, [Kel] (see the note to Theorem 1 of Sect. 3.4.9) unless we replace Cauchy sequences by Cauchy filters.

Recall that if E is any set, then a non-empty collection $\mathcal{F}$ of non-empty subsets of E is called a filter if it satisfies:
F1: $A \in \mathcal{F}, B \in \mathcal{F} \Longrightarrow A \cap B \in \mathcal{F}$,
F2: $A \in \mathcal{F}, A \subseteq B \Longrightarrow B \in \mathcal{F}$.

A non-empty collection $\mathcal{B}$ of non-empty subsets of E is called a filter base if it satisfies:

FB: $A \in \mathcal{B}, B \in \mathcal{B} \Longrightarrow \exists C \in \mathcal{B}, C \subseteq A \cap B$.

The set $\mathcal{F}$ of all sets containing a set of $\mathcal{B}$ is then a filter, called the filter generated by $\mathcal{B}$. If E is a topological space. The filter $\mathcal{F}$ is said to converge to a, written $\mathcal{F} \to a$, if every neighborhood of a contains some set of $\mathcal{F}$. In a separated space a filter can not converge to more than one point.

If $\{x_n\}$ is a sequence, then the filter generated by the filter base $\{X_n\}$, $X_n = \{x_j : j \geq n\}$ is called the elementary filter. A sequence converges to a if and only if the associated elementary filter converges to a.

In a locally convex topological vector space E a filter $\mathcal{F}$ is called a Cauchy filter if for each neighborhood U, $\mathcal{F}$ contains a set small of order U. The measure of smallness (size) in a metrizable space is given by the diameter. In the general case a set A is called small of order U if $x - y \in U$ for all $x, y \in U$.

If every Cauchy filter in E is convergent, then E is said to be complete.

Theorem 2 *Let $\mathcal{G}$ be a $\bar{\mathbb{C}}$-module. $L(\mathcal{G}, \bar{\mathbb{C}})$ is complete for the weak topology $\sigma(L(\mathcal{G}, \bar{\mathbb{C}}), \mathcal{G})$.*

Proof Let $\mathcal{F}$ be a Cauchy filter on $L(\mathcal{G}, \bar{\mathbb{C}})$. For each $u \in \mathcal{G}$ and $\epsilon > 0$, there is some $A \in \mathcal{F}$ for which $b(u, A)$ has diameter less than ϵ. Thus $b(u, \mathcal{F})$ is the base of a Cauchy filter of scalars. The scalar field being complete, there is some $f(u)$ with

$$b(u, \mathcal{F}) \longrightarrow f(u). \quad (*)$$

We show that $f(u)$ is a linear form on $\mathcal{G}$. For each $u_1, u_2 \in \mathcal{G}$ and $\epsilon > 0$ there are sets $A_1, A_2 \in \mathcal{F}$ with $|b(u_1, v_1) - f(u_1)| < \epsilon/2$ for all $v_1 \in A_1$ and $|b(u_2, v_2) - f(u_2)| < \epsilon/2$ for all $v_2 \in A_2$. Then letting $B = A_1 \cap A_2 \in \mathcal{F}$ we have $|b(u_1 + u_2, v) - (f(u_1) + f(u_2)| < \epsilon$ for all $v \in B$. But then by $(*)$ this should mean $b(u_1 + u_2, \mathcal{F}) \longrightarrow f(u_1) + f(u_2)$, i.e. $f(u_1 + u_2) = f(u_1) + f(u_2)$. Similarly $f(\lambda u) = \lambda f(u)$ so that $f \in L(\mathcal{G}, \bar{\mathbb{C}})$. Hence there is some $v \in L\mathcal{G}, \bar{\mathbb{C}})$ such that by $(*)$ $b(u, \mathcal{F}) \longrightarrow b(u, v)$ for all $u \in \mathcal{G}$. Thus $\mathcal{F} \longrightarrow v$ in $\sigma(L(\mathcal{G}, \bar{\mathbb{C}}), \mathcal{G})$. $\square$

Polar Topologies, Strong Topology

The definition of polar sets, polar topologies in $\bar{\mathbb{C}}$-topological modules follow closely those in the classical theory of topological vector spaces and locally convex spaces.

Let $\mathcal{G}$ and $\mathcal{H}$ be two $\bar{\mathbb{C}}$-modules in duality through the bilinear form b. The polar of a subset A of $\mathcal{G}$ is the subset A^0 of $\mathcal{H}$ consisting of those $v \in \mathcal{H}$ such that $|b(u, v)| \leq 1$ for all $u \in A$.

Clearly $A_1 \subseteq A_2 \Longrightarrow A_2^0 \subseteq A_1^0$ and $(\bigcup_{A_i \in I})^0 = \bigcap_{i \in I} A_i^0$. Also for all invertible $\lambda \in \bar{\mathbb{C}}$, $(\lambda A))^0 = \lambda^{-1} A^0$. Other properties of polar sets like being absolutely convex and $\sigma(\mathcal{H}, , \mathcal{G})$-closed are proved like in the classical texts, (e.g. [Rob-R], Chap. II, Proposition 9). Furthermore A^0 is absorbent if and only if A is bounded in the topology $\sigma(\mathcal{G}, \mathcal{H})$. Hence by the Example following Definition 3 $\mathcal{P}_{A^0} = e^{-v_{A^0}(v)}$ (the gauge of A^0) defines an ultra pseudo-seminorm on $\mathcal{H}$.

Definition 5 Let $(\mathcal{G}, \mathcal{H})$ be a dual pair of $\bar{\mathbb{C}}$-modules with the bilinear form b. A topology on $\mathcal{H}$ is said to be polar if it is determined by the family of ultra pseudo-seminorms $\{\mathcal{P}_{A^0}\}_{A\in\mathcal{A}}$, where $\mathcal{A}$ is a collection of all $\sigma(\mathcal{G}, \mathcal{H})$-bounded subsets of $\mathcal{G}$. Then the corresponding polar topology is called the <u>strong topology</u> and denoted by $\beta(\mathcal{H}, \mathcal{G})$.

<u>Link to Generalized Functions</u>

Definition 6 . Let E be a locally convex topological vector space topologized by the family of seminorms $\{p_i\}_{i\in I}$. Define the set of E-moderate elements by:

$$\{\mathcal{M}_E \doteq \{(u_\epsilon)_\epsilon \in E^{(0,1]} : \forall i \in I, \exists N \in \mathbb{N} \ \ p_i(u_\epsilon) = O(\epsilon^{-N}) \text{ as } \epsilon \to 0\}$$

and the set of E-negligible elements by

$$\mathcal{N}_E \doteq \{(u_\epsilon)_\epsilon \in E^{(0,1]} : \forall i; \in I, \ \forall q \in \mathbb{N}, \ p_i(u_\epsilon) = O(\epsilon^q) \text{ as } \epsilon \to 0\}$$

Then we call the factor space $\mathcal{G}_E = \mathcal{M}_E/\mathcal{N}_E$ the space of generalized functions based on E. Use the notation $[(u_\epsilon)_\epsilon]$ for the class u of $(u_\epsilon)_\epsilon$ in $\mathcal{G}_E$ and we imbed E into $\mathcal{G}_E$ via the constant imbedding $f \longrightarrow [(f_\epsilon)_\epsilon]$.

Multiplication of a generalized constant by an element of $\mathcal{G}_E$ is defined via the map $\bar{\mathbb{C}} \times \mathcal{G}_E \longrightarrow \mathcal{G}_E : ([(\lambda_\epsilon)_\epsilon], [(u_\epsilon)_\epsilon]) \longrightarrow [(\lambda_\epsilon u_\epsilon)_\epsilon]$, which equips $\mathcal{G}_E$ with the structure of a $\bar{\mathbb{C}}$- module.

As the growth in ϵ of an E-moderate net is estimated in terms of seminorms p_i of E, the valuation (called p_i-valuation) is naturally given in terms of seminorms of E:

$$v_{p_\epsilon} ((u_\epsilon)_\epsilon) = \sup\{b \in \mathbb{R} : \ p_i(u_\epsilon) = O(\epsilon^b) \text{ as } \epsilon \to 0\} \quad (5)$$

Note that $v_{p_\epsilon} ((u_\epsilon)_\epsilon) = v ((p_i(u_\epsilon))_\epsilon)$ where v is the valuation on $\bar{\mathbb{C}}$ given by (1) v_{p_i} has properties in parallel to those of the valuation on $\mathcal{G}$ given in Definition 3:
(i) $v_{p_i}((u_\epsilon)_\epsilon) = +\infty$ for all $i \in I$ if and only if $(u_\epsilon)_\epsilon \in \mathcal{N}_E$,
(ii) $v_{p_i} ((\lambda_\epsilon)_\epsilon) \geq v ((\lambda_\epsilon)_\epsilon) + v_{p_i} ((u_\epsilon)_\epsilon)$, for all $(\lambda_\epsilon)_\epsilon \in \mathcal{E}_M$ and $(u_\epsilon)_\epsilon \in \mathcal{M}_E$,
(ii)' $v_{p_i} ((\lambda_\epsilon)_\epsilon) = v ((\lambda_\epsilon)_\epsilon)$ for all $(\lambda_\epsilon)_\epsilon = (ce^b)_\epsilon$, $c \in \mathbb{C}$, $b \in \mathbb{R}$,
(iii) $v_{p_i} ((u_\epsilon)_\epsilon + (w_\epsilon)_\epsilon) \geq \min\{v_{p_i} ((u_\epsilon)_\epsilon), v_{p_i} ((w_\epsilon)_\epsilon)\}$.

In a similar fashion $\mathcal{P}_i = e^{-v_{p_i}}$ is an ultra pseudo-seminorm on the $\bar{\mathbb{C}}$-module $\mathcal{G}_E$. $\mathcal{G}_E$ endowed with the topology of ultra pseudo-seminorms $\{\mathcal{P}_{\rangle}\}_{i\in <I}$ is a locally convex topological $\bar{\mathbb{C}}$-module. We denote thus created sharp topology by $\tau_\sharp$. Then $(\mathcal{G}_E, \tau_\sharp)$ is a separated locally convex topological $\bar{\mathbb{C}}$-module.

Theorem 3 *If E is a locally convex topological space with a countable base of neighborhoods of the origin, then $\mathcal{G}_E$ with the sharp topology $\tau_\sharp$ is complete.*

Proof Under the assumptions $\mathcal{G}_E$ is metrizable (c.f. [Gar-1], Theorem 3.3. (ii)). Then it is sufficient to show that every Cauchy sequence converges to a point of the space.

The proof of this fact resembles to the completeness proof of $\bar{\mathbb{C}}$ under the sharp topology as in Theorem 1 (details in [Gar-1], Proposition 3.4).

Examples of Colombeau algebras obtained as $\bar{\mathbb{C}}$-modules $\mathcal{G}_E$.

(I) For $E = \mathbb{C}$, we obtain $\mathcal{G}_E = \bar{\mathbb{C}}$, which is also a topological $\bar{\mathbb{C}}$-algebra.

(II) Let Ω be an open subset of $\mathbb{R}^n$ and $E = C^\infty(\Omega)$. It is topologized through the family of seminorms $p_{K_i,j}(f) = \sup_{x \in K_i, |\alpha| \leq j} |D^\alpha f(x)|$ where $K_0 \subset K_1 \subset \cdots \cdots \subset K_i \subset \cdots$ is a countable and exhaustive sequence of compact subsets of Ω. In this way we have $\mathcal{G}_E = \mathcal{G}(\Omega)$.

A more general form of the simplified Colombeau algebras endowed with the sharp topology determined by the seminorms $\{p_{K_i,j}\}_{i,j \in \mathbb{N}}$ is a Fréchet $\bar{\mathbb{C}}$-module and a topological $\bar{\mathbb{C}}$-algebra.

(III) Colombeau algebras of compactly supported generalized functions:

For $K \subset\subset \Omega$ we denote by $\mathcal{G}_K(\Omega)$ the space of all generalized functions in $\mathcal{G}(\Omega)$ with support contained in K. $\mathcal{G}_K(\Omega)$ is contained in $\mathcal{G}_{\mathcal{D}_{K'}(\Omega)}$ for all compact subsets K' of Ω such that $K \subseteq int(K')$ where $\mathcal{G}_{\mathcal{D}_{K'}(\Omega)}$ is the space of all smooth functions f with support in K'. Then $\mathcal{G}_K(\Omega) \longrightarrow \mathcal{G}_{\mathcal{D}_{K'}(\Omega)} : u \longrightarrow (u_\epsilon)_\epsilon + \mathcal{N}_{\mathcal{D}_{K'}(\Omega)}$ is a well-defined and injective $\bar{\mathbb{C}}$-linear map.

Also $\mathcal{G}_{\mathcal{D}_{K'}(\Omega)}$ is naturally imbedded into $\mathcal{G}(\Omega)$ via

$$\mathcal{G}_{\mathcal{D}_{K'}(\Omega)} \longrightarrow \mathcal{G}(\Omega) : (u_\epsilon)_\epsilon + \mathcal{N}_{\mathcal{D}_{K'}(\Omega)} \longrightarrow (u_\epsilon)_\epsilon + \mathcal{N}(\Omega).$$

In $\mathcal{G}(\Omega)$, the $p_{K,n}$-valuation where $p_{K,n}(f) = \sup_{x \in K, \alpha| \leq n} |D^\alpha f(x)|$ is obtained as the valuation of the complex generalized number $\sup_{x \in K, |\alpha| \leq n} |D^\alpha u(x)| \doteq (\sup_{x \in K, |\alpha| \leq n} |D^\alpha u_\epsilon(x)|)_\epsilon + \mathcal{N}$. Hence for $K \subseteq int K'$ we have $v_{K,n} = v_{p_{K'},n}(u)$. as a valuation on $\mathcal{G}_K(\Omega)$. Furthermore this does not depend on K', i.e. for any K_1', K_2' containing K in their interiors $v_{K_1',n} = v_{K_2',n}$.

$\mathcal{G}_K(\Omega)$ with the topology induced by the ultra pseudo seminorms $\{\mathcal{P}_{\mathcal{G}_K(\Omega),n}(u) = e^{-v_{K,n}(u)}\}_{n \in \mathbb{N}}$ is a locally convex topological $\bar{\mathbb{C}}$-module.

This topology is separated, metrizable and induced by any $\mathcal{G}_{\mathcal{D}_{K'}(\Omega)}$ with $K \subseteq int(K')$. Also $\mathcal{G}_K(\Omega)$ is closed in $\mathcal{G}_{\mathcal{D}_{K'}(\Omega)}$ and since $\mathcal{D}_{K'(\Omega)}$ is complete in sharp topology, $(\mathcal{G}_K(\Omega), \{\mathcal{P}_{\mathcal{G}_K(\Omega),n}\}$ is a Fréchet $\bar{\mathbb{C}}$- module. Then letting $\{K_n\}_{n \in \mathbb{N}}$ be an exhaustive sequence of Ω as in I), we clearly have $\mathcal{G}_c(\Omega)$ of the Colombeau algebras of compactly supported generalized functions will be given by $\mathcal{G}_c(\Omega) = \bigcup_{n \in \mathbb{N}} \mathcal{G}_{K_n}(\Omega)$. Each $\mathcal{G}_{K_n}(\Omega)$ is a Fréchet $\bar{\mathbb{C}}$-module and $\mathcal{G}_c(\Omega)$ will be endowed with the strict inductive limit topology.

(IV) The algebra of tempered generalized functions $\mathcal{G}_\tau(\mathbb{R}^n)$:

Inspired by the fact that the tempered distributions generalize slowly increasing functions as the members of the dual of rapidly decreasing functions, $\mathcal{G}_\tau(\mathbb{R}^n)$ is constructed as the factor algebra $\mathcal{E}_\tau(\mathbb{R}^n)/\mathcal{N}_\tau(\mathbb{R}^n)$. Here $\mathcal{E}_\tau(\mathbb{R}^n) = \mathcal{O}_M(\mathbb{R})^{(0,1]}$ where $\mathcal{O}_M$ is the set of all functions which are smooth and polynomially bounded along with all of their derivatives. More precisely $\mathcal{E}_\tau$ consists of all τ- moderate nets $(u_\epsilon)_\epsilon$ where

$$\forall \alpha \in \mathbb{N}^n, \quad \sup_{x \in \mathbb{R}^n} (1 + |x|^2)^{-N} |D^\alpha u_\epsilon(x)| = O(\epsilon^{-N}) \text{ as } \epsilon \to 0,$$

and $\mathcal{N}_\tau(\mathbb{R}^n)$ is the ideal of all negligible nets $(u_\epsilon)_\epsilon \in \mathcal{E}_\tau(\mathbb{R}^n)$ such that

$$\forall \alpha \in \mathbb{N}^n, \exists N \in \mathbb{N}, \forall q \in \mathbb{N} \quad \sup_{x \in \mathbb{R}^n}(1 + |x|^2)^{-N}|D^\alpha u_\epsilon| = O(\epsilon^q) \quad \text{as } \epsilon \to 0.$$

A locally convex $\bar{\mathbb{C}}$-linear topology on $\mathcal{G}_\tau(\mathbb{R}^n)$ is given whose construction involves a countable family of different algebras of generalized functions (for the details c.f. [Gar-1], Example 3.9).

(V) Regular generalized functions based on E. for any locally convex topological vector space $(E, \{p_i\}_{i \in I})$ the set

$$\mathcal{M}_E^\infty \doteq \{(u_\epsilon)_\epsilon \in E^{(0,1]} : \exists N \in \mathbb{N}, \forall i \in I \ p_i(u_\epsilon) = O(\epsilon^{-N}) \text{ as } \epsilon \to 0\}$$

is a subspace of the set $\mathcal{M}_E$ of E-moderate nets. Therefore the corresponding factor space $\mathcal{G}_E^\infty \doteq \mathcal{M}_E^\infty/\mathcal{N}_E$ is a subspace of $\mathcal{G}_E$ whose elements are called regular generalized functions based on E. When E is in addition a topological algebra, $\mathcal{G}_E$ and $\mathcal{G}_E^\infty$ are both algebras. Firstly since $\mathcal{G}_E^\infty \subset \mathcal{G}_E$, the sharp topology induced by $\mathcal{G}_E$ on $\mathcal{G}_E^\infty$ gives the structure of a separated locally convex topology on $\bar{\mathbb{C}}$-module to $\mathcal{G}_E^\infty$. The ultra pseudo-seminorms obtained in this way are the original $\mathcal{P}_i$ on $\mathcal{G}_E$ restricted to $\mathcal{G}_E^\infty$.

The moderateness properties of $\mathcal{M}_E^\infty$ allow us to define the valuation $\nu_E^\infty : \mathcal{M}_E^\infty \to (-\infty, +\infty]$ as

$$\nu_E^\infty((u_\epsilon)_\epsilon) = \sup\{b \in \mathbb{R} : \forall i \in I \ p_i(u_\epsilon) = O(\epsilon^p) \text{ as } \epsilon \to 0\}$$

which can be extended to $\mathcal{G}_E^\infty$. This yields the existence of the ultra pseudo-seminorm $\mathcal{P}_E^\infty(u) \doteq e^{-\nu_E^\infty(u)}$ on $\mathcal{G}_E^\infty$. Since for all $i \in I$ and $u \in \mathcal{G}_E^\infty$, $\nu_{p_i}(u) \geq \nu_E^\infty$, the topology $\tau_\natural^\infty$ determined by $\mathcal{P}_E^\infty$ on $\mathcal{G}_E^\infty$ is finer than the topology induced by $\mathcal{G}_E$.

When E is a locally convex topological vector space with a countable base of neighborhoods of the origin, $\mathcal{G}_E^\infty$, with the topology $\tau_\natural^\infty$ is complete.

(V') The Colombeau algebra of S-regular generalized functions $\mathcal{G}_S^\infty(\mathbb{R}^n)$ is precisely $\mathcal{G}_E^\infty$ with $E = S(\mathbb{R}^n)$.

(VI) The Colombeau algebra $\mathcal{G}^\infty(\Omega)$.

All $u \in \mathcal{G}(\Omega)$ having a representative $(u_\epsilon)_\epsilon$ satisfying the following condition

$$\forall K \subset\subset \Omega \ \exists N \in \mathbb{N} \ \forall \alpha \in \mathbb{N}^n \quad \sup_{x \in K}|D^\alpha u_\epsilon(x)| = O(\epsilon^{-N}) \text{ as } \epsilon \to 0.$$

defines the subset $\mathcal{E}_M^\infty(\Omega)$ of the set of moderate nets $\mathcal{E}(\Omega)$ and determines $\mathcal{G}^\infty(\Omega)$ as the factor $\mathcal{E}_M(\Omega)/\mathcal{N}(\Omega)$. $\mathcal{G}^\infty(\Omega)$ can be seen as the intersection $\bigcap_{K \subset\subset} \mathcal{G}^\infty(K)$ where $\mathcal{G}^\infty(K)$ is the space of all $u \in \mathcal{G}(\Omega)$ such that there exists a representative $(u_\epsilon)_\epsilon$ satisfying the condition

$$\exists N \in \mathbb{N}^n \quad \sup_{x \in K}|D^\alpha u_\epsilon(x)| = O(\epsilon^{-N}) \text{ as } \epsilon \to o.$$

We choose an exhausting sequence $K = K_0 \subset K_1 \subset \cdots$ of compact subsets of Ω. We equip $\mathcal{G}^\infty(K)$ with the locally convex $\bar{\mathbb{C}}$-linear topology determined by the usual ultra pseudo-seminorms $\mathcal{P}_i(u) = e^{-v_{p_i}(u)}$ on $\mathcal{G}(\Omega)$ where $p_i(u_\epsilon) = \sup_{x \in K_i, |\alpha| \le i} |D^\alpha u_\epsilon(x)|$, and the $\mathcal{G}^\infty(K)$-ultra pseudo-seminorm $\mathcal{P}_{\mathcal{G}^\infty(K)}$ defined via the valuation

$$v_{\mathcal{G}^\infty(K)}(u) = \sup\{b \in \mathbb{R} \forall \alpha \in \mathbb{N}^n \ \ \sup_{x \in K} |D^\alpha u_\epsilon(x)| = O(\epsilon^b)\}$$

with the topology $\mathcal{G}^\infty(K)$ is separated and metrizable, and also complete as proved in [Gar-2]. Example 3.12.

By definition $\mathcal{G}(K')$ is a $\bar{\mathbb{C}}$-submodule of $\mathcal{G}^\infty(K)$ when $K \subset K'$ and noting that $\mathcal{P}_{\mathcal{G}^\infty(K)}(u) \le \mathcal{P}_{\mathcal{G}^\infty(K')}(u)$ for all $u \in \mathcal{G}^\infty(K')$, the topology on $\mathcal{G}^\infty(K')$ is finer than the topology induced by $\mathcal{G}^\infty(K)$ on $\mathcal{G}^\infty(K')$. Hence $\mathcal{G}^\infty(\Omega)$ equipped with the initial topology for the injections $\mathcal{G}^\infty(\Omega) \to \mathcal{G}^\infty(K)$ is a complete locally convex topological $\bar{\mathbb{C}}$-module.

(VII) The Colombeau algebra $\mathcal{G}_c^\infty(\Omega)$ is the algebra of generalized functions in $\mathcal{G}^\infty(\Omega)$ which have compact support. $\mathcal{G}_{\mathcal{D}_K(\Omega)}^\infty$, $K \subset\subset \Omega$ is a complete ultra pseudo-seminormed $\bar{\mathbb{C}}$-mmodule with valuation

$$v_{\mathcal{D}_K(\Omega)}^\infty(u) = \sup\{b \in \mathbb{R} : \ \forall \alpha \in \mathbb{N}^n \ \ \sup_{x \in K} |D^\alpha u_\epsilon(x)| = O(\epsilon^b)\}.$$

We note that $v_{\mathcal{G}^\infty(K)}$ defined in the previous example and $v_{\mathcal{D}_K(\Omega)}^\infty$ coincide on $\mathcal{G}_{\mathcal{D}_K(\Omega)}^\infty$. $\mathcal{G}_K^\infty(\Omega)$, the space of all generalized functions in $\mathcal{G}^\infty(\Omega)$ with support contained in K, is contained in $\mathcal{G}_{\mathcal{D}_K'(\Omega)}$, for every compact K' containing K in its interior. Again choosing an exhausting sequence $K_0 \subset K_1 \subset K_2 \subset \cdots$ of compact sets, we have $\mathcal{G}_c^\infty(\Omega) = \bigcup_{n \in \mathbb{N}} \mathcal{G}_{K_n}(\Omega)$ equipped with strict inductive limit is a complete and separated $\bar{\mathbb{C}}$-linear topology.

3.4.11 *Pseudodifferential Operators and Wave Front Set*

Consider the basic formulae of the Fourier transform and the inverse Fourier transform for any $f \in L^1(\mathbb{R}^n)$:

Fourier transform

$$\hat{f}(\xi) = (2\pi)^{-\frac{n}{2}} \int_{\mathbb{R}^n} e^{-i<\xi,x>} f(x)dx$$

Inverse Fourier transform

$$\tilde{f}(\xi) = (2\pi)^{-\frac{n}{2}} \int_{\mathbb{R}^n} e^{i<\xi,x>} f(x)dx$$

In this Subsection for the convenience of notations we take the character of the Fourier transform as $e_\xi(x) = (2\pi)^{-\frac{1}{n}} e^{i<\xi,x>}$ in accordance with the Important Remark (3), Sect. 1.8.1.

By the Riemann-Lebesgue lemma Fourier transform is continuous and vanishes at infinity.

<u>Notes.</u>

(i) By the Fourier inversion theorem if $f \in L^2(\mathbb{R}^n)$, then $\hat{\tilde{f}} = f$ and $\tilde{\hat{f}} = f$ a.e.

(ii) Plancherel's formula: if $f \in L^2(\mathbb{R}^n)$, then $\|f\|_2 = \|\hat{f}\|_2 = \|\tilde{f}\|_2$.

(iii) If $f \in \mathcal{S}(\mathbb{R}^n)$, then $\hat{f} \in \mathcal{S}(\mathbb{R}^n)$ and $\tilde{f} \in \mathcal{S}(\mathbb{R}^n)$.

<u>Motivation.</u> By Note i) and the inversion formula we have

$$f(x) = (2\pi)^{-\frac{n}{2}} \int_{\mathbb{R}^n} e^{i<x,\xi>} \hat{f}(\xi)d\xi.$$

If $f \in \mathcal{S}$ the integral is uniformly continuous and differentiation can be put under the integral sign. Using the multi-index notation $\alpha = (\alpha_1, \cdots, \alpha_n)$, $D^\alpha = D_1^{\alpha_1} \cdots D_n^{\alpha_n}$, where $D_j = \frac{\partial}{\partial x_j}$

$$D^\alpha f(x) = (2\pi)^{-\frac{n}{2}} \int_{\mathbb{R}^n} \xi^\alpha e^{i<x,\xi>} \hat{f}(\xi)d\xi; \quad \xi^\alpha = \xi_1^{\alpha_1} \cdots \xi_n^{\alpha_n}$$

Therefore if the differential operator is $p(x, D) = \sum_{|\alpha| \leq k} a_\alpha D^\alpha$; $a_\alpha \in \mathcal{C}^\infty(\mathbb{R}^n)$:

$$p(x, D)f(x) = (2\pi)^{-\frac{n}{2}} \int_{\mathbb{R}^n} p(x, \xi) e^{i<x,\xi>} \hat{f}(\xi)d\xi \quad (1)$$

We can view (1) as an operator acting on f. To generalize it to lead to the definition of pseudo-differential operators we have to take the polynomial $p(x, \xi)$ belonging to a large class of functions called <u>symbols</u>.

Definition 1 For any $m \in \mathbb{R}$, the space S^m is defined to be the set of functions $\sigma \in \mathcal{C}^\infty(\mathbb{R}^n \times \mathbb{R}^n)$ such that for all multi-indices α and β there exists a constant $C_{\alpha,\beta}$ such that

$$|D_x^\alpha D_\xi^\beta \sigma(x, \xi)| \leq C_{\alpha,\beta}(1 + |\xi|)^{m-|\beta|}; \quad x, \xi \in \mathbb{R}^n$$

A function in $\bigcup_{m \in \mathbb{R}} S^m$ is called a symbol.

<u>Notes.</u> It is not difficult to verify that

(i) S^m is a vector space,

(ii) If $\sigma \in S^{m_1}$, $\tau \in S^{m_2}$, then $\sigma\tau \in S^{m_1+m_2}$,

(iii) If $m_1 \leq m_2$, then $S^{m_1} \subseteq S^{m_2}$. ([Tor], p. 18).

Definition 1'. A more generalized symbol: Let Ω be an open subset of $\mathbb{R}^n$, $m \in \mathbb{R}$ and $\rho, \delta \in [0, 1]$. Then $S^m_{\rho,\delta}$ denotes the set of all symbols $\sigma \in \Omega \times \mathbb{R}^n$ of order m and type (ρ, δ), if for any $K \subset\subset \Omega$, for any multi-indices α and β there exists a constant $C_{K,\alpha,\beta}$ such that

$$|D^\alpha_x D^\beta_\xi \sigma(x, \xi) \le C_{K,\alpha,\beta}(1 + |\xi|)^{m-\rho|\alpha|+\delta|\beta|}, \ \forall x \in K, \forall \xi \in \mathbb{R}^n.$$

For $\rho = 1$ and $\delta = 0$ the definition reduces to the previous definition and then the subscript $[\rho, \delta)$ is omitted. We also use $S^1_{hg}(\Omega \times \mathbb{R}^n \setminus 0)$ of all $\sigma \in S^1 \in (\Omega \times \mathbb{R}^n \setminus 0$ homogeneous of degree 1 in ξ.

Definition 2 (Elliptic symbols) A symbol called m-elliptic if $\sigma \in S^m$ and there exist constants $C > 0$, $R \ge 0$ such that

$$|\sigma(x, \xi)| \ge C(1 + |\xi|)^m, \ \ |\xi| \ge R.$$

Elliptic symbols are important in the theory partial differential equations. General symbols are bounded from above while elliptic symbols are bounded from below.

Proposition 1 *Let* $p(x, \xi) = \sum_{|\alpha| \le m} a_\alpha(x)\xi^\alpha$; $x, \xi \in \mathbb{R}^n$, *where* $m \in \mathbb{N}$ *and all the coefficients* $a_\alpha \in C^\infty(\mathbb{R}^n)$ *and the derivatives* $D^\beta a_\alpha$ *are bounded for all multi-indices* β. *Then* $p \in S^m$.

Proof For multi-indices $|\alpha| \le m$ and β, let $C_{\alpha,\beta} = \sup_{x \in \mathbb{R}^n}\{D^\beta a_\alpha(x)\}$. Then using the multi-index formula

$$D^\beta \xi^\alpha = \begin{cases} \beta!\binom{\alpha}{\beta}x^{\alpha-\beta} & \text{if } \beta \le \alpha \\ 0 & \text{otherwise} \end{cases}$$

we have

$$|D^\beta_x D^\gamma_\xi p(x, \xi)| \le \sum_{|\alpha| \le m} |D^\beta_x D^\gamma_\xi (a_\alpha(x)\xi^\alpha)| = \sum_{|\alpha| \le m} |D^\beta a_\alpha(x)| |D^\gamma \xi^\alpha|$$

$$\le \sum_{|\alpha| \le m, \alpha \ge \gamma} \gamma!\binom{\alpha}{\gamma} C_{\alpha,\beta} |\xi|^{|\alpha-\gamma|} \le \sum_{|\alpha| \le m, \alpha \ge \gamma} \gamma!\binom{\alpha}{\gamma} C_{\alpha,\beta} |\xi|^{|\alpha|-|\gamma|}$$

$$\le \sum_{|\alpha| \le m, \alpha \ge \gamma} \gamma!\binom{\alpha}{\gamma} \alpha, \beta |\xi|^{m-|\gamma|} \le C'_{\beta,\gamma}(1 + |\xi|)^{m-|\gamma|}$$

where $C'_{\beta,\gamma} = \sum_{|\alpha| \le m, \alpha \ge \gamma} \gamma! \binom{\alpha}{\gamma} C_{\alpha,\beta}$, it the follows that $p \in S^m$. $\qquad \square$

Definition 3 (Pseudo-differential operator) For any symbol σ, the pseudo-differential operator $T_\sigma : \mathcal{S}_n \longrightarrow \mathcal{S}_n$ is defined to be the linear operator such that for any $\phi \in \mathcal{S}_n$:

$$(T_\sigma\phi)(x) = (2\pi)^{-\frac{n}{2}} \int_{\mathbb{R}^n} e^{i<x,\xi>} \sigma(x,\xi)\hat{\phi}(\xi)d\xi, \quad x \in \mathbb{R}^n.$$

$(\mathcal{S}_n(\mathbb{R}^n)$ is the Schwartz space of rapidly decreasing functions).

As by Proposition 1 $p(x,\xi)$ is a symbol, (1) defines a pseudo-differential operator.
The class of pseudo-differential operators is larger than the class of differential operators as the following proposition shows:

Proposition 2 *Let $p(x,D) = \sum_{|\alpha|\leq m} a_\alpha(x)D^\alpha$ be a partial differential operator where all the coefficients $a_\alpha \in C^\infty(\mathbb{R}^n)$ and the derivatives $D^\beta a_\alpha$ are bounded for all multi-indices β. Then $p(x,D)$ is also a pseudo-differential operator with symbol $p(x,\xi) = \sum_{|\alpha|\leq m} \alpha(x)\xi^\alpha$.*

Proof It was proved in Proposition 2 that $p(x,\xi)$ is a symbol. By the Fourier inversion formula $(T_p\phi)(x) = (2\pi)^{-\frac{n}{2}} \int_{\mathbb{R}^n} e^{i<x,\xi>} p(x,\xi)\phi(\hat{\xi})\,d\xi = (2\pi)^{-\frac{n}{2}} \sum_{|\alpha|\leq m} \int_{\mathbb{R}^n} e^{i<x,\xi>} \widehat{(D^\alpha\phi)}(\xi)\,d\xi = \sum_{|\alpha|\leq m} D^\alpha\phi(x) = p(x,D)\,\phi(x)$ which proves that $p(x,D)$ is a pseudo-differential operator with symbol $p(x,\xi)$.

Fourier integral operators. Operators closely related to pseudo-differential operators are the Fourier integral operators. They have become important tools in the theory of partial differential equations. The class of Fourier integral operators contain the differential operators as well as classical integral operators as special cases.
A Fourier integral operator Q is given by

$$(Qf)(x) = \int_{\mathbb{R}^n} e^{i\Phi(x,\xi)}\sigma(x,\xi)\hat{f}(\xi)d\xi,$$

where $\hat{f}$ denotes the Fourier transform of f, $\sigma(x,\xi)$ is a standard symbol which is compactly supported in x and Φ is real-valued and homogeneous of degree 1 in ξ, it is also necessary to require that $Det\frac{\partial^2\Phi}{\partial x i \xi_j} \neq 0$ on the support of σ. If σ is of order zero, then it is shown that Q defines a bounded operator from L^2 into L^2.

Oscillatory integrals. They are integrals depending on a real parameter ϵ of the type

$$\int_{\mathbb{R}^n} e^{i<\omega,x>} a_\epsilon dx$$

with phase function ω and amplitude $a_\epsilon \in \mathbb{R}^n$.
The phase function $\omega \in C^\infty \setminus 0$ is real-valued and order $k > 0$, i.e. $\omega(tx) = t^k\omega(x)$ and $\Delta\omega(x) \neq 0$ for $x \neq 0$.

Wave Front Set
In analysis, more precisely in microlocal analysis the wave front set $WF(u)$ characterizes the singularities of a linear distribution or generalized function u, not only

in space but also with respect to its Fourier transform at each point (the term wave front was coined by Lars Hörmander).

Definition 4 In Euclidean space, the wave front set of a distribution or generalized function u is defined as

$$WF(u) = \{(x, \xi) \in \Omega \times \mathbb{R}^n : \xi \in \sum_x (u)\}, \ \Omega \subset \mathbb{R}^n$$

where $\sum_x (u)$ is the singular fibre of u at x.

The singular fibre is defined to be the complement of all directions ξ such that the Fourier transform of u, localized at x, is sufficiently regular when restricted to an open cone containing ξ (microlocal regularity). More precisely a direction ψ is the complement of $\sum_x (u)$, if there is a compactly supported smooth function ϕ with $\phi(x) \neq 0$ and an open cone Γ containing ψ such that the following estimate holds for each integer N:

$$\widehat{(\phi u)}(\xi) \leq C_N (1 + |\xi|)^{-N} \ \text{ for all } \ \xi \in \Gamma.$$

(i.e. $\widehat{(\phi)}$ is fast decreasing on Γ).

Once such an estimate holds for a particular cut-off function ϕ at x, it also holds for all cut-off functions with smaller support, possibly for a different open cone containing v.

Notes.

(1) Observe that the wave front set is conical, i.e. if $(x, v) \in WF(\omega)$, then $(x, \lambda v) \in WF(\omega)$ for all $\lambda > 0$.

(2) In the example of the curve delta function (Example 1 c), 1.2) the wave front set is the set-theoretic complement of the image of the tangent bundle of the curve inside the tangent bundle of the plane.

(3) Since the definition involves cut-off by a compactly supported function, the notion of a wave front set can be transported to any differential manifold M.

In the more general situation, the wave front set is a closed conical subset of the cotangent bundle $T^*(M)$, since the ξ variable naturally localizes to a covector rather than a vector. In the case of a differentiable manifold M, using the local coordinates x, ξ on the cotangent bundle, the wave front set $WF(u)$ of a distribution u can be defined as $WF(u) = \{(x, \xi) \in T^*(X) : \xi \in \Sigma_x (u)\}$, where the singular fibre $\Sigma_x (u)$ is again the complement of all directions ξ such that the Fourier transform of u, localized at x, is sufficiently regular when restricted to a conical neighborhood of ξ. The regularity estimate transforms well under diffeomorphism, and so the notion of regularity is independent of the choise of local coordinates.

(4) The projection of $WF(u)$ on M is the <u>singular support</u> of the function (see also 1.9.1).

The distributional wave front set can be described ([Gar-4] (in fact was originally defined in [Hörm-3]) in terms of characteristic sets of pseudodifferential operators:

If $m \in \mathbb{R}$ arbitrary, then for any $v \in \mathcal{D}'$

$$WF(v) = \bigcap_{Av \in C^\infty} \text{Char}\,(A)$$

where Char(A) is the characteristic set of $A = a(x, D)$ $(a \in S^m$ the space of smooth symbols of order m and type $(0, 1)$, i.e., the complement of those points (x_1, ξ_1), where a is (micro)-elliptic in the sense that an estimata $|a(x, \xi) \geq C(1 + |\xi|)^m$ holds for x near x_1 and ξ with $|\xi| \geq 1$ in some neighborhood of ξ_1.

3.4.12 *Pseudo-Differential and Fourier Integral Operators on Generalized Functions*

As this section is closely related to Garetto's work ([Gar-2] and [Gar-3]), we stick to her notation.

In 3.4.10 the Colombeau algebra of tempered generalized functions was defined as the factor

$$\mathcal{G}_\tau(\mathbb{R}^n) = \mathcal{E}_\tau(\mathbb{R}^n)/\mathcal{N}_\tau(\mathbb{R}^n).$$

It is a differential algebra containing the space of distributions $S'(\mathbb{R}^n)$, where the derivatives extend the usual ones on $S'(\mathbb{R}^n)$.

Definition 1 We denote by $\mathcal{G}_{\tau,S}(\mathbb{R}^n)$ the factor space $\mathcal{E}_\tau(\mathbb{R}^n)/\mathcal{N}_S(\mathbb{R}^n)$, where $\mathcal{N}_S$ is the set of all $(u_\epsilon)_\epsilon \in \mathcal{E}(\mathbb{R}^n)$ fulfilling the condition
$\forall \alpha, \beta \in \mathbb{N}^n$, $\forall q \in \mathbb{N}$, $\sup_{\epsilon \in (0,1]} \epsilon^{-q} \|x^\alpha D^\beta u_\epsilon\|_{L^\infty(\mathbb{R}^n} < \infty$.
Since $\mathcal{N}_S(\mathbb{R}^n) \subset \mathcal{N}_\tau(\mathbb{R}^n)$, S' is a subspace of $\mathcal{G}_{\tau,S}(\mathbb{R}^n)$ and for all $f \in S(\mathbb{R}^n)(f - f \star \phi_\epsilon)_\epsilon \in \mathcal{N}_S(\mathbb{R}^n)$ implies the imbedding as subalgebra of $S(\mathbb{R}^n)$ into $\mathcal{G}_{\tau,S}(\mathbb{R}^n)$.

Definition 2 Let $m, l \in \mathbb{R}$. Denote by $\mathcal{A}_l^m(\mathbb{R}^n)$ the set of all generalized amplitudes $(a_\epsilon)_\epsilon \in \mathcal{E}(\mathbb{R}^n)$ satisfying the requirement: for all $\alpha \in \mathbb{N}^n$ there exists $N \in \mathbb{N}$ such that
$\sup_{\epsilon \in (0,1]} \epsilon^N \|(x)^{l|\alpha|-m} D^\alpha a_\epsilon\|_{L^\infty(\mathbb{R}^n)} < \infty$, where $(x) = (1 + |x|^2)^{1/2}$.
$\mathcal{A}_l^m$ is a linear space with the following easily verified properties:
(i) $m \leq m'$ and $l \leq l' \implies \mathcal{A}_l^m \subset \mathcal{A}_{l'}^{m'}$,
(ii) $(a_\epsilon)_\epsilon \in \mathcal{A}_l^m$, $(b_\epsilon)_\epsilon \in \mathcal{A}_{l'}^{m'} \implies (a_\epsilon b_\epsilon)_\epsilon \in \mathcal{A}_{\min l, l'}^{m+m'}$,
(iii) $(a_\epsilon)_\epsilon \in \mathcal{A}_l^m \implies \forall \alpha \in \mathbb{N}^n, (D^\alpha(a_\epsilon)_\epsilon \in \mathcal{A}_l^{m-l|\alpha|}$.

Definition 3 For $N \in \mathbb{N}$, $\mathcal{A}_{l,N}^m(\mathbb{R}^n)$ is the set of generalized functions such that $\forall \alpha \in \mathbb{N}^n$, $\sup_{\epsilon \in (0,1]} \epsilon^N \|(x)^{l|\alpha|-m} D^\alpha u_\epsilon\|_{L^\infty(\mathbb{R}^n)} < \infty$ and $\bigcup_N \mathcal{A}_{l,N}^m$ is the set of regular generalized amplitudes.

Symbols and Amplitudes.
In comparison to the case of classical functions, the symbols have some extra features.

For f and g real functions on $\mathbb{R}^p$, $f(z) \prec g(z), z \in A \subseteq \mathbb{R}^p$ will mean: there exists a positive constant c, such that $f(z) \leq cg(z)$ for all $z \in A$.

Definition 4 A continuous real function $\Lambda(z)$ on $\mathbb{R}^{2n}$ is a <u>weight function</u> if and only if

(i) $\exists \mu > 0$ such that $(z)^{\mu} \prec \Lambda(z) \prec (z)$ on $\mathbb{R}^{2n}$,

(ii) $\Lambda(z) \sim \Lambda(\xi)$ on $A = \{(z, \zeta) : |\zeta - z\| \leq \mu \Lambda(z)\}$

(iii) $\forall t \in \mathbb{R}^{2n}$, $\Lambda(tz) \prec \Lambda(z)$ on $\mathbb{R}^{2n}$

(where $tz = (t_1 z_1, t_2 z_2, \cdots, t_{2n} z_{2n})$ and $\sim$ means $f \prec g$ and $g \prec f$).

Definition 5 Let $m \in \mathbb{R}$, $\rho \in (0, 1]$ and $\Lambda(z)$ be a weight function and $z = (x, \xi) \in \mathbb{R}^{2n}$.

Denote by $S^m_{\Lambda, \rho}(\mathbb{R}^{2n}$ the set of generalized symbols $(a_\epsilon)_\epsilon \in \mathcal{E}(\mathbb{R}^{2n})$ satisfying $\forall \alpha \in \mathbb{N}^{2n}$, $\exists c > 0 : \forall \epsilon \in (0, 1], \forall z \in \mathbb{R}^{2n}$, $|D^\alpha a_\epsilon(z)| \leq c \Lambda(z)^{m - \rho|\alpha|} \epsilon^{-N}$.

Definition 6 Let $m \in \mathbb{R}$, $\rho \in (0, 1]$, Λ be a weight function and $N \in \mathbb{N}$. We denote by $S^m_{\Lambda, \rho, N}$, the subset of the elements of $S^m_{\Lambda, \rho}$ satisfying the property:

$\forall \alpha \in \mathbb{N}^n$, $\exists c > 0 : \forall \epsilon \in (0, 1], \forall z \in \mathbb{R}^2$ $|D^\alpha a_\epsilon(z)| \leq c \Lambda(z)^{m - \rho|\alpha|} \epsilon^{-N}$.

The symbols of $\bigcup_N S^m_{\Lambda, \rho N}$ are called <u>regular</u>.

Definition 7 Let $m \in \mathbb{R}$, $\rho \in (0, 1]$ and Λ be a weight function. We denote by $\bar{S}^m_{\Lambda, \rho}(\mathbb{R}^{3n})$ the set of all amplitudes $(a_\epsilon(x, y, \xi))_\epsilon \in \mathcal{E}(\mathbb{R}^{3n})$ fulfilling the condition

$\forall \alpha, \beta, \gamma \in \mathbb{N}^n$, $\exists N \in \mathbb{N}$, $\exists c > 0 :$

$\forall \epsilon \in (0, 1], \forall (x, y, \xi \in \mathbb{R}^{3n}$

$|D^\alpha_\xi D^\beta_x D^\gamma_y a_\epsilon(x, y, \xi)| \leq c \lambda_{m, m', \alpha, \beta, \gamma}(x, y, \xi) \epsilon^{-N}$ for a particular $\lambda_{m, m', \alpha, \beta, \gamma}$.

Taking the same $N \in \mathbb{N}$ for all $\alpha, \beta, \gamma \in \mathbb{N}^n$ we define the subset $\bar{S}^m_{\Lambda, \rho, N}$ of $\bar{S}^m_{\Lambda, \rho}$.

The amplitudes of $\bigcup_N \bar{S}^m_{\Lambda, \rho, N}$ are called <u>regular</u>.

<u>Pseudo-differential Operators.</u>

In the classical theory, an integral

$$Au(x) = \int_{\mathbb{R}^{2n}} e^{i(x-y)\xi} a(x, y, \xi) u(y) dy d\xi \quad (*)$$

where a is an amplitude and $u \in S(\mathbb{R}^n)$ defines a continuous map ftom $S(\mathbb{R}^n)$ to $S(\mathbb{R}^n)$ which can be extended to a continuous map from $S'(\mathbb{R}^n)$ to $S'(\mathbb{R}^n)$.

Finally we give a formal definition of a pseudo-differential operator acting on $\mathcal{G}_{\tau, S}(\mathbb{R}^n)$ as follows:

Definition 8 (Pseudo-differential operator on generalized functions)

This is defined by means of $(*)$, where a and u are replaced by a_ϵ and $(u_\epsilon)(\hat{\phi}_\epsilon(y)$ respectively and $(a_\epsilon)_\epsilon \in \bar{S}^m_{\Lambda, \rho}$, $(u_\epsilon)_\epsilon \in \mathcal{E}_\tau(\mathbb{R}^n)$ and ϕ is a fixed mollifier, (c.f. [Gar-2], pp. 142).

Garetto in [Gar-3] also gives the definition of generalized forms of oscillatory integrals and Fourier integral operators. Recalling the symbol $S^m_{\delta, \rho}$ denote by $S^1_{hg}(\Omega \times \mathbb{R}^p \setminus 0$ of all $a \in S^1(\Omega \times \mathbb{R}^p \setminus 0)$, homogeneous of degree 1 in $\xi \in \mathbb{R}^p \setminus 0$.

$\phi(y, \xi)$ is a phase function on $\Omega \times \mathbb{R}^p$ (Ω open $\subset \mathbb{R}^n$) if it is a smooth function on $\Omega \times \mathbb{R}^p \setminus 0$, real valued, positively homogeneous of degree 1 in ξ) with $\nabla_{y,\xi}\phi(y, \xi) \neq 0$ for all $y \in \Omega$ and $\xi \in \mathbb{R}^p$. We denote the set of all phase functions on $\Omega \times \mathbb{R}^p$ by $\Phi(\Omega \times \mathbb{R}^p)$ and the set of al nets in $\Phi(\Omega \times \mathbb{R}^p)^{(0,1]}$ by $[\Omega \times \mathbb{R}^p]$. An element of $\mathcal{M}_\Phi(\Omega \times \mathbb{R}^p)$, is a net $(\phi_\epsilon)_\epsilon \in \Phi(\Omega \times \mathbb{R}^p)$ satisfying

(i) $(\phi_\epsilon)_\epsilon \in \mathcal{M}_{S^1_{hg}(\Omega \times \mathbb{R}^p \setminus 0}$,
(ii) For all $K \subset\subset \Omega$, the net,

$$\left(\inf_{y \in K, \xi \in \mathbb{R}^p \setminus 0} |\nabla \phi_\epsilon(y, \frac{\xi}{|\xi|}|^2 \right)_\epsilon$$

is strictly non-zero.

On $\mathcal{M}_\Phi(\Omega \times \mathbb{R}^p)$ an equivalence relation $\sim$ is introduced as follows: $(\phi_\epsilon)_\epsilon \sim (\omega_\epsilon)_\epsilon \Leftrightarrow (\phi_\epsilon - \omega_\epsilon) \in \mathcal{N}_{S^1_{hg}(\Omega \times \mathbb{R}^p \setminus 0}$. The elements of the factor space

$$\tilde{\Phi}(\Omega \times \mathbb{R}^p) \doteq \mathcal{M}_\Phi(\Omega \times \mathbb{R}^p)/\sim$$

are called generalized phase functions.

As a consequence of Proposition 3.3, [Gar-3] any generalized phase function $\phi \in \tilde{\Phi}(\Omega \times \mathbb{R}^p)$ defines a generalized partial differential operator

$$L_\phi(y, \xi, \partial_y, \partial_\xi) = \sum_{j=1}^{p} a_j(y, \xi) \frac{\partial}{\partial \xi_j} + \sum_{k=1}^{n} b_k(y, \xi) \frac{\partial}{\partial y_k} + c(y, \xi)$$

where coefficients $\{a_j\}_{j=1}^{p}$ and $\{b_k\}_{k=1}^{n}$ are generalized symbols specified in Proposition 3.3.

By construction, L_ϕ maps $\tilde{S}^m_{\rho,\delta}$ continuously into $\tilde{S}^{m-s}_{\rho,\delta}$ where $s = \min\{\rho, 1 - \delta\}$. Hence L_ϕ^k is continuous from $\tilde{S}^m_{\rho,\delta}(\Omega) \times \mathbb{R}^p$ to $\tilde{S}^{m-ks}_{\rho \times \delta}(\Omega \times \mathbb{R}^p)$.

Finally we have Definition 3.6 of [Gar-3];

Let $\phi \in \tilde{\Phi}(\Omega \times \mathbb{R}^p)$, $a \in \tilde{S}^m_{\rho,\delta}$ and $u \in \mathcal{G}_c(\Omega)$ (c.f. 3.4.10) the general oscillatory integral $\int_{\Omega \times \mathbb{R}^p} e^{i\phi(y,\xi)}a(y, \xi)u(y)dyd\xi$. is defined as $\int_{\Omega \times \mathbb{R}^p} e^{i\phi(y,\xi)}L_\phi^k(a(y, \xi)u(y))dyd\xi$, where k is chosen such that $m - ks + 1 < -p$.

The functional $I_\phi(a): \mathcal{G}_c(\Omega) \to \bar{C}: u \to \int_{\Omega \times \mathbb{R}^p} e^{i\phi(y,\xi)}a(y, \xi)u(y)dyd\xi$ belongs to the dual $\mathcal{L}(\mathcal{G}_c(\Omega), \bar{C})$. For defining the generalized Fourier operators an additional parameter x, varying in an open set Ω' of $\mathbb{R}^{n'}$ is considered. It appears in the phase function ϕ and in the symbol a of oscillatory integral.

The dependence on x is investigated in the Colombeau context. Denote by $\tilde{\Phi}[\Omega'; \Omega \times \mathbb{R}^p]$ the set of all nets $(\phi_\epsilon)_{\epsilon \in (0,1]}$ of continuous functions on $\Omega' \times \Omega \times \mathbb{R}^p$ which are smooth on $\Omega' \times \Omega \times \mathbb{R}^p \setminus \{0\}$ and such that $\phi_\epsilon(x, ., .) \in \Phi[\Omega'; \Omega \times \mathbb{R}^p]$ for all $x \in \Omega'$.

An element of $\mathcal{M}_\Phi(\Omega'; \Omega \times \mathbb{R}^p)$ is a net $(\phi_\epsilon)_\epsilon \in \Phi[\Omega'; \Omega \times \mathbb{R}^p]$ satisfying the conditions

(i) $(\phi_\epsilon)_\epsilon \in \mathcal{M}_{S^1_{hg}(\Omega' \times \Omega \times \mathbb{R}^p \setminus 0)}$,

(ii) for all $K' \subset\subset \Omega'$ and $K \subset\subset \Omega$ the net

$$\left(\inf_{x \in K', y \in K, \xi \in \mathbb{R}^p \setminus 0} |\nabla_{y,\xi} \phi_\epsilon(x, y, \frac{\xi}{|\xi|}|^2 \right)_\epsilon$$

is strictly nonzero.

on $\mathcal{M}_\Phi(\Omega'; \Omega \times \mathbb{R}^p)$ an equivalence reation $\sim$ is defined as follows:
$(\phi_\epsilon)_\epsilon \sim (\omega_\epsilon)_\epsilon \Leftrightarrow (\phi_\epsilon - \omega_\epsilon)_\epsilon \in \mathcal{N}_{S'_{hg}(\Omega' \times \Omega \times \mathbb{R}^p \setminus 0)}$.
The elements of the factor space $\tilde{\Phi}(\Omega'; \Omega \times \mathbb{R}^p) \doteq \mathcal{M}_\Phi(\Omega'; \Omega \times \mathbb{R}^p)/\sim$ are called generalized phase functions with respect to the random variables in $\Omega \times \mathbb{R}^p$.

Generalized Fourier integral operator
***

Let $\phi \in \tilde{\Phi}(\Omega'; \Omega \times \mathbb{R}^p)$, $a \in \tilde{S}^m_{\rho,\delta}(\Omega' \times \Omega \times \mathbb{R}^p)$ and $u \in \mathcal{G}_c$. The generalized oscillatory integral

$$I_\phi(a)(u)(x) = \int_{\Omega \times \mathbb{R}^p} e^{i\phi(x,y,\xi)} a(x, y, \xi) u(y) dy d\xi$$

defines a generalized function in $\mathcal{G}(\Omega')$ and the map

$$A : \mathcal{G}_c(\Omega) \to \mathcal{G}(\Omega') : u \to I_\phi(a)(u) \quad (*)$$

is continuous.

The operator A defined in $(*)$ is called generalized Fourier integral operator with amplitude $a \in \tilde{S}_{\rho,\delta}(\Omega' \times \Omega \times \mathbb{R}^p)$ and phase function $\phi \in \tilde{\Phi}(\Omega'; \Omega \times \mathbb{R}^p)$.

3.5   Other Theories of Non-linear Distributions

3.5.1   *Delcroix and Scarpalezos' Asymptotic Algebras of Generalized Functions*

In Colombeau's theory the polynomial growth of the non-linear maps is an essential feature. As a consequence many singular problems with nonlinearities growing faster than any polynomial remain outside the scope of this theory. Spaces of generalized functions associated to a given asymptotic scale and stable by nonlinear maps growing faster than poynomials have been introduced by Delcroix and Scarpalezos, [Del-Scar1, Del-Scar2].

Of course using scales stronger than polynomial ones implies some limitations like the lack of canonical imbedding of spaces of distributions respecting the product of smooth functions contrary to the polynomial case.

Asymptotic Extensions

Let H be locally convex, metrizable topological vector space the uniform structure of which is given by an increasing countable family of semi-norms $\{p_n\}_{n\in\mathbb{N}}$. The construction done below depends only on the uniform structure but not on the sequence of semi-norms. Denote by $\mathcal{X}(H)$, the set of all mappings from $(0,1]$ to H, the elements of it are represented by $(f_\epsilon)_{\epsilon\in(0,1]}$, or in short by $(f_\epsilon)_\epsilon$, $\mathcal{X}(H)$ is in its natural vector space structure.

Asymptotic scale and asymptotic extensions

Definition 1 A sequence $\{a_n\}_{n\in\mathbb{Z}}$ of increasing functions from $(0,1]$ to $(0,+\infty)$ is called an asymptotic scale if for each $n \in \mathbb{N}$ (except possibly zero) $\lim_{\epsilon\to 0} a_n(\epsilon) = 0$ and if

(i) $\forall n \in \mathbb{N}\backslash\{0\}$, $a_{-n} = \dfrac{1}{a_n}$,

(ii) $\forall n \in \mathbb{N}, a_{n+1} = o(a_n)$ as $\epsilon \to 0$,

(iii) $\forall n \in \mathbb{N}, \forall m \in \mathbb{N}, \exists N \in \mathbb{N}, a_N(\epsilon) = O(a_n(\epsilon)a_m(\epsilon))$ as $\epsilon \to 0$.

Definition 2 Let $a = \{a_n\}_{n\in\mathbb{Z}}$ be an asymptotic scale.

1. An element $\phi \in \mathcal{X}(H)$ is said to be of moderate growth with respect to the scale $\{a_n\}_{n\in\mathbb{Z}}$ (briefly a-moderate) if
$\forall n \in \mathbb{N}, \exists q \in \mathbb{Z}, p_n(\phi(\epsilon)) = o(a_q(\epsilon))$ as $\epsilon \to 0$.

2. An element ϕ is said to be rapidly decreasing with respect to the scale $\{a_n\}_{n\in\mathbb{N}}$ (briefly a-negligible) if
$\forall n \in \mathbb{N}, \forall q \in \mathbb{Z}, p_n(\phi(\epsilon)) = o(a_q(\epsilon))$ as $\epsilon \to 0$.

3. Denote by $\mathcal{X}_{M,a}$ the set of all a-moderate elements of $\mathcal{X}(H)$ and by $\mathcal{N}_a(H)$ the set of all a-negligible elements.

Then $\mathcal{X}_{M,a}(H)$ is a vector subspace of $\mathcal{X}(H)$ and $\mathcal{N}_a(H)$ is a vector subspace of $\mathcal{X}_{M,a}(H)$.

The quotient vector space

$$\mathcal{G}_a(H) = \mathcal{X}_{M,a}/\mathcal{N}_a(H)$$

is called the a-asymptotic extension of H or the Colombeau$a$-extension.

Note. The space H is canonically imbedded in $\mathcal{G}_a(H)$, by associating any $\phi \in H$ with the class of the net $(f_\epsilon)_\epsilon$ defined by $f_\epsilon = \phi$ for $\epsilon \in (0,1]$. We denote the class of a net $(f_\epsilon)_\epsilon$ by $[f(\epsilon)]$.

Examples

I) For $H = \mathcal{C}^\infty(\Omega)$ with Ω an open subset of $\mathbb{R}^n$ and the polynomial scale $a = \{a_n\}_{n\in\mathbb{Z}} \, \forall n \in \mathbb{Z}, \forall \epsilon \in (0,1]$, $a_n(\epsilon) = \epsilon^n$, we recover the space $\mathcal{G}_s(\Omega)$ of simplified Colombeau generalized functions.

A natural family of increasing semi-norms on $\mathcal{C}^\infty(\Omega)$ i given by

$$p_n(\phi) = \sup_{|\alpha| \leq n} \left(\sup_{K_n} |D^\alpha \phi| \right)$$

where $\alpha \in \mathbb{N}^n$, $\{K_n\}_{n \in \mathbb{N}}$ is an increasing family of compact subsets of Ω with $\bigcup_{n \in \mathbb{N}} K_n = \Omega$. If $f \in \mathcal{D}'(\Omega)$ (or $f \in \mathcal{C}(\Omega)$) take $(f_\epsilon)_\epsilon = f * \phi$, $\phi \in \mathcal{D}(\Omega)$, then $f * \phi \in \mathcal{C}^\infty(\Omega)$, but then there will be no privileged inclusion. (See also Sect. 3.4.8).

(II) For any given asymptotic scale a, the space $\mathcal{G}_a(\Omega)$ of a-Colombeau generalized function on an open subset of $\mathbb{R}^n$ is the set $\mathcal{G}_a(\mathcal{C}^\infty(\Omega))$. For example an algebra stable under exponentiation is obtained by considering the exponential scale defined by $\forall n \in \mathbb{N}$, $a_n(\epsilon) = 1/Exp_n(1/\epsilon)$, $\forall n \in \mathbb{N} \backslash \{0\}$, $a_{-n}(\epsilon) = Exp_n(1/\epsilon)$, where $(Exp)_{n \in \mathbb{N}}$ is given by $\forall \xi \in \mathbb{R}$, $Exp_0(\xi) = \xi$, $\forall n \in \mathbb{N}$, $\forall \xi \in \mathbb{R}$, $Exp_{n+1}(\xi) = \exp(Exp_n(\xi))$.

Hence

$$a_1(\epsilon) = e^{-\frac{1}{\epsilon}}$$

$$a_2(\epsilon) = e^{-e^{\frac{1}{\epsilon}}}$$

$$a_3(\epsilon) = \exp(-e^{e^{\frac{1}{\epsilon}}})$$

It is easily verified that these $\{a_n(\epsilon)\}_{n \in \mathbb{Z}}$ satisfy the conditions of Definition 1.

(III) Taking $H = \mathbb{C}$ with the topological structure given by the modulus we obtain the set $\mathcal{G}_a(\mathbb{C})$, $= \tilde{\mathbb{C}}$ of the generalized constants associated with the scale a. This set is a ring and equivalent to Colombeau's generalized constants of the polynomial case.

Application: A well-posed Dirichlet problem

Let $\{a_n\}_n \in \mathbb{N}$ be an asymptotic scale. An increasing function $g : \mathbb{R}_+ \backslash \{0\} \to \mathbb{R}_+ \backslash \{0\}$ is said to be a-moderate if for each $p \in \mathbb{Z}$ there exists $m(p)$ such that $g(a_p(\epsilon)) = o(a_{m(p)}(\epsilon))$. Let Ω be an open bounded subset of $\mathbb{R}^n$ with a smooth boundary $\delta\Omega$. Let $\Theta : \bar{\Omega} \times \mathbb{R} \to \mathbb{R}$ be a strictly increasing function with respect to the second variable, which is $\mathcal{C}^\infty$ on both variables up to the boundary with respect to the first variable. following Dirichlet problem

$$\begin{cases} -\Delta u + \theta(x, u) = f \\ u|_{\delta\Omega} = g \end{cases}$$

It is well-known that under some technical hypotheses on θ, the Dirichlet problem stated is well-posed with solutions in $\mathcal{C}^\infty(\Omega)$ when the data (f, g) belong to $\mathcal{C}^\infty(\bar{\Omega}) \times \mathcal{C}^\infty(\delta\Omega)$. However if the data have some singularities, weak solutions are sought in distribution spaces or the spaces of H^s type. In general this Dirichlet problem can not be solved in spaces of distributions.

Nevertheless the present paper ([Del-Scar2] provides a solution under some assumptions in the framework of a-asypmtotic extension as stated in the following theorem:

Theorem *Let $a = \{a_n\}_{n \in \mathbb{N}}$ be an asymptotic scale and θ a function belonging to $\mathcal{C}^\infty(\bar{\Omega} \times \mathbb{R})$ strictly increasing with respect to the second variable and such that for all $\alpha \in \mathbb{N}^{n+1}$ there exists an a-moderate function h_a with*

$$\forall r \in \mathbb{R}\backslash\{0\} \quad \sup_{\bar{\Omega}\times[-R,R]} |\partial^\alpha \theta(x,t)| \le h_a(R).$$

Then for each $f \in \mathcal{G}_a(\mathcal{C}^\infty(\Omega))$ and $g \in \mathcal{G}_a(\mathcal{C}^\infty(\delta(\Omega)))$, there exists a unique solution $U_{f,g} \in \mathcal{G}_a(\mathcal{C}^\infty(\Omega))$ such that

$$\begin{cases} -\Delta U_{f,g} + \theta(x, U_{f,g}) = f \in \Omega \\ U_{f,g} = g \ \text{on} \ \delta\Omega \end{cases}$$

Moreover the mapping

$$\mathcal{G}_a(\bar{\mathcal{C}}^\infty(\Omega)) \times \mathcal{G}_a(\mathcal{C}(\delta\Omega)) \to \mathcal{G}_a\mathcal{C}^\infty(\Omega)) : \quad (f,g) \to U_{f,g}$$

is continuous for the respective sharp topologies, (for the sharp topologies see Sect. 5.12).

Proof [Del-Scar2]
In the paper some growth estimates of the solution $U_{f,g}$ are given.

3.5.2 *Egorov's Theory of Generalized Functions*

Egorov starts by stating that the linear distribution theory does not enable one either to find or even to define a solution for example to Cauchy problem of the Klein-Gordon equation

$$\frac{\partial^2}{\partial t^2} - \Delta u + m^2 u + gu^3 = 0$$

or the Kolmogorov-Petrovvskii-Piskunov equation

$$\frac{\partial u}{\partial t} \Delta u + F(u) = 0$$

in the case where the initial values are Schwartz distributions.

Egorov in his work "'A contribution to the theory of generalized functions'" [Ego], has a different approach inspired by Mikusinski visualizing distributions as uniform limits of sequences of smooth functions. (We refer to the nonlinear distributions as generalized functions and preserve the term 'distribution' for the Schwartz distributions, i.e. members of $\mathcal{D}'$).

Let Ω be a domain in $\mathbb{R}^n$, we consider the sets of sequences of functions $\{f_k(x)\}$ which are infinitely differentiable in x for $x \in \Omega$.

We will call two sequences $\{f_k(x)\}$ and $\{g_k(x)\}$ equivalent if for each compact set $K \subset\subset \Omega$ there exists $N > 0$ such that $f_k(x) = g_k(x)$ for $k \geq N$, $x \in K$. The set of classes of equivalent functions is denoted by $\mathcal{G}(\Omega)$ and its elements are called generalized functions (in Egorov sense).

Clearly the space of generalized functions is an algebra. Here the product $f(x)g(x)$ of two generalized functions is a class of sequences equivalent to $\{f_k(x)g_k(x)\}$, where $\{f_k(x)\}$ is any representative of the class $f(x)$ and $\{g_k(x)\}$ is any representative of the class $g(x)$. Let $F \in \mathbb{C}(\mathcal{C}^p)$ for some $p \geq 1$, and let

$$F_1(x_1, \cdots, x_{2p}) = F(x_1 + ix_2, \cdots, x_{2p-1} + ix_{2p}) \in \mathcal{C}^\infty(\mathbb{R}^{2p})$$

If $f_1, \cdots, f_p \in \mathcal{G}(\Omega)$, then a generalized function $f = F(f_1, \cdots, f_p)$ is well defined. Moreover a generalized function $F(f_1, \cdots, f_p)$ can be defined even in the case where $F_1 \in \mathcal{G}(\mathbb{R}^{2p})$.

Theorem 1 $\mathcal{D}'(\Omega) \subset \mathcal{G}(\Omega)$, *i.e. every distribution is a generalized function.*

Proof For the proof we choose an averaging kernel $\omega(x)$. This is a function from $\mathcal{C}_0^\infty$ such that $\omega(x) \geq 0$, $\omega(x) = 0$ for $|x| \geq 1$, $\int \omega(x)dx = 1$. By $\omega_k(x)$ we will denote the averaging kernel $k^n\omega(kx)$ for $k \geq 1$.

First let us consider the case $\Omega = \mathbb{R}^n$ and $T \in \mathcal{D}'(\mathbb{R}^n)$. Then we know from Subsection 3.4.1 that replacing k by $1/\epsilon$, $\omega_k(x)$ converges in distribution to the Dirac delta function. Thus $f_k = T * \omega_k \longrightarrow T$ in $\mathcal{D}'$.

If Ω is an arbitrary domain in $\mathbb{R}^n$ and $T \in \mathcal{D}'(\Omega)$ we need to introduce a factor χ_k, given by $\chi_k \in \mathcal{C}_0^\infty$ with $\chi_k(x) = 1$ if $dist(x, \delta\Omega) > 1/k$ and $dist(x, 0) < k$. Then if $T_k = \chi_k T$ and $f_k(x) = (T_k * \omega)_k(x)$, we easily show that $f_k \in \mathcal{C}^\infty(\Omega)$ and $f_k \longrightarrow T$ in $\mathcal{D}'(\Omega)$. $\square$

In general a generalized function f is a distribution if for every function $\phi \in \mathcal{D}(\Omega)$ there exists a limit $\lim_{k\to\infty} \int f_k(x)\phi(x)dx$, where $f_k(x)$ is an element from the class f.

Similarly a generalized function f determines an element from $\mathcal{C}(\Omega)$, $L^p(\Omega)$, $\mathcal{C}^\infty(\Omega)$, and so on if $\{f_k\}$ converges as $k \to \infty$ to an element of the corresponding space in the norm of this space.

However due to the choice of different averaging kernels in the above construction, infinitely many different generalized functions may correspond to a single distribution.

For instance the limit as $k \to \infty$ of $\rho_k(x) = k^n\rho(kx)$ given above is the Dirac delta function $\delta(x)$. But also $\delta(x)$ is the limit of the sequence of functions $f_k(x) = \dfrac{k}{\pi(1 + k^2x^2)}$. In [Vlad], Vladimirov demonstrates that $\int_{-\infty}^{\infty} f_k^2\phi(x)dx \longrightarrow 0$ as $k \to \infty$ if $\phi \in \mathcal{D}(\mathbb{R})$, $\phi(0) = 0$ and bases his proof of the equation $\delta^2(x) = C\delta(x)$, $x \in \mathbb{R}$, C arbitrary, on this demonstration.

Generalized complex numbers.

The extended complex plane $\tilde{C}$ is introduced by using a construction typical for a non-standard analysis. We denote by $\bar{C}$ the complex plane compactified at a point at infinity. Consider the set of sequences $\{f_k\}$ where $f_k \in \bar{C}$ We call two sequences $\{f_k\}$ and $\{g_k\}$ from this set equivalent if there is an $\mathbb{N}$, such that $f_k = g_k$ for $k > N$. The set $\tilde{C}$ of equivalence classes will be called the extended complex plane and any of the classes a generalized complex number.

$\bar{C} \subset \tilde{C}$ by associating with any ordinary complex number $c \in \bar{C}$, a class containing a sequence with $f_k = c$. Clearly $\tilde{C}$ is an algebra. The product $c_1 c_2$ of two generalized complex numbers is the class of those sequences eqivalent to the sequence $\{c_{1k} c_{2k}\}$ where $\{c_{ik}\}$ is an arbitrary representative of the class c_i, $i = 1, 2$.

If $F \in C(\mathbb{C}^p)$, $p \geq 1$ and $c_1, c_2, \cdots, c_p \in \tilde{C}$, then $F(c_1, \cdots, c_p)$ is a well-defined element of $\tilde{C}$.

One can also define the extended real line $\tilde{\mathbb{R}}$ in a similar fashion. Neither $\tilde{\mathbb{R}}$ nor $\tilde{C}$ is a field.

Topology.

A topology in $\tilde{C}$ is determined by a collection of open sets $U_{c,\delta}$, where $c \in \tilde{C}$, $\delta > 0$, such that $c' \in U_{c,\delta}$ if there exist representatives c_k, c'_k with $|c_k - c'_k| < \delta$ for $k \geq 1$. By a neighborhood of the point at infinity we mean a set of generalized complex numbers $c \in \tilde{C}$ for which we can find representatives satisfying $|c_k| > \delta$, for $\delta > 0$ A sequence $\{c^{(m)}\}$ converges to $c \in \tilde{C}$, if for every $\delta > 0$, there exists a N such that for $m > N$ one can find representatives $c_k^{(m)}$ and c_k such that $|c_k^{(m)} - c_k| < \delta$ for $k \geq 1$.

This topology is not seperated.

Topology in $\mathcal{G}(\Omega)$.

A topology parallel to the one by Biagioni ([Bia]) is given as follows:

Define the open sets $U_{f,M,\delta,K}$ of the topology where $f \in \mathcal{G}(\Omega), M \in \mathbb{N}, \delta > 0, K \subset\subset \Omega$ such that $g \in U_{f,M,\delta,K}$ if there are representatives f_k, g_k with $\sup_{x \in K} |D_x^\alpha [f_k(x) - g_k(x)]| < \delta$ for $|\alpha| \leq M, k \geq 1$.

In particular a sequence $\{f^{(m)}(x)\} \subset \mathcal{G}(\Omega)$ converges to $f \in \mathcal{G}(\Omega)$, if for every $\delta > 0$, $K \subset\subset \Omega$, $M \in \mathbb{N}$ there exists $N \in \mathbb{N}$ and representatives $f_k^{(m)}(x)$ and $f_k(x)$ such that

$$\sup_{x \in K} |D_x^\alpha [f_k(x) - f_k^{(m)}]| < \delta \text{ for } |\alpha| \leq M, k \geq 1, m > N$$

It is not difficult to verify that both spaces $\tilde{C}$ and $\mathcal{G}(\Omega)$ are complete in the indicated topologies.

Properties of generalized functions

Localization. If Ω and Ω' are open sets in $\mathbb{R}^n$ and $\Omega' \subset \Omega$, $g \in \mathcal{G}(\Omega)$, then the restriction $g|_{\Omega'}$ is determined by the canonical restriction of representatives $g_k(x)$ to Ω' and it is an element of $\mathcal{G}(\Omega')$. The support of a generalized function $g \in \mathcal{G}(\Omega)$ is the complement of the largest open subset of Ω on which $g = 0$.

Theorem 2 *Let $\{\Omega_i, i \in I\}$ be a family of open subsets of $\mathbb{R}^n$ and $g_i \in \mathcal{G}(\Omega_i)$. Suppose that for every pair of indices i, j with $\Omega_i \cap \Omega_j \neq \emptyset$ the restrictions $g_i|_{\Omega_i \cap \Omega_j}$ and $g_j|_{\Omega_i \cap \Omega_j}$ coincide. Then there exists a unique generalized function g on $\Omega = \cup \Omega_i$ of which the restriction to Ω_i is g_i.*

Proof The proof is achieved by a standard application of the Partition of Unity.

Restriction to a subspace. Values at a point.

The definition of generalized functions yields immediately the possibility of restricting them to a subspace of $\mathbb{R}^n$. If Ω' is a smooth submanifold of dimension $m, 0 \leq m \leq n - 1$, and $\Omega' \subset \Omega$, then for every generalized function $f \in \mathcal{G}(\Omega)$ there is defined a restriction $f|_{\Omega'} \in \mathcal{G}(\Omega')$. In particular one can speak about the value $f(x) \in \tilde{C}$ at any point $x \in \Omega$.

If $f \in \mathcal{G}(\bar{\Omega})$ and the boundary $\delta\Omega$ of the domain Ω is a smooth manifold, then the restriction $f|_{\delta\Omega}$ is defined as an element of the subspace $\mathcal{G}(\delta\Omega)$.

As an example consider the function $x\omega(kx)$, where $\omega \in C_0^\infty(\mathbb{R})$, $\omega(0) = 1$, determines a generalized function f, the support of which consists of a single point at $x = 0$ and in addition $f(0) = 0$, however $f \neq 0$ in any neighborhood of this point.

Differentiation of generalized functions.

If a representative $f_k(x)$ corresponds to a generalized function f, then for every α the function $D_x^\alpha f_k(x)$ determines a generalized function called a derivative $D^\alpha f$.

This definition agrees with ordinary differentiation in $\mathcal{D}'(\Omega)$. If a generalized function f determines a distribution g, that is, for every function ϕ from $\mathcal{D}(\Omega)$ there exists a finite limit $\lim_{k \to \infty} \int f_k(x)\phi(x)dx = g(\phi)$, then a generalized function $D^\alpha f$ determines a distribution $D^\alpha g$.

The rules of differentiation of classical analysis also hold for generalized functions, e.g. the Leibniz formula for the derivative of a product of two generalized functions is valid:

$$D^\alpha(f.g) = \sum_{\beta \leq \alpha} c_{\alpha\beta}(D^{\alpha-\beta} f)(D^\beta g)$$

where positive integers $c_{\alpha\beta}$ are well-known in analysis.

Integration of generalized functions.

Let $g \in \mathcal{G}(\Omega)$ and let $K \subset\subset \Omega$ be a compact subset of Ω. Let $\{f_k(x)\}$ be a representative of g. The sequence $F_k = \int_K f_k(x)dx$ determines an element of $\tilde{C}$ which does not depend on the choice of $\{f_k\}$. Thus an element of $\int g(x)dx \in \tilde{C}$ is defined.

Example. Let $\delta(x)$ be the Dirac function on the real line, $\int_{-1}^1 \delta(x)dx = 1$. However the integral $\int_0^1 \delta(x)dx$ does not make sense for δ as a distribution. But if we consider

δ as an element of $\mathcal{G}(\mathbb{R})$, then depending on the choice of the representatives this integral may take an arbitrary value from $\mathbb{R}$ (even from $\tilde{C}$).

If $g \in \mathcal{G}(\Omega)$ and $supp\, g = K$ is a compact subset of Ω, then for any compact subsets K_1 and K_2 of Ω containing an open neighborhood of the compact K we have $\int_{K_1} g(x)dx = \int_{K_2} g(x)dx$. Therefore one may denote the above value by $\int_{\Omega} g(x)dx$. For example $\int_{\mathbb{R}} \delta(x)dx = 1$.

Weak equality.

Two generalized complex numbers b and c are called weakly equal, $b \approx c$, if there exist representatives of them, $\{b_k\}$ and $\{c_k\}$, such that $\overline{\lim}_{k\to\infty}(b_k - c_k) = 0$. For example, the sequence $b_k = \frac{1}{k}$ determines a class weakly equal to 0, (here $\lim b_k$ need not exist).

Similarly two generalized functions f and g are called weakly equal $f \approx g$, if there exist representatives of them $f_k(x)$ and $g_k(x)$, such that $\lim_{k\to\infty} \int ([f_k(x) - g_k(x)]\phi(x) = 0$ for every function ϕ from $\mathcal{D}(\Omega)$.

The distinctions between strong and weak equalities are important in the deduction of differenetial equations from physical laws.

3.5.3   A Global Theory of Generalized Functions

Grosser, Kunzinger, Steinbauer and Vickers ([Gross-et-al]) present a geometric approach to defining an algebra $\widehat{\mathcal{G}}(M)$, the Colombeau algebra of generalized functions on a smooth manifold M containing the space $\mathcal{D}'(M)$ of distributions on M.

As a summary of their work, they achieve an intrinsic construction of $\widehat{\mathcal{G}}(M)$ based on a differential calculus in convenient vector spaces. Its elements possess Lie derivative with respect to arbitrary smooth vector fields.They also construct a canonical linear imbedding of $\mathcal{D}'(M)$ into $\widehat{\mathcal{G}}$ that renders $C^\infty(M)$ as a faithful subalgebra. Furthermore their imbedding commutes with the Lie derivatives.

It is difficult to give a full account of this rather involved work. However to give some details in summary form, throughout the paper M denotes an oriented paracompact C^∞ manifold with an atlas $\mathcal{U} = (\{U_\alpha, \psi_\alpha\}, \alpha \in I\}$, I is an index set of a family of a coordinate neighborhoods.

In their setting test objects no longer have a function character, but rather n forms, i.e. alternating covariant tensor fields of order n, (or exterior differential form of order n), (for a review of some basic differential geometry concepts see below).

For $\Omega \subset \mathbb{R}^n$, Λ_c^n will denote the space of compactly supported (smooth) n-forms on M. Locally for coordinates $y^1, \cdots, y^n$ on U_α, elements of $\Lambda_c^n(\psi_\alpha(U_\alpha))$ will be written as $\phi\, d^n y \doteq \phi\, dy^1 \wedge \cdots \wedge dy^n$. The pullback of any $\omega \in \Lambda_c^n((\psi(U_\alpha))$ under ψ_α is written as $\psi_\alpha^*(\omega)$. Then

$$\int_M \psi_\alpha^*(\phi\, d^n y) = \int_{\psi_\alpha(U_\alpha)} \phi\, d^n y, \quad \text{for all}\ \ \phi\, d^n y \in \Lambda_c^n(\psi_\alpha(U_\alpha)).$$

For a diffeomorphism $\mu : \tilde{\Omega} \to \Omega$, the pullback of any $u \in \mathcal{D}'(\Omega$ under μ is defined by $< \mu^*(u, \phi) >=< u(y), \phi^*(\mu^{-1}y) > |Det D\mu^{-1}(y)|$.

The space of distributions on M is defined as $\mathcal{D}'(M) = \Lambda_c^n(M)'$.

Operations on distributions are defined as (sequentially) continuous extensions of classical operations on smooth functions. In particular $X \in \mathfrak{X}(M)$, (the space of smooth vector fields on M) and $u \in \mathcal{D}'(M)$, the Lie derivative of u with respect to X is given by $< \mathcal{L}_X u, \omega >= - < u, \mathcal{L}_X \omega >$.

If $u \in \mathcal{D}'(M)$, $(U_\alpha, \psi_\alpha) \in \mathcal{U}$, then the local representation of u on U_α is the element $(\psi_\alpha^{-1}) * (u) \in \mathcal{D}'(\psi_\alpha(U_\alpha))$ defined by

$$< (\psi_\alpha^{-1}) * (u), \phi >=< u|_{U_\alpha}, \psi_\alpha^* n(\phi d^n y) >, \quad \forall u \in \mathcal{D}(\psi_\alpha(U_\alpha)).$$

For the diffeomorphism invariant Colombeau theory we consider for $\Omega \subset \mathbb{R}^n$

$$\mathcal{A}_0(\Omega) = \{\phi \in \mathcal{D}(\Omega) : \int \phi(\xi)d\xi = 1\}$$

$$\mathcal{A}_q(\mathbb{R}^n) = \left\{\phi \in cal A_0(\mathbb{R}^n) : \int \phi(\xi)\xi^\alpha d\xi = 0, 1 \leq \alpha| \leq q, \alpha \in \mathbb{N}_0^n\right\} \quad (q \in \mathbb{N})$$

(Compare with 3.4.1 Definition 1 of $\mathcal{A}_m$).

The basic space of the diffeomorphism ivariant local theory is defined to be $\mathcal{E}(\Omega) = \mathcal{C}^\infty(\mathcal{A}_0(\Omega) \times \Omega)$ (compare with $(\mathcal{A}(\mathbb{R}^n)^\Phi$ of Subsection 3.4.1).

For $x \in \mathbb{R}^n, \epsilon \in I = (0, 1]$, define translation (resp.scaling) operators by $T_x \mathcal{D}(\mathbb{R}^n) \to \mathcal{D}(\mathbb{R}^n)$, $T_x(\phi) = \phi(. - x)$ and $S_\epsilon : \mathcal{D}(\mathbb{R}^n) \to \mathcal{D}(\mathbb{R}^n)$, $S_\epsilon = \epsilon^{-1}\phi(\frac{.}{\epsilon})$.

The subspaces of moderate elements of $\mathcal{E}(\Omega)$ are defined by

$$\{\mathcal{E}_m(\Omega) = \{R \in \mathcal{E}(\Omega)| \forall K \subset\subset \Omega \, \forall \alpha \in \mathbb{N}_0^n \, \exists N \in \mathbb{N} \, \forall \phi \in \mathcal{C}_b^\infty(I \times \Omega, \mathcal{A}_0(\mathbb{R}^n)) :$$

$$\sup_{x \in K} |\partial^\alpha(R(T_x S_\epsilon \phi(\epsilon, x))| = O(\epsilon^{-N}) \, (\epsilon \to 0)\}$$

where $\mathcal{C}_b^\infty(I \times \Omega, \mathcal{A}_0(\mathbb{R}^n))$ is the space of smooth maps $\phi : I \times \Omega \to \mathcal{A}_0(\mathbb{R}^n)$ such that for each $K \subset\subset \Omega$ and for any $\alpha \in \mathbb{N}_0^n$, the set $\{\partial_x^\alpha \phi(\epsilon.x) \, | \, \epsilon \in (0, 1], x \in K\}$ is bounded in $\mathcal{D}(\mathbb{R}^n)$ and negligible elements of $\mathcal{N}(\Omega)$:

$$\{\mathcal{N}(\Omega) = \{R \in \mathcal{E}(\Omega) \, | \, \forall K \subset\subset \Omega, \forall \alpha \mathbb{N}_0^n \, \forall r \in \mathbb{N} \exists m \in \mathbb{N} \, \forall \phi \in \mathcal{C}_b^\infty(I \times \Omega, \mathcal{A}_m(\mathbb{R}^n)) :$$

$$\sup_{x \in K} |\partial^\alpha(R(T_x S_\epsilon \phi(\epsilon, x), x))| = O(\epsilon^r) \, (\epsilon \to 0)\}$$

The diffeomorphism invariant Colombeau algebra $\mathcal{G}(\Omega)$ on Ω is the quotient algebra $\mathcal{E}_m(\Omega)/\mathcal{N}(\Omega)$.

Partial derivatives in $\mathcal{G}(\Omega)$ are defined as

$$(D_i R)(\phi, x) = -((d_1 R)(\phi, x))(\partial_i \phi) + (\partial_i R)(\phi, x) =$$

where d_1 and ∂_i denote differentiation with respect to ϕ and x_i ,respectively.

The basic definitions and operations needed for an intrinsic definition of algebras of generalized functions on manifolds are as follows:

Definition 1 $\hat{\mathcal{A}}_0(M) \doteq \omega \in \Omega_c^n(M) : \int \omega = 1.$
The basic space for the forthcoming definition of the Colombeau algebra on M is:

Definition 2 $\widehat{\mathcal{E}}(M) = \mathcal{C}^\infty(\hat{\mathcal{A}}_0(M) \times M).$

Then locally we have $\widehat{\mathcal{E}}(\psi_\alpha(U_\alpha)) \cong \mathcal{E}(\psi_\alpha(U_\alpha))$, the isomorphism being given by $\widehat{\mathcal{E}}(\psi_\alpha(U_\alpha)) \ni R \to (\phi d^n y, x).$

We now review some basic concepts from differentiable manifolds.

Consider a vector space V, a tensor Φ on V is by definition a multilinear map $\Phi; \underbrace{V \times \cdots \times V}_{r} \times \underbrace{V_1^* \times \cdots \times V^*}_{s}$ (V^* denotes the dual space to V (covectors)), r its covariant order and s is contravariant order. For a fixed (r, s) we let $\mathfrak{J}_s^r(V)$ be the collection of all tensors on V of covariant order r and contravariant order s which has a natural vector space structure.

A vector field X on a manifold M is a function assigning to each point p of M an element X_p of $T_p(M)$ (tangent space to M at p, c.f. 2.4 Definitions 2, 3, 5).

In a similar fashion a $\mathcal{C}^\infty$-covariant tensor field of order r on a $\mathcal{C}^\infty$-manifold M is a function assigning to each $p \in M$ an element of $\mathfrak{J}^r(T_p(M))$, (For the objectives of the paper only covariant tensor field is used).

$\Phi \in \mathfrak{J}^r$ is symmetric if for each pair i, j , $1 \le i \le\le r$ we have $\Phi(v_1, v_2, \cdots, v_i \cdots v_j, \cdots v_r) = \Phi(v_1, \cdots, v_j, \cdots v_i \cdots v_r).$

Similarly if interchanging the i-th and j-th variables we have $\Phi(v_1, \cdots v_i, \cdots v_j, \cdots v_r) = -\Phi(v_1, \cdots, v_j, \cdots v_i, \cdots v_r)$ we say Φ is a skew or antisymmetric covariant tensor. Alternating covariant tensors are often called exterior forms. We define two linear transformation on the vector space $\mathfrak{J}^r V$):

A symmetrizing mapping $\mathcal{S} : \mathfrak{J}^r(V) \to \mathfrak{J}^r(V)$ and an alternating mapping are defined $\mathcal{A} : \mathfrak{J}^r(V) \to \mathfrak{J}^r(V)$ by the formulas:

1. $(\mathcal{S}\Phi(v_1, \cdots, v_r) = \frac{1}{r!} \sum_\sigma \Phi(v_{\sigma(1)}, \cdots, v_{\sigma r})$
2. $(\mathcal{A})\Phi)(v_1, \cdots, v_r) = \frac{1}{r!} \sum_\sigma sgn\, \sigma \,\, \Phi(v_{\sigma(1)}, \cdots, v_{\sigma(r)})$

the summations being over all $\sigma \in \mathfrak{G}_r$, the group of all permutations of r letters.
An alternating covariant tensor field of order r on M will be called <u>exterior differential form of order</u> r (or sometimes simply an r form).
For each $r > 0$, $\Lambda^r(V) \subset \mathfrak{J}^r(V)$; $\Lambda^0(V) = \mathfrak{J}^0(V) = \mathbb{R}$ and $\Lambda^1(V) = \mathfrak{J}^1(V) = V^*$, but $\Lambda^r(V)$ is properly contained in $\mathfrak{J}^r$ for $r > 1$.

$\mathcal{A}(\mathfrak{J}^r(V)) = \Lambda^r(V)$ and Φ is alternating if and only if $\mathcal{A}\Phi = \Phi$

The mapping from $\Lambda^r(V) \times \Lambda^s(V) \to \Lambda^{r+s}(V)$ is defined by $(\phi, \psi) \mapsto \frac{(r+s)!}{r!s!} \mathcal{A}(\phi \otimes \psi)$ is called the <u>exterior product</u> (or wedge product) of ϕ and ψ and is denoted by $\phi \wedge \psi$.

Given a vector field X on a manifold M, we say a curve $t \to F(t)$ on an open interval $J \subset \mathbb{R}$ is an integral curve of X if $\frac{dF}{dt} = X_{F(t)}$ on J.

Let X be a smooth vector field on a smooth manifold M of dimension n and let $p \in M$. Then there exists a neighborhood U of p on which the integral curve of X exists. Let ϕ_t be the mapping taking each point p of M a parameter distance t along the integral curve through the point The following properties hold for ϕ_t :

(i) ϕ_0 is the identity,
(ii) There exists $\epsilon > 0$ such that $\phi_s \circ \psi_t = \psi_t \circ \phi_s = \phi_{s+t}$ for $|s| < \epsilon, |t| < \epsilon, |s + t| < \epsilon$,
(iii) $\phi_t^{-1} = \phi_{-t}$.

We denote the image of $x \in M$ under the mapping ϕ_t by $\phi(x, t)$.

ϕ_t induces a mapping $\phi_{t*} : TM_x \to TM_{\phi(x,t)}$ and the pullback mapping $\phi_t^* : TM_{\phi(x,t)}^* \to TM_x^*$, (the dual of the tangent space TM_x).

The Lie derivative $\mathcal{L}_X Y$ of a C^r-covariant vector field Y with respect to X is defined at each $p \in M$ as

$(\mathcal{L}_X Y)_p = \lim_{t \to 0} \frac{1}{t}(\phi_t^* Y_{\phi(x,t)} - Y_p)$,

$\mathcal{L}_X Y$ is a vector field.

Grosser-et-al uses two forms of the Lie derivative as applied to an alternating covariant tensor field of order n, i.e. to $\Lambda^n(M)$ (n-forms).

First they define a function f on $M \times M$ such that for each fixed $p \in M$, $f \doteq (q \to f(p, q))$ represents a member of $\Lambda^n(M)$.

Given a smooth vector field X on M:

$\mathcal{L}_X' f(p, q) = \frac{d}{dt}|_0 f(\phi_t p, q)$

$\mathcal{L}_X f(p, q) = \frac{d}{dt}|_0 (\phi_t^* f_p, q)$.

The second one is the usual Lie derivative.

In terms of these two versions of the Lie derivative they define the space of <u>smoothing kernels</u> in their sense, (Grosser-et-al, Definition 3.3). The space of smoothing kernels on M is denoted by $\tilde{\mathcal{A}}_0(M)$.

Also for each $m \in \mathbb{N}$ by $\tilde{\mathcal{A}}_m(M)$ is denoted the set of all $\Phi \in \tilde{\mathcal{A}}_0(M)$ such that $\forall f \in C^\infty(M)$ and $\forall K \subset\subset M$

$$s \sup_{p \in K} |f(p) - \int_M f(q)\Phi(\epsilon, p)(q)| = O(\epsilon^{m+1})$$

We introduce the appropriate notion of a Lie derivative for elements of $\widehat{\mathcal{E}}(M)$:

For any $R \in \widehat{\mathcal{E}}(M)$ and any $X \in \mathfrak{X}(M)$ set

$\widehat{\mathcal{L}}(R))(\omega, p) \doteq -dR(\omega, p)\mathcal{L}_X(\omega) + \mathcal{L}_X(R(\omega, \cdot))p$.

Here $dR(\omega, x)$ denotes the derivative of $\omega \to R(\omega, x)$.

Definition 3　$R \in \widehat{\mathcal{E}}(M)$ is moderate if $\forall K \subset\subset M$, $\forall k \in \mathbb{N}$, $\exists N \in \mathbb{N}$ such that

$$\forall X_1, \cdots, X_k \in \mathfrak{X}(M), \forall \Phi \in \widehat{\mathcal{A}}_0(M) \ \sup_{p \in K} |\mathcal{L}_{X_1}, \cdots \mathcal{L}_{X_k}(R(\Phi(\epsilon, p), p)| = O(\epsilon^{-N}).$$

The subset of moderate elements of $\widehat{\mathcal{E}}(M)$ is denoted by $\widehat{\mathcal{E}}_m(M)$, (Grosser-et-al Definition 3.10).

Definition 4　$R \in \widehat{\mathcal{E}}(M)$ is called negligible if it satisfies: $\forall K \subset\subset M$, $\forall k, l \in \mathbb{N}_0$, $\exists m \in \mathbb{N}$ such that,

$$\forall X_1, \cdots, X_k \in \mathfrak{X}(M), \forall \Phi \in \tilde{\mathcal{A}}_0 \ \sup |p \in K \mathcal{L}_{X_1}, \cdots \mathcal{L}_{X_k}(R(\Phi(\epsilon, p), p))| = O(\epsilon^l).$$

The set of negligible elements of $\widehat{\mathcal{E}}_m(M)$ is denoted by $\widehat{\mathcal{N}}(M)$, Grosser-et-al Definition 3.11.

Similar to the original theory of Colombeau algebras, $\widehat{\mathcal{E}}_m(M)$ is a subalgebra of $\widehat{\mathcal{E}}(M)$ and $\widehat{\mathcal{N}}(M)$ is an ideal in $\widehat{\mathcal{E}}_m(M)$ and again $\widehat{\mathcal{G}}(M) = \widehat{\mathcal{E}}_m(M)/\widehat{\mathcal{N}}(M)$ is the Colombeau algebra on manifold M.

By construction $\widehat{\mathcal{G}}(M)$ is a sheaf of differential algebras on M.
The stability of $\widehat{\mathcal{E}}_m(M)$ and $\widehat{\mathcal{N}}(M)$ under Lie derivative also follows from the local description $\widehat{\mathcal{L}}_X \widehat{\mathcal{E}}_m \subset \widehat{\mathcal{E}}(M)$, $\widehat{\mathcal{L}}_X \widehat{\mathcal{N}}(M) \subset \widehat{\mathcal{N}}(M)$.

By construction, every $\widehat{\mathcal{L}}$ induces a Lie derivative (again denoted by $\widehat{\mathcal{L}}$) on $\widehat{\mathcal{G}}(M)$, so that $\widehat{\mathcal{L}}(M)$ becomes a differential algebra.

3.5.4　*Optimal Imbedding of Ultradistributions into Differential Algebras*

Hyperfunctions and ultradistributions are among the linear theories of generalized functions. Therefore some deficiencies and pathological features similar to those in Schwartz' distributions also apply to them. One way of overcoming such limitations is to imbed them into differential algebras which in the case of Schwartz distributions was achieved in an optimal manner by Colombeau.

However in the more general case of imbedding hyperfunctions into differetial algebras was not an easy task although it was posed in the 90 s by Oberguggenberger, it was one of the longest standing problems in the nonlinear theory of generalized functions. It was not solved until 2019.

In this Subsection we confine ourselves to the more restricted problem of imbedding problem for non-quasianalytic ultradistributions that was achieved by Debrouwere, Vernaeve and Vindas in [Deb-Ver-Vin].

The notation and preliminaries are as in Sect. 1.9.2. the duals $\mathcal{D}^{(M_p)'}(\Omega)$ and $\mathcal{D}^{\{M_p\}'}(\Omega)$ are ultradistribution spaces of class (M_p) (Beurling type) and class $\{M_p\}$ (Roumieu type) respectively. As customary we use the notation $* = (M_p)$ or $\{M_p\}$ to treat both cases.
The dual spaces $\mathcal{E}^{(M_p)'}(\Omega)$ and $\mathcal{E}^{\{M_p\}'}(\Omega)$ correspond to the compactly supported

ultradistributions.

Impossibility Result.

There is an analogue of Schwartz impossibility result as discussed in Sect. 3.2.

Recall that in the case of Schwartz' distributions this result asserts that it is impossible to imbed $\mathcal{D}'(\Omega)$ into an associative and commutative differential algebra preserving differentiation (like Leibniz rule), the unit function and at the same time the pointwise multiplication of continuous functions while possessing some natural properties.

The role of continuous functions in the present impossibility result is played by a space $\mathcal{E}^{\dagger}$ with slightly less regular than $\mathcal{E}^*(\Omega)$. We assume that N_p is a weight function that satisfies $(M.1)$ and stability under differential operators $(M.2)'$, namely

$$N_{p+1} \leq AH^p N_p, \ p \in \mathbb{N} \ |; \text{for} \ \text{some} A, H \geq 1.$$

We set $\dagger = (N_p)$ or $\{N_p\}$. When imbedding $\mathcal{D}^{*'}$ into some associative and commutative algebra $(\mathcal{A}^{*,\dagger}, +, \circ)$, the following requirements appear to be natural, [Deb-Ver-Vin];

$(P.1) \mathcal{D}^{*'}(\Omega)$ is linearly imbedded into $\mathcal{A}^{*,\dagger}(\Omega)$ and $f(x) \equiv 1$ is the unity in $\mathcal{A}^{*,\dagger}(\Omega)$.

$(P.2)$ For each $*$-ultradifferential operator $P(D)$ there is a linear operator $P(D)$: $\mathcal{A}^{*,\dagger}(\Omega) \to \mathcal{A}^{*,\dagger}(\Omega)$ satisfying

$$P(D)(q \circ f) = \sum_{\beta \leq degq} \frac{1}{\beta!} D^{\beta} q \circ (D^{\beta} P)(Df), \ \forall f \in \mathcal{A}^{*,\dagger}(\Omega), \ (\bullet)$$

and every polynomial q. Moreover, $P(D)|_{\mathcal{D}^{*'}(\Omega)}$ coincides with the usual action of $P(D$ on $*$-ultradistributions.

$(P.3) \circ |_{\mathcal{E}^{\dagger}(\Omega) \times \mathcal{E}^{\dagger}(\Omega)}$ coincides with the pointwise product of functions.

The following result imposes a limitation on the possibility of constructing such an algebra.

Theorem 1 (Impossibility Result) *Let $M_p \prec N_p$. Then, there is no associative and commutative algebra $\mathcal{A}^{*,\dagger}$ satisfying $(P.1) - (P.3)$ Moreover, in the Beurling case of $*$, there does not exist an algebra $\mathcal{A}^{(M_p),\{M_p\}}(\Omega)$ either satisfying $(P.1) - (P.3)$. Then Debrouwere-et-al settle for constructing the algebra of generalized functions of class (M_p) and $\{M_p\}$. Their construction of the algebra $\mathcal{G}^*(\Omega)$ of generalized functions consisting of nets of ultradifferentiable functions is very much analogous to that of Colombeau distributions, (see also 3.4.10 Garetto's work)*

We first introduce a new symbol:

Let $\Omega \subset \mathbb{R}^d$ be open. For $K \subset\subset \Omega$ and $h > 0$, $\mathcal{E}^{\{M_p\}h}$ consists of all $\phi \in \mathcal{C}^{\infty}(\Omega)$ such that

$$\|\phi\|_{\mathcal{E}^{\{M_p\},h}(K)} \doteq \sup_{x \in K, \alpha \in \mathbb{N}^d} \frac{|\phi^{\alpha}(x)|}{h^{|\alpha|} M_{|\alpha|}} < \infty$$

where $\mathcal{E}^{\{M_p\}}(\Omega) = \lim_{K \subset\subset \Omega} \lim_{h \to \infty} \mathcal{E}^{\{M_p\},h}(K)$.
($\lim_{\leftarrow}$ denotes the projective limit).

We introduce the following spaces of $*$-moderate nets of ultradifferentiable functions,

$$\mathcal{E}_M^{(M_p)}(\Omega) = \{(f_\epsilon)_\epsilon \in \mathcal{E}^{(M_p)}(\Omega)^{(0,1]} : (\forall K \subset\subset \Omega)(\forall h > 0)(\exists k > 0)$$

$$\|f_\epsilon\|_{\mathcal{E}^{(M_p),h}(K)} = O(e^{M(k/\epsilon)}) \text{ as } \epsilon \to 0^+\}$$

$$\mathcal{E}_M^{\{M_p\}}(\Omega) = \{(f_\epsilon)_\epsilon \in \mathcal{E}^{\{M_p\}}(\Omega)^{(0,1]} : (\forall K \subset\subset \Omega)(\forall k > 0)(\exists h > 0)$$

$$\|f_\epsilon\|_{\mathcal{E}^{\{M_p\},h}(K)} = O(e^{M(k/\epsilon)}) \text{ as } \epsilon \to 0^+\}$$

and the spaces of $*$-negligible elements

$$\mathcal{E}_N^{(M_p)}(\Omega) = \{(f_\epsilon)_\epsilon \in \mathcal{E}^{(M_p)}(\Omega)^{(0,1]} : (\forall K \subset\subset \Omega(\forall h > 0)(\forall k > 0)$$

$$\|f_\epsilon\|_{\mathcal{E}^{\{M_p\},h}(K)} = O(e^{-M(k/\epsilon)}) \text{ as } \epsilon \to 0^+\}$$

$$\mathcal{E}_N^{\{M_p\}}(\Omega) = \{(f_\epsilon)_\epsilon \in \mathcal{E}^{\{M_p\}}(\Omega)^{(0,1]} : (\forall K \subset\subset \Omega(\exists k > 0)(\exists h > 0)$$

$$\|f_\epsilon\|_{\mathcal{E}^{\{M_p\},h}(K)} = O(e^{-M(k/\epsilon)}) \text{ as } \epsilon \to 0^+\}$$

Note. M is the associate function of M_p as given in 1.9.2.
The conditions $(M.1)$ and $(M.2)$ ensure that $\mathcal{E}_M^*(\Omega)$ is an algebra under pointwise multiplication of nets and that $\mathcal{E}_N^*$ is an ideal of $\mathcal{E}_M^*(\Omega)$. Hence we can define the algebra $\mathcal{G}^*(\Omega)$ of generalized functions of class $*$ as the factor algebra

$$\mathcal{G}^*(\Omega) = \mathcal{E}_M^*(\Omega)/\mathcal{E}_N^*(\Omega).$$

We denote by $[(f_\epsilon)_\epsilon]$ the equivalence class of $(f_\epsilon)_\epsilon \in \mathcal{E}_M^*(\Omega)$. Observe that $\mathcal{E}^*(\Omega)$ can be regarded as a subalgebra of $\mathcal{G}^*(\Omega)$ via the constant imbedding $\sigma(f) \doteq [f]$, $f \in \mathcal{E}^*(\Omega)$.

We also remark that $\mathcal{G}^*(\Omega)$ can be endowed with a canonical action of $*$-ultradifferential operators. In fact since $*$-ultradifferential operators act continuously on $\mathcal{E}^*(\Omega)$, we have that $\mathcal{E}_M^*(\Omega)$ and $\mathcal{E}_N^*(\Omega)$ are closed under $*$-ultradifferential operators if we define their actions on nets as $P(D)((f_\epsilon)_\epsilon) \doteq (P(D)f_\epsilon)_\epsilon$. Consequently, every $*$-ultradifferential operator $P(D)$ conveniently induces a linear operator

$$P(D) : \mathcal{G}^*(\Omega) \to \mathcal{G}^*(\Omega),$$

which satisfies he generalized Leibniz rule (see ($\bullet$) following $(P.2)$) for any $f \in \mathcal{G}^*(\Omega)$ and the polynomial q seen as $\sigma(q)$. $\mathcal{G}^*\Omega)$ has also the sheaf property as given in [Deb-Ver-Vin], Proposition 4.3:

The mapping $\Omega \to \mathcal{G}^*(\Omega)$ is a fine and supple sheaf of differential algebras. Every $*$-ultradifferential operator $P(D) : \mathcal{G}^* \to \mathcal{G}^*$ is a sheaf morphism. Moreover the sheaf $\mathcal{G}^{(M_p)}$ is not flabby

(For the flabby sheaf see 1.9.1, for the terms fine, supple sheafs see the GLOSSARY).

Generalized Point-values

The ring of $*$-generalized numbers are introduced to regard $*$-generalized functions as poinwise objects. We define the generalized number $\tilde{\mathbb{C}}^*$ as

$$\mathbb{C}_M^{(M_p)} = \{(z_\epsilon)_\epsilon \in \mathbb{C}^{(0,1]} : (\exists k > 0)(|z_\epsilon| = O(e^{M(k/\epsilon)}))\}$$

$$\mathbb{C}^{M_p} = \{(z_\epsilon)_\epsilon \in \mathbb{C}^{(0,1]} : (\forall k > 0)(|z_\epsilon| = O(e^{M(k/\epsilon)}))\}$$

and the ideals

$$\mathbb{C}^{\{M_p\}} = \{z_\epsilon)_\epsilon \in \mathbb{C}^{(0,1]} : (\forall k > 0)(|z_\epsilon| = O(e^{-M(k/\epsilon)}))\}$$

$$\mathbb{C}_N^{\{M_p\}} = \{z_\epsilon)\epsilon \in \mathbb{C}^{(0,1]} : (\exists k > 0)(|z_\epsilon| = O(e^{-M(k/\epsilon)}))\}$$

(Compare with the contruction of generalized complex numbers in 3.4.10.
<u>Note.</u> We have $f = 0$ in $\mathcal{G}^*(\Omega)$ if and only if $f(x) = 0$ in $\tilde{\mathbb{C}}^*$ for all x.
We denote by $\mathcal{G}_c^*(\Omega) = \{f \in \mathcal{G}^*(\Omega) : \ suppf \ is \ compact \}$ the ideal of compactly supported $*$-generaized functions on Ω.

Imbedding ofM_p-ultradistributions

$\mathcal{D}^{*'}(\Omega)$ is imbedded into $\mathcal{G}^*(\Omega)$ in such a way that the multiplication of functions from $\mathcal{E}^*(\Omega)$ is preserved.

Debrouwere-et-al do this in more than one step. They first construct an imbedding of $\mathcal{E}^{*'}$ into the ideal $\mathcal{G}_c^*(\Omega)$ by means of convolution with a suitable mollifier and then use the sheaf-theoretical properties of $\mathcal{G}^*(\Omega)$ (as indicated in their Proposition 4.3) to extend the imbedding to the whole space $\mathcal{D}^{*'}(\Omega)$. For such a suitable mollifier:

Let N_p be a weight sequence with associated function N. We assume that N_p fulfills $(M.1)$ and $(M.3)'$ of 1.9.2. We fix an even function $\phi \in \mathcal{F}^{-1} D^{(N_P)}(\mathbb{R}^d)$ with the following properties $\hat{\phi}(x) = 1$ for $|x| \leq 1$ and $\hat{\phi}(x) = 0$ for $|x| \geq 2$. The net $\phi(x) = \frac{1}{\epsilon^d}\phi(\frac{x}{\epsilon})$, $\epsilon \in (0, 1]$ is called a (N_p)-net mollifiers. We also impose the following assumption N_p, relating the growth of M and N. For each $l > 0$

$$2M(t) \leq N(lt) + C, \quad \forall t \geq 0$$

for some $C > 0$. By lemma 5.1 ([Deb-Ver-Vin]) it is always possible to select a weight sequence N_p with the properties $(M.1)$, $(M_3)'$and the last inequality.

Notice that ϕ_ϵ is an entire function, thus the convolution $f * \phi_\epsilon$ is well-defined for $f \in \mathcal{E}^{*'}(\Omega)$.

Proposition *The mapping*

$$\iota_c : \mathcal{E}^{*'}(\Omega) \to \mathcal{G}_c^*(\Omega) : f \to \iota_c(f) = [((f * \phi_\epsilon)|_\Omega)_\epsilon],$$

is a linear mapping. Furthermore, $\iota_c|_{\mathcal{D}^*}(\Omega) = \sigma$, where σ is the constant mapping (i.e. $\sigma(f) = [(f)_\epsilon]$), [Deb-Ver-Vin], Proposition 5.6).

Finally

Theorem 2 *There is a linear mapping $\iota = \iota_\Omega : \mathcal{D}^{*'}(\Omega) \to \mathcal{G}^*(\Omega)$ with the following properties:*

(i) $\iota|_{\mathcal{E}^{*'}} = \iota_c$,
(ii) ι *commutes with $*$-ultradifferential operators, that is, for all $*$-differential operators $P(D)$*
        $$P(D)\iota(f) = \iota(P(D)f) \quad f \in \mathcal{D}^{*'}.$$
(iii) $\iota|_{\mathcal{E}^*(\Omega)}$ *coincides with the constant mapping σ. Consequently $\iota(fg) = \iota(f)\iota(g)$, $f, g \in \mathcal{E}^*(\Omega)$.*

Moreover, the entirety of all $\iota_\Omega : \mathcal{D}^{*'}(\Omega) \to \mathcal{G}^*(\Omega)$ is a sheaf monomorphism $\mathcal{D}^{*'} \to \mathcal{G}^*$ on any open subset of $\mathbb{R}^d$. ([Deb-Ver-Vin] Theorem 5.6).

3.6   Applications to Partial Differential Equations

This section is closely related to Sect. 3.4.5 of the coupled calculus in Colombeau algebras.

Suppose given a m-th order nonlinear partial differential equation

$$T(D)U(x) = 0, \ x \in \Omega \subset \mathbb{R}^n$$

as for instance the first order shock wave equation
$U_t + U_x.U = 0, \ (t, x) \in (0, \infty) \times \mathbb{R}$, or a nonlinear wave equation
$(\frac{\partial^2}{\partial t^2} - \Delta) = F(U)$ in $\mathcal{G}(\mathbb{R}^4)$, $F : \mathbb{R} \to \mathbb{R}$, C^∞-smooth.
We differentiate between different cases:
A) When $U \in C^\infty(\Omega)$, there is a perfect identity between $T(D)U$ computed in C^∞ or in $\mathcal{G}(\Omega)$. Therefore we have $T(D)U = 0$ both in C^∞ and $\mathcal{G}(\Omega)$.
B) We may have a second case of classical solutions $U \in C^m \setminus C^\infty$. Then we still have $T(D)U = 0$ in C^0, but $T(D)U$ computed in $\mathcal{G}(\Omega)$ will in general be no longer identical with $T(D)U$ computed in C^0 and then will only be associated by the equivalence relation $\approx$, that is
$$T(D)U|_{\mathcal{G}(\Omega)} \approx T(D)U|_{C^0}, \ (\dagger)$$

Then as $T(D)U = 0$, ($\dagger$) implies $T(D)U| \approx 0$ since $\approx$ is an equivalence relation. Conversely if ($\dagger$) holds and $T(D)U|_{\mathcal{G}(t)} \approx 0$ this will give $T(D)U|_{C^0} \approx 0$. But left of this equivalence relation is in $\mathcal{D}'$ since it is in $\mathcal{C}^0$. Then this will yield $T(D)U|_{\mathcal{C}^0} = 0$ according to the Theorem a) 2) of 3.4.5.

Thus we conclude that in case B) U is a classical solution if and only if it is a solution of $T(D)U \approx 0$ in $\mathcal{G}(\Omega)$.

C) In this case we have non-classical generalized solution $U \in \mathcal{G}(\Omega)$. Then for $U \in \mathcal{G}(\Omega) \setminus \mathcal{D}'(\Omega)$ for linear and nonlinear partial differential equations the solution can not be obtained within the theory of Schwartz distributions. However in some particular initial value and/or boundary conditions,the generalized solution of $T(D)U|_{\mathcal{G}(\Omega)} \approx 0$, even $T(D)U \sim 0 \in \mathcal{G}(\Omega)$ may be closely related to their distributional even classical solution.

<u>EXAMPLE</u> Shock wave equations and its solution:

For simplicity we consider the initial value problem in one dimension

$$U_t + U_x.U = 0, \quad t > 0, x \in \mathbb{R}$$

$$U(0, x) = u(x), \quad x \in \mathbb{R}$$

$$u(x) = \begin{cases} \alpha \text{ if } x < 0 \\ \beta \text{ if } x > 0 \end{cases}$$

where $-\infty < \beta < \alpha < \infty$ are given. It is well known that (c.f. [Lax]) that the unique, physically meaning solution of the above initial value problem is given by the shock wave

$$U(t.x) = \begin{cases} \alpha \text{ if } x < \frac{\alpha+\beta}{2}t \\ \beta \text{ if } x > \frac{\alpha+\beta}{2}t \end{cases}$$

which satisfies the Rankine- Hugoniot condition (on the propagation of the movements in the bodies specially in the perfect gases).

However there are many other possible solutions of this initial value problem.

For instance consider $\Gamma : [0, \infty) \to \mathbb{R}$, $\mathcal{C}^\infty$ smooth function such that $\beta r \leq \Gamma(t) \leq \alpha t, ; t \in (0, \infty)$ and define

$$U_\Gamma(t, x) = \begin{cases} \alpha \text{ if } x < \Gamma(t) \\ \beta \text{ if } x > \Gamma(t) \quad (\ddagger) \end{cases}$$

Then U_Γ satisfies the differential equation given above along with the initial conditions.

It is shown that U_Γ is a solution of $U_t + U_x.U \approx 0$ if and only if $\Gamma(t) = \frac{\alpha+\beta}{2}$, $t > 0$, that is if and only if U_Γ is the solution given in ($\ddagger$).

For this purpose U_Γ is expressed with the help of the Heaviside function $H = f + J(\mathbb{R}) \in \mathcal{G}(\mathbb{R})$.

Recall that (c.f. 3.4.2) $f(\psi, y) = \int_y^\infty \psi(z)dz, \psi \in \Phi(\mathbb{R}), y \in \mathbb{R}$ and $\delta = DH$, $\delta(\psi, y) = \psi(-y), \psi \in \Phi(\mathbb{R}), y \in \mathbb{R}$.

Therefore $U_\Gamma = \alpha + (\beta - \alpha)H \circ g \in \mathcal{G}((0, \infty) \times \mathbb{R}$ where $g : (0, \infty) \times \mathbb{R} \to \mathbb{R}$ is given by $g(t, x) = x - \Gamma(t), t > 0, x \in \mathbb{R}$.

Since g is C^∞-smooth $H \circ g$ is well-defined:
$H \circ g = h + J(\Lambda) \in \mathcal{G}(\Lambda)$ where $\Lambda = \Phi(\mathbb{R}^2) \times \Delta$, $\Delta = (0, \infty) \times \mathbb{R}$ and for $(\phi, x.t) \in \Lambda)$ we have $h(\phi, t, x) = f(\psi, x - \Gamma(t))$ with $\psi(y) = \int_{\mathbb{R}} \phi(y, \xi) \, d\xi$, $y \in \mathbb{R}$.

After some calculations (c.f. in detail, [Ros], pp. 150) we find $(U_\Gamma)_t + (U_\Gamma)_x \approx (\beta - \alpha)(\frac{\alpha+\beta}{2} - \Gamma')(\delta \circ g)$. Therefore U_Γ satisfies $U_t + U_x.U \approx 0$ if and only if $(\beta - \alpha)(\frac{\alpha+\beta}{2} - \Gamma')(\delta \circ g) \approx 0$ which is equivalent with $\Gamma(t) = \frac{\alpha+\beta}{2}t$.

Chapter 4
Random Distributions

4.1 Some Preliminaries

In this chapter we assume that the reader has a senior or graduate first year level background in probability theory and stochastic processes. However we review some of the basic concepts and definitions.

A probability space is a triplet $(\Omega, \mathcal{F}, P)$, where Ω is an abstract space (can be a vector space or a topological space), $\mathcal{F}$ is a σ-algebra of subsets of Ω and P is a non-negative countably additive set function on $\mathcal{F}$ with $P(\Omega) = 1$.

If Ω is a metric space and $\mathcal{F} = \mathcal{B}(\Omega)$, the Borel σ-algebra generated by the open subsets of Ω, then P is called Radon if it satisfies for every $B \in \mathcal{B}(\Omega)$, $P(B) = \sup\{P(K), K \text{ compact } \subset B\}$.

A measurable map $X : \Omega \to \mathbb{R}$ (or $\mathbb{C}$), is called a a random variable (r.v.). It induces a probability measure μ_X on $(\mathbb{R}, \mathcal{B}(\mathbb{R}))$ defined by $\mu_X(B) = P(X^{-1}(B)) \equiv P(X \in B)$.

μ_X is completely determined by the distribution function of X defined by $F_X(x) = P(X \leq x) = P(X^{-1}(-\infty, x])$, $-\infty < x < \infty$. If F_X has a Radon-Nikodym derivative f_X, called the probability density function, (p.d.f.) of X, then $\mu_X(B) = \int_B f_X(x)dx$, (a Lebesgue integral), $\int_{\mathbb{R}} f_X(x)dx = 1$.

The members of $\mathcal{F}$ are called events. The probability space $(\Omega, \mathcal{F}, P)$ is called complete if every subset of an event with probability 0 (a null event) is also an event (hence also has probability 0).

The expected value (or mean) of a r.v. X denoted by $E(X)$ or m_X is given by the integral $E(X) = \int_\Omega X dP(\omega)$, whenever the integral exists.

If X and Y are two r.v.s in $L^2(\Omega, \mathcal{F}, P)$ with the expected values $E(X)$ and $E(Y)$ respectively, then the covariance between X and Y is defined by

$$Cov(X, Y) = \int_\Omega (X - E(X))(Y - E(Y)dP(\omega) = E(XY) - E(X)E(Y).$$

© The Author(s), under exclusive license to Springer Nature Switzerland AG 2026

U. Çapar, *A Guide to Generalized Functions*,

https://doi.org/10.1007/978-3-032-09184-0_4

For the special case $X = Y$, then $Cov(X, X)$ (is the <u>variance</u>) of X denoted by $V(X) = \sigma_X^2 = E(X^2) - (E(X))^2$. Here X^2 is also a r.v. defined by $X^2(\omega) \doteq [X(\omega)]^2$, $\omega \in \Omega$. Similarly $\phi(X)$, where ϕ is a continuous (or measurable) function is again a r.v.

For $\phi(x) = x^r$, $r \in \mathbb{N}$, $E(X^r)$ is called the <u>r th moment</u> of X.

When we have more than one r.v.s, the algebra of random variables yields more r.v.s. In particular a n-tuple of r.v.s $\vec{X} = (X_1, \cdots, X_n)$ is called a <u>random vector</u>. It induces a probability measure on $(\mathbb{R}^n, \mathcal{B}(\mathbb{R}^n))$ by $\mu_{\vec{X}}(B) = P(\vec{X}^{-1}(B))$, where $B \in \mathcal{B}(\mathbb{R}^n)$ is a Borel subset in n dimensions. If this measure is absolutely continuous with respect to the n dimensional Lebesgue measure then it is given by $\mu_{\vec{X}}(B) = \int_B f_{\vec{X}}(x_1, \cdots, x_n)dx_1 \cdots dx_n$. $f_{\vec{X}}$ is an n dimensional density function. n random variables are said to be <u>mutually independent</u> if for all collections of Borel sets $B_i \in \mathcal{B}(\mathbb{R})$, $(i = 1, \cdots, n)$, the events $\{X_1 \in B_1\}, \cdots \{X_n \in B_n\}$ are independent, i.e. $P(\bigcap_{i=1}^{n}\{X_i \in B_i\}) = \prod_{i=1}^{n} P(X_i^{-1}(B_i))$.

A powerful tool in probability theory is the Fourier-Stieltjes transform of r.v.s, random vectors or probability measures called <u>characteristic function</u>, <u>characteristic functional</u> depending on the particular set-up they appear.

The characteristic function Φ_X of a r.v. X is given by $\Phi_X(t) = \int_{-\infty}^{\infty} e^{itx}dF_X(x)$, $(i^2 = -1)$. It has the following properties:

(i) $\Phi_X(0) = 1$,

(ii) Φ_X uniformly continuous,

(iii) Φ_x, is positive definite, i.e. for any finite set $\{t_1, \cdots, t_m\} \subset \mathbb{R}$ and $\{\alpha_1, \cdots, \alpha_m\} \subset \mathbb{C}$, we have

$$\sum_{j,k=1}^{m} \alpha_j \bar{\alpha}_k \Phi_X(t_1, \cdots, t_m) \geq 0. \quad (1)$$

A random variable of utmost importance in probability theory is the <u>normal random variable</u> which has the p.d.f. $f_X(x) = \dfrac{1}{\sqrt{2\pi}\sigma_X} e^{-\frac{(x-m_X)^2}{2\sigma_X^2}}$. The probability measure it induces is called one dimensional <u>Gaussian measure.</u> Its characteristic function is given by $\Phi_X(t) = e^{m_X t - \frac{\sigma_X^2 t^2}{2}}$.

If $(\Omega, \mathcal{F}, P)$ is a probability space, a mapping $X : \Omega \times T \to \mathbb{R}$ with properties to be specified in proper order is called a <u>stochastic process</u> (s.p.). Thus it is a function of two variables of the form $X(\omega, t)$, $\omega \in \Omega$, $t \in T$, where T is usually interpreted as time. If T is one of the sets $(-\infty, \infty)$, $[0, \infty)$ or $[a, b]$ then the process is called a continuous time s.p., otherwise when it is $\mathbb{N}$, $\mathbb{N}_+$, $\mathbb{Z}$ or $\mathbb{Z}_+$, we have a discrete time process. (In this chapter we consider only continuous time stochastic processes).

There are two ways of conceiving a s.p.:

(i) For fixed $t \in T$, $X_t(\omega)$ is assumed to be a r.v. on $(\Omega, \mathcal{F})$. Thus with this point of view a s.p is a collection $\{X_t(\omega)\}_{t \in T}$ of r.v.s.

(ii) For fixed $\omega \in \Omega$, as $t \in T$ varies we trace a <u>sample path</u> (or trajectory) $X_\omega(t)$. In this way a s.p. is a <u>random function</u>.

The expected value function of a s.p. is $X(t) = m_{X_t} \equiv E(X_t)$ and the
covariance function is $c_X(s, t) = Cov(X_s, X_t)$.
If $X : \Omega \times T \to \mathbb{R}^d$, then we have a d-dimensional s.p.
Finite dimensional distributions of a stochastic process, (fidis)
If $I = \{t_1, \cdots, t_m\} \subset T$, $(t_i, \ i = 1, \cdots, m$ are ordered), we denote by P_I the
probability measure induced by the random vector $\vec{X} = \{X_{t_1}, \cdots, X_{t_m}\}$, namely
$P_I(B) = P(\vec{X}^{-1}(B))$, $B \in \mathcal{B}(\mathbb{R}^m)$. Let $\mathcal{I}$ be the collection of all finite, ordered subsets of T Then the family $\{P_I\}_{I \in \mathcal{I}}$ is called the finite dimensional distributions (fidis)
of the s.p. This family satisfies the following compatibility (consistency) property:
Let $I = \{t_1, \cdots, t_m\}$, $J = \{t_1, \cdots, t_m, t_{m+1}, \cdots, t_{m+k}\}$. Then $P_J|_I = P_I$. More
explicitly

$$P_I(B) = P(\vec{X}_I^{-1}(B)) = P(\vec{X}_J^{-1}(B \times \mathbb{R}^k)) = P_J(B \times \mathbb{R}^k); \quad B \in \mathcal{B}(\mathbb{R}^m).$$

Cylindrical sets
Let $S(X) = \bigcup_{\omega \in \Omega}\{X_\omega(t), \ t \in T\} \subset R^T$ be the sample path space of the s.p. X. A
cylinder set with base $I \in \mathcal{I}$ is a subset of $S(X)$, if it is of the form
$\{X_\omega(t) : X_\omega(t_1) \in B_1, \cdots, X_\omega(t_m) \in B_m\}$ for   fixed $B_j \in \mathcal{B}(\mathbb{R})$ $(j =
1, \cdots, m)$, $I = \{t_1, \cdots, t_m\}$ or equivalently if it is of the form:

$$\{X_\omega(t) : \vec{X}_I(\omega) \in B, \ B \in \mathcal{B}(\mathbb{R}^m)\} \quad (*).$$

Let $\mathcal{A}_I$ be the set of all cylinder sets with base I and $\mathcal{A} = \bigcup_{I \in \mathcal{I}} \mathcal{A}_I$. Then $\mathcal{A}$ is an
algebra but not a σ-algebra in general.
   $\mathcal{A}_I$ can be probabilized by the finite dimensional distribution P_I. In this way the
consistent family $\{P_I\}_{I \in \mathcal{I}}$ defines a finitely additive set function Q on $\mathcal{A}$. Then the
following measure theoretical question arises: Can this finitely additive set function
be extended to a unique probability measure μ on the σ- algebra $\sigma(\mathcal{A})$ generated
by $\mathcal{A}$? (The cylinder sets given by (*) generate the same σ-algebra). The answer
is affirmative and takes the name of Kolmogorov or Daniel-Kolmogorov extension
theorem, (c.f. also [Bad]), μ can be looked upon as the probability measure of the
stochastic process. In the proof first the countable additivity of Q is shown and then
the Caratheodory extension theorem is utilized, (c.f. [Kar-Shr], Theorem. 2.2). μ has
the family, $\{P_I\}$ as its finite dimensional projections.

4.2   Brownian Motion (Wiener Process) and a Random Tempered Distribution

One of the most important stochastic processes which plays a central role in probability theory, physics, finance etc. was first observed by the biologist Robert Brown
as of 1820 studying very irregular motions of pollin particles dissolved in a liquid.

Its mathematical theory was developed later by Norbert Wiener, Louis Bachelier and also by Albert Einstein.

The Brownian motion is a s.p. $\{W_t\}_{t \in T}$, $T = [0, \infty)$ on a probability space $(\Omega, \mathcal{F}, P)$ satisfying the following axioms:

(i) $W_0 = 0$ almost surely (a.s.) (i.e. with probability one),

(ii) It has stationary, independent increments, i.e. $W_t - W_s$ and $W_{t+h} - W_{s+h}$ has the same probability distribution with $h \geq 0$ and $t + h, s + h \in T$ and also for $I = \{t_1, \cdots, t_m\} \in \mathcal{I}$, $W_{t_2} - W_{t_1}, \cdots, W_{t_m} - W_{t_{m-1}}$ are independent r.v.s,

(iii) For every $t > 0$, W_t is a normal r.v. with mean 0 and variance t,

(iv) The sample paths are a.s. continuous.

As a result of these axioms the r.v.s $W_t - W_s$ and W_{t-s} $(t > s)$ are r.v.s with mean zero and variance $t - s$.

Besides Brownian sample paths are <u>a.s. nowhere differentiable</u> which can be expected from the very irregular motions of a Brownian particle in rapidly changing directions. For the proof of this fact see for instance [Hida-2] Theorem 2.2 or [Kar-Shr] Theorem 1.8.

(Note. The notation $\{B_t\}_{t \in T}$ is also frequently used in the literature to denote the Brownian motion). The axioms (i)-(iii) suffice to determine the expectation and covariance functions of the process as $m_W(t) = 0$, $c_W(s, t) = \min(s, t)$ also allow us to obtain the fidis of the process by $P_I(B) = \int_B f_I(x_1, \cdots, x_m) dx_1 \cdots dx_m$, $I \in \mathcal{I}$, $B \in \mathcal{B}(\mathbb{R}^m)$, where f_I is the multivariate Gaussian probability density given by

$$f_I(x_1, \cdots, x_m) = (2\pi)^{-m/2}[t_1, t_2 - t_1, \cdots, t_m - t_{m-1}]^{-1/2} e^{-\frac{1}{2}\left\{\frac{x_1^2}{t_1} + \frac{(x_2 - x_1)^2}{t_2 - t_1} + \cdots + \frac{(x_m - x_{m-1})^2}{t_m - t_{m-1}}\right\}} \quad (1)$$

The unique probability measure extending this family of consistent family $\{P_I\}_{I \in \mathcal{I}}$ is called the <u>Wiener measure</u>.

A more general Brownian motion in which $T = \mathbb{R}$ instead of $[0, \infty)$ allows us to construct the relationship to the tempered distributions. When T is assumed to be the whole real line $\mathbb{R}$, the axioms i)-iv) remain the same, e.g. still we have $W_0 = 0, a.s..$ The set function created by the finite dimensional distributions again extend uniquely to $\sigma(\mathcal{A})$, still called the Wiener measure.

Let Ω_0 denote the space of all continuous functions on $\mathbb{R}$, null at 0 equipped with the topology of uniform convergence on bounded sets. Then Ω_0 is a Fréchet space. The Borel σ-algebra $\mathcal{B}(\Omega_0)$ coincides with $\mathcal{A} = \sigma(\bigcup_{I \in \mathcal{I}} \mathcal{A}_I)$. (c.f. [Kar-Shr], p. 60). Let μ be the Wiener measure on $(\Omega_0, \mathcal{B}(\Omega_0))$ created by the Brownian motion $W_t(\omega) = \omega(t)$, $\omega \in \Omega_0, t \in \mathbb{R}$.

The Brownian motion process $W(t, \omega)$ and the process $\{t W(\frac{1}{t}), \omega, t \neq 0\}$. have the same probability distribution. We have $\lim_{|t| \to 0, t \neq 0} t W(\frac{1}{t}, \omega) = 0, a.s.$ and replacing t by $\frac{1}{t}$ we find $\lim_{|t| \to \infty} t^{-1}|W(t, \omega)| = 0$. This also implies $\lim_{|t| \to \infty} \frac{|\omega(t)|}{\sqrt{1 + t^2}} = 0$. Define

$$\Omega_0^- = \{\omega \in \Omega_0 : \lim_{|t| \to \infty} (1 + t^2)^{-1/2}|\omega(t)| = 0.\} \quad (2)$$

From the sample path properties of Brownian motion $\mu(\Omega_0^-) = 1$. Let $\mathcal{F} = \Omega_0^- \cap \mathcal{B}(\Omega_0)$. Then $\{W_t, \ t \in \mathbb{R}\}$ is still a Brownian motion when restricted to $(\Omega_0^-, \mathcal{F}, \mu)$. From Sect. 1.5 we know that $\Omega_0^- \subset S'(\mathbb{R})$, (the tempered distributions space of Schwartz). Thus the identity map i on the sample paths of W, defines a random tempered distribution. Alternatively we may regard Brownian motion W and its generalized derivative $\dot{W}$ of its sample paths as $S'(\mathbb{R})$-valued random elements.

$\dot{W}$ is the so called <u>white noise</u>, (see also Sects. 4.4 and 5.7). Its probability distribution on the sample path space $(S'(\mathbb{R}), \mathcal{B}(S'(\mathbb{R}))$ is called the <u>white noise measure</u>. In symbols $< W(\omega), \phi > = < \omega, \phi >$ and $< \dot{W}(\omega), \phi > = - < \omega, \phi' >$ $\phi \in S(\mathbb{R})$ (the space of rapidly decreasing functions), where $< \ . \ >$ is the bilinear form on $(S'(\mathbb{R}) \times S(\mathbb{R}))$.

We further observe that $< \dot{W}(\omega), \phi >$ can also be regarded as a generalized stochastic process where the parameter space T is replaced by $S(\mathbb{R})$, (see also the next Section). The expected value and covariance functionals of this generalized process are computed as follows:
$E_{\dot{W}}(\phi) = E(< \dot{W}, \phi >) = E(-\dot{\phi}(t)W(t)) = \int \dot{\phi} \ E(W(t)dt = 0, \ m_{\dot{W}}(\phi) \equiv 0; \ \phi \in S(\mathbb{R}).$
$Cov_{\dot{W}}(\phi, \psi) = E(< \dot{W}, \phi >< \dot{W}, \psi >) = E(\int_{-\infty}^{\infty} W(t)\dot{\phi}(t)dt \int_{-\infty}^{\infty} W(s)\dot{\psi}(s)ds = \int_{-\infty}^{\infty} E(W_t W_s)\dot{\phi}(t)\dot{\psi}(s)dtds = \int_{-\infty}^{\infty} \min(t,s)\dot{\phi}(t)\dot{\psi}(s)dtds = \int_{t=-\infty}^{\infty} \dot{\phi}(t)(\int_{s=-\infty}^{t} s\dot{\psi}(s)ds)dt + \int_{s=-\infty}^{\infty} \dot{\psi}(s)(\int_{t=-\infty}^{s} \int_{t=-\infty}^{s} t\dot{\phi}(t)dt)ds, \ \phi, \ \psi \in S(\mathbb{R})$
After some manipulations (c.f. [Gel-Vil], 5.2): $Cov_{\dot{W}}(\phi, \psi) = \int_{-\infty}^{\infty} \phi(t)\psi(s)dtds = \int_{-\infty}^{\infty} \int_{-\infty}^{\infty} \delta(t-s)\phi(t)\psi(s)dtds.$ Hence the covariance functional of the generalized process is the generalized function $\delta(t-s)$.

4.3   Random Generalized Functions and Generalized Stochastic Processes

Definition 1 A $\mathcal{F} \leftrightarrow \mathcal{B}$ measurable map $\mathcal{U}$ from a probability space into a space $\mathcal{K}$ of generalized functions is called a <u>random distribution</u>. $\mathcal{B}$ is the Borel σ-algebra of $\mathcal{K}$.
$\mathcal{K}$ may be $\mathcal{D}'$, S' or else. The map $\mathcal{U}$ probabilizes $\mathcal{K}$, hence creating random elements in $\mathcal{K}$. In the case of S' we have a random tempered distribution, an example of which was given in the previous Section (the identity map i on Ω_0^- is obviously measurable).

For the next definition we introduce the notion of a Gelfand triplet $E \subset H \subset E^*$. Here E is a countably Hilbertian nuclear space (for the detailed definition see Subsection 5.2.3), H is a separable Hilbert space, E^* is the continuous dual of E and both inclusions are dense and continuous. The canonical bilinear form that links E and E^* is denoted by $< x, \xi >$, $x \in E^*, \xi \in E$. In particular if $x \in H$, then $< x, \xi >$ coincides with the inner product in H.

Definition 2 Consider a Gelfand triplet $E \subset H \subset E^*$. A <u>generalized stochastic process</u> is a system of random variables $\{X(\xi, \omega) : \xi \in E\}$ with parameter space

E defined on a probability space $(\Omega, \mathcal{F}, P)$ in such a way that $X(\xi, \omega)$ is linear and continuous in ξ.

For the special Gelfand triplet $S(\mathbb{R}) \subset L^2(\mathbb{R}) \subset S'(\mathbb{R})$ and the probability space $(\Omega_0^-, \mathcal{F}, \mu)$ we had an example of such a generalized process, namely $< \dot{W}, \omega >$. The joint probability distribution of the random vector $(X(\xi_1), X(\xi_2), \cdots, X(\xi_n))$ is completely and uniquely determined by the characteristic function $\int_\Omega \exp[i \sum_j z_j X(\xi_j, \omega)] \, dP(\omega)$, $z_j \in \mathbb{R}$.

Because of the linearity of X in ξ this turns out to be the characteristic function of $X(\sum_j z_j \xi_j)$. Since $\sum_j z_j \xi_j \in E$, it is concluded that the distribution of the generalized process is completely determined by

$$C_X(\xi) = \int_\Omega e^{iX(\xi, \omega)} \, dP(\omega), \quad \xi \in E \quad (1)$$

which is called the <u>characteristic functional</u> of the generalized process. It has the same properties as given in 4.1, (1).

In parallel to the definition of Sect. 4.1, a <u>cylinder set</u> in E^* is defined as

$\{x \in E^* : (x, \xi_1) \in B_1, \cdots, (x, \xi_n) \in B_n\}$, B_j are fixed Borel subsets in $\mathcal{B}(\mathbb{R})$, $\xi_j \in E$, $(j = 1, \cdots, n)$.

As examples of cylinder sets consider the half spaces in E^* defined by $(x, \xi) \leq a$, $x \in E^*$, $\xi \in E$ and also sets of a more general type: strips, defined by $a_k \leq (x, \xi_j) \leq b_k$, $x \in E^*$, $\xi_k \in E$, $k = 1, \cdots, n$.

For a more convenient concept of a cylinder set consider a fixed n-dimensional subspace N of E. The <u>annihilator</u> $N^\perp$ of N is the subspace of E^* defined by

$$N^\perp = \{x \in E^* :< x, \xi >= 0 \text{ for all } \xi \in N\}. \quad (2)$$

The quotient space $E^*/N^\perp$ is isomorphic to $N^*(\cong N)$ and in particular n-dimensional. The cosets in $E^*/N^\perp$ consist of functionals taking on same values on N. That means if the bilinear form $< ., . >_N$ connecting $E^*/N^\perp$ and N is defined as $< \bar{x}, \xi >_N =< x, \xi >$, $\bar{x} \in E^*/N^\perp$, $\xi \in N$ then this value does not depend on the representative $x \in E^*$ of $\bar{x} \in E^*/N^\perp$.

Let $A \subset E^*/N^\perp$, then $Z = \{x \in E^* : x + N^\perp \in A\}$, is called a cylinder set with base A and generating subspace $N^\perp$.

All cylinder sets with a fixed generating subspace $N^\perp$ is a σ-algebra $\mathcal{A}_N$. The union $\mathcal{A} = \bigcup_{N \subset E} \mathcal{A}_N$ taken over all finite dimensional subspaces N of E is only an algebra of subsets of E^*, Then we consider $\mathcal{B} = \sigma(\mathcal{A})$, the smallest σ-algebra containing $\mathcal{A}$.

The objective is to create a measure μ, on the measurable space $(E^*, \mathcal{B})$ which can be called, if it exists, the probability distribution of the generalized stochastic process $\{X(\xi, \omega)\}$.

The problem is finding a probability measure μ such that $C_X(\xi) = \int_{E^*} e^{i<x, \xi>} \, d\mu(x)$

is satisfied when C_X is a given functional satisfying properties i)-iii) of a characteristic functional given in 4.1 This is done in three steps:

(1) Construct a measure space $(E^*, \mathcal{A}_N, \nu_N)$,
(2) Let N vary to obtain a system of measure spaces which determine a finitely additive measure space $(E^*, \mathcal{A}, \nu)$,
(3) Extend $(E^*, \mathcal{A}, \nu)$ to a countably additive measure space $(E^*, \mathcal{B} = \sigma(\mathcal{A}), \mu)$.

For the first step, the restriction $C_N(\xi)$ of $C_X(\xi)$ to N may be viewed as a characteristic function on $\mathbb{R}^n$ and so the Bochner theorem yields a unique probability measure $\tilde{\nu}_N$ on E^*/N such that

$$C_N(\xi) = \int_{E^*/N} e^{i<x,\xi>} d\tilde{\nu}_N. \quad (3)$$

(See the GLOSSARY for Bochner theorem). Thus we have a probability space $(E^*/N^\perp), N, \tilde{\nu}_N)$, where $\mathcal{B}_N$ is the Borel subsets of $E^*/N^\perp$.

Let $\pi_N : E^* \to E^*/N^\perp :$ $E^* \ni x \to \bar{x} = x + N^\perp$,   With   this   notation $\pi_N^{-1}(\mathcal{B}_N) = \mathcal{A}_N$. If $B \in \mathcal{B}_N$, then $A = \pi_N^{-1}(B) \in \mathcal{A}_N$. In this way we define $\nu_N(A) = \tilde{\nu}_N(B)$, obtaining a measure space $(E^*, \mathcal{A}_N, \nu_N)$.

Let N_1 and N_2 be two finite-dimensional subspaces of E with $N_1 \subset N_2$. Denote the canonical projection of N_2 on N_1 by $\pi_{N_1 N_2} : N_2 \to N_1$. If $\mathcal{N}$ denote the family of all finite dimensional subspaces of E, the family $\{\nu_N\}_{N \in \mathcal{N}}$ satisfy the compatibility condition $\nu_{N_1} = \pi_{N_1 N_2} \nu_{N_2}$ $(N_1 \subset N_2)$. If $N_1, N_2 \in \mathcal{N}$, the annihilator of the subspace generated by N_1 and N_2 is the intersection $N_1^\perp \cap N_2^\perp$.

This can be generalized to any finite set N_i, $(i = 1, \cdots, r)$ of finite dimensional subspaces of E and the cylinder sets Z_i, with generating subspaces $N_i^\perp$, $(i = 1, \cdots, r)$. If N is generated by the subspaces $N_1, \cdots, N_r$, then using C_N we produce measures ν_N, hence we obtain a finitely additive set function ν on $\mathcal{A}$ which can be suitably called <u>cylinder set measure</u> satisfying the properties:

(1) $0 \le \nu(Z) \le 1$ for all Z,
(2) $\nu(E^*) = 1$,
(3) If a set Z is the union of a sequence $Z_1, Z_2, \cdots$ of non-intersecting cylinder sets having Borel bases and a common generating subspace $N^\perp$, then $\nu(Z) = \sum_{n=1}^{\infty} \nu(Z_n)$.

If the condition of common generating subspace is not satisfied then ν is only additive, i.e. if $Z_1, \cdots, Z_n$ is a finite system of disjoint cylinder sets in E^*, then $\nu(\bigcup_{j=1}^{n} Z_j) = \sum_{j=1}^{n} \nu(Z_j)$. For extending the measure ν to the σ-algebra $\mathcal{B} = \sigma(\bigcup_{I \in \mathcal{I}} \mathcal{A}_I)$ generated by all cylinder sets with Borel bases as a countably additive measure, we have the following result:

Proposition *A necessary and sufficient condition for a finitely additive measure ν to be extendable to a σ-additive measure μ on $(E^*, \mathcal{B})$, is for any $\epsilon > 0$ there exists a natural number n and a ball $S_n = \{x \in E^* : |x|_{-n} \le \gamma_n\}$ such that for any $A \in \mathcal{A}$ disjoint from S_n we have $\mu(A) < \epsilon$.*

Proof Necessity: Suppose that ν has an extension μ. Choose a sequence S_n of balls with increasing radii γ_n such that $\gamma_n \to \infty$. Then $\bigcup_n S_n = E^*$ and so as μ is countably additive $\mu(S_n^c) < \epsilon$ must hold for sufficiently large n. The required inequality for $\mu(A)$ follows.

For sufficiency we use the method of proof by reducing to contradiction. Suppose that $\{A_n\}$ is a sequence of pairwise disjoint elements of $\mathcal{A}$ such that $\sum_n A_n = E^*$. Since ν is finitely additive $\nu(\sum_1^n A_j) = \sum_1^n \nu(A_j) \leq 1$.

Suppose that this last inequality strict. Then there exists $\epsilon > 0$ such that $\sum_1^\infty \nu(A_n) = 1 - 3\epsilon < 1$. (†).

For each A_n we can find an open cylinder set A_n' (i.e. the base is open) such that $A_n' \supset A_n$ and $\nu(A_n' \backslash A_n) < \frac{\epsilon}{2^n}$. Clearly $\bigcup_j A_j' \supseteq S_n.$, But in a countably Hilbert nuclear space all closed, bounded sets are weakly compact, (c.f. [Hu-Yan], Theorem 3.14). Therefore we can choose a finite number $A_1', \cdots, A_k'$ of the A_j' which cover S_n. Setting $A' = \bigcup_1^k A_j'$, we have $A' \in \mathcal{A}$ and $\nu(A') \leq \sum_1^k \nu(A_j')$. We have then $1 = \nu(A' + A'^c) = \nu(A') + \nu(A'^c)$. As $A'^c \subset S_n^c$, $\nu(A'^c) < \epsilon$. This implies $1 \leq \nu(A') + \epsilon$ and combining with the inequality above

$$1 \leq \sum_j \nu(A_j') + \epsilon = \sum_j \nu(A_j) + \sum_n \nu(A_N' \backslash A_n) + \epsilon \overset{(\dagger)}{<} 1 - 3\epsilon + \sum_n \frac{\epsilon}{2^n} + \epsilon = $$
$1 - \epsilon$, a contradiction. $\qquad\qquad\qquad\qquad\qquad\qquad\qquad\qquad\qquad\qquad\qquad\qquad\square$

On the other hand as a more direct approach to the extension problem we recall that a if and only if condition of extending ν to $(E^*, \mathcal{B})$ as a unique σ-additive measure is that it should be so on $\mathcal{A}$, i.e. if $Z_1, Z_2, \cdots$ are pairwise disjoint cylinder sets such that $\bigcup_{j=1}^\infty Z_j \in \mathcal{A}$, then $\nu(\bigcup_{j=1}^\infty Z_j) = \sum_{j=1}^\infty \nu(Z_j)$. Using the above Proposition and a technical lemma due to Minlos ([Hida-2], Lemma 3.1), we arrive at the following result:

Theorem 1 *Let $C(\xi)$ be a functional on E which is*

(1) Continuous in the norm $|.|_p$ for some p,
(2) Positive definite,
(3) $C(0) = 1$.

(i.e. $C(\xi)$ is a characteristic functional)
If for some $n > p$ the injection $E_n \to E_p$ is of Hilbert-Schmidt type, (c.f. 5.2.3) then there exists a unique countably additive extension μ of ν to $(E^*, \mathcal{B})$ and μ is supported by E_n^*.
<u>*Note.*</u> *In the topology of countably normed Hilbert space E, condition 1 implies the continuity of $C(\xi)$ on E.*
With minor modifications Theorem 1 implies the following well-known Theorem.

Theorem 2 *(Minlos). A complex-valued function C on , E (a countably Hilbert nuclear space) is the characteristic function of a unique probability measure μ on E^*, (hence the characteristic functional of the generalized process indexed by E), i.e.*

$$C(\xi) = \int_{E^*} e^{i<x,\xi>} d\mu(x), \ \xi \in E$$

if and only if (1) $C(\xi)$ is continuous in ξ, (2) $C(\xi)$ is positive definite, that is for any set $z_1, \cdots, z_n \in \mathbb{C}$ of complex numbers and $\xi_1, \cdots, \xi_n \in E$, $\sum_{j,k=1}^{n} z_j \bar{z}_k C(\xi_j - \xi_k) \geq 0$, (3) $C(0) = 1$.

Remark Minlos Theorem will be taken up again in 5.2.3.

<u>APPLICATION</u>. Considering the Gelfand triplet $E \hookrightarrow H \hookrightarrow E^*$, the functional on E given by $\exp\{-\frac{1}{2}|\xi|^2\}$ is easily seen to satisfy (1)–(3) of the Theorem 2, thus there exists a unique probability measure μ (Gaussian measure) on $(E^*, \mathcal{B}(E^*))$ satisfying $\int_{E^*} e^{<x,\xi>} d\mu(x) = \exp\{-\frac{1}{2}|\xi|^2\}$, $|\,.\,|$ is the norm in H.

4.4 Random Colombeau Distributions and Their Applications to the Non-linear Stochastic Differential Equations

The concept of a random Colombeau distribution was initiated by Çapar in 1993 (c.f. [Çap-8]). Later on the idea was also taken up by Oberguggenberger and Russo (c.f. [Ober-3], [Ober-Rus]), their main concern being the solution of non-linear stochastic p.d.e.

Given a probability space $(\Omega, \mathcal{F}, \mu)$, a random Colombeau distribution is still a measurable map $\mathcal{U} : \Omega \to \mathcal{G}$. But because of additional features of the Colombeau algebra it is natural that there will be some extra conditions in comparison to Definition 1 of Sect. 4.3.

Definition 1 By a Colombeau random generalized function (C.r.g.f.) on a probability space $(\Omega, \mathcal{F}, \mu)$ is meant a map $\mathcal{U} : \Omega \to \mathcal{G}(O)$ (O an open subset in $\mathbb{R}^n$) such that there is a representing function $R_{\mathcal{U}} : \mathcal{A}(\mathbb{R}^n) \times O \times \Omega \to \mathbb{R}$ with the properties

 (i) For fixed $\phi \in \mathcal{A}(\mathbb{R}^n)$, $(x, \omega) \to R_{\mathcal{U}}(\phi, x, \omega)$ is jointly measurable on $O \times \Omega$,
(ii) Almost surely in $\omega \in \Omega$, $\phi \to R_{\mathcal{U}}(\phi, .., \omega)$ belongs to $\mathcal{E}_M(O)$ and it is a representative of $\mathcal{U}(\omega)$.

(For $\mathcal{A}(\mathbb{R}^n)$ and $\mathcal{E}_M(O)$ see the Sect. 3.4.1 Definitions 1 and 2).

The algebra of random Colombeau generalized functions is denoted by $\mathcal{G}_\Omega(O)$.

According to Definition 1 of Sect. 4.3 a generalized random function (distribution) is a weakly measurable map $V : \Omega \to \mathcal{D}'(\mathbb{R}^n)$ (Notation: $V \in \mathcal{D}'_\Omega(\mathbb{R}^n)$).

If $\phi \in \mathcal{A}(\mathbb{R}^n)$, then $V(\omega) * \phi(x) = < V(\omega), \phi(x - .) >$ is measurable with respect to $\omega \in \Omega$ and smooth with respect to $x \in \mathbb{R}^n$, hence jointly measurable.

Thus $R_V(\phi, x, \omega) = V(\omega) * \phi(x)$ qualifies as a representing function for an element of $\mathcal{G}_\Omega(\mathbb{R}^n)$. In this way one obtains an imbedding $\mathcal{D}'_\Omega(\mathbb{R}^n) \to \mathcal{G}_\Omega(\mathbb{R}^n)$.

Example 1 Denote by $S(T) \doteq S(\mathbb{R}^{n+1}|T)$, the space of rapidly decreasing smooth functions on $T = \mathbb{R}^n \times [0, \infty)$ $(.|T$ meaning restriction $)$. Let $\Omega = S'$, $\mathcal{F}$ the Borel

σ-algebra generated by the weak topology. By the Minlos Theorem (Sect. 4.3, Theorem 2) there is a unique probability measure μ on $(\Omega, \mathcal{F})$ such that

$$\int_{S^*} e^{i<\omega,\phi>} \, d\mu(\omega) = e^{-\frac{1}{2}\|\phi\|^2_{L^2(T)}}, \quad \phi \in S(T), \quad \text{specific Gelfand triplet being } S(T) \subset L^2(T) \subset S'(T).$$

White noise with support in T (can be interpreted as time in the standard stochastic process) is the random generalized function $\dot{W} : \Omega \to \mathcal{D}'(\mathbb{R}^{n+1}) : < \dot{W}(\omega), \phi >=< \omega, \phi|T)$. (For white noise in different set-ups see Sects. 4.3, 5.2.3 and 5.2.4 (Example 3)).

In fact it is a generalized Gaussian process with mean zero and variance $E(\dot{W}^2) = \|\phi|T\|^2_{L^2(T)}$ for $\phi \in \mathcal{D}(\mathbb{R}^{n+1})$. Viewed as a Colombeau random generalized function it has a representing function $R_{\dot{W}}(\phi, x, t, \omega) = \langle \omega, \phi(x - .t - .)|T \rangle$, (denoting the variables on $\mathbb{R}^{n+1}$ by (x, t)) which vanishes when t is less than minus the diameter of the support of ϕ; thus $\dot{W}$ is zero on $\mathbb{R}^n \times (-\infty, 0)$ in $\mathcal{G}_\Omega(\mathbb{R}^{n+1}$. Its variance is the generalized constant

$$E(R_{\dot{W}}(\phi, x, t)^2) = \int_{\mathbb{R}^n} |\phi(y, t - x)|^2 dy ds == \int_{\mathbb{R}^n} \int_{-\infty}^{t} |\phi(y, s)|^2 dy ds.$$

Application to the Solutions of Stochastic Partial Differential Equations

Example 2 The nonlinear stochastic wave equation ([Ober-Rus]).

Consider the semi-linear stochastic wave equation

$$\Box U = F(U) + H \; ; \text{on } \mathbb{R}^{n+1}; \quad U|\{t < 0\} = 0 \quad (1)$$

where F is smooth and globally Lipschitz, $\Box$ denotes the d'Alembertian $\partial_t^2 - \partial_{x_1}^2 - \cdots - \partial_{x_n}^2$ and H is a a generalized stochastic process with support in the half space $T = \mathbb{R}^n \times [0, \infty)$.

It is well-known that the solution to the linear wave equation

$$\Box V = H \quad \text{on } \mathbb{R}^{n+1} \quad V|\{t < 0\} = 0$$

is usually a generalized stochastic process, for example when H is a space-time white noise on $T.(n \geq 2)$. But considering (1), because of $F(U)$ term we are faced with non-linear operations on white noise or Schwartz distribution. Hence it is preferable to visualize (1) with H being a random Colombeau generalized function.

We also accept that $F \in \mathcal{O}_M(\mathbb{R})$, that is F is smooth with all dervatives growing at most polynomially at infinity. Further we require that F is globally Lipschitz and satisfies $F(0) = 0$.

Theorem 1 *Let F be described as above, $H \in \mathcal{G}_\Omega(\mathbb{R}^{n+1})$ with $supp H \subset T$. Then there is an a.s. unique solution $U \in \mathcal{G}_\Omega(\mathbb{R}^{n+1})$ to problem (1).*

Sketch of the proof. By fixing $\eta > 0$ we can choose a representative $R_H^\eta(\phi, x, t, \omega)$ of H such that $R_H^\eta(\phi, x, t, \omega) \equiv 0$ when $t < -\eta$, for all $\phi \in \mathcal{A}(\mathbb{R}^{n+1})$, $x \in \mathbb{R}^n$ and almost all $\omega \in \Omega$. Define $R_U^\eta(\phi, x, t, \omega)$ to be the classical smooth solution $u \in C^\infty(\mathbb{R}^{n+1})$ to the problem

$$\Box u(x, t) = F(u(x, t)) + R_H^\eta(\phi, x, t, \omega); \quad u(x, t) = 0 \ \textit{for} \ t - \eta, x \, n\mathbb{R}^n.$$

This solution is obtained by Picard iteration to the corresponding integral equation using the Lipschitz property of F. For example in the case $n = 3$, Kirchoff's formula gives

$$u(x, t) = \frac{1}{4\pi} \int_{-\eta}^{t} \frac{1}{t - s} \int_{|x-y|=t-s} (F(u(y, s) + R_H^\eta(\phi, y, s, \omega)) d\sigma(y) ds$$

for $t \geq -\eta$; in the case $n = 1, 2$, d'Dalembert's and Poisson's formula apply respectively. The solution can be smoothly continued by zero on $t < -\eta$ due to the assumption $F(0) = 0$.

It is clear that $R_U^\eta(\phi, xit, \omega)$ is jointly measurable in (x, t, ω). In the rest of the paper the authors show that $R_U^\eta(\phi, ., \omega)$ has the $\mathcal{E}_M$ property a.s. Thus its class $U(\omega)$ in $\mathcal{G}_\Omega(\mathbb{R}^{n+1})$ will define a solution to

$$\Box U = F(U) + H; \quad supp \, U \subset \mathbb{R}^n \times [-\eta, \infty).$$

Oberguggenberger and Russo also study the pathwise behavior of the Colombaeu generalized solution to the nonlinear wave equation with white noise excitation

$$\Box U = F(U) + \dot{W} \ \text{ on } \ \mathbb{R}^{n+1}; \quad U|\{t < 0\} = 0$$

where $\dot{W}$ is the white noise with support in $T = \mathbb{R}^n \times [0, \infty)$ viewed as an element of $\mathcal{G}_\Omega(\mathbb{R}^{n+1})$, ([Ober-Rus]).

Example 3 Colombeau solutions of a nonlinear stochastic predator-prey equation Deterministic semi-linear hyperbolic system (Lotka-Volterra) in two variables is:

$$D_1 u_1(x, t) = (\partial_t + c_1 \partial_x) u_1(x, t) = \lambda_1 u_1(x, t) u_2(x, t)$$

$$D_2 u_2(x, t) = (\partial_t + c_2 \partial_x) u_2(x, t) = \lambda_2 u_1(x, t) u_2(x, t)$$

$$u_j(x, 0) = \gamma_j(x); \quad j = 1, 2, \ \lambda_1 \lambda_2 < 0, t \geq 0, \ x \in \mathbb{R} \quad (2)$$

As a special case $\gamma_1(x) = \Delta_1(x - \xi_1)$, $\gamma_2(x) = \Delta_2(x - \xi_2)$ and $\lambda_1 = 1 = -\lambda_2$ respectively, represent a predator-prey system where the initial masses Δ_1 and Δ_2 of predators and preys have been concentrated at ξ_1 and ξ_2 respectively and they move in opposite directions towards each other with velocities c_1 and c_2.

Çapar considers in [Çap-3] the following stochastic version of the system where u_1 and u_2 undergo white noise excitations

$$D_1 u_1 = \lambda_1 (u_1 + \dot{W}_1)(u_2 + \dot{W}_2)$$

$$D_2 u_2 = \lambda_2 (u_1 + \dot{W}_1)(u_2 + \dot{W}_2) \quad (3)$$

We note that it is more realistic to attach different white noise excitations to the predator and prey populations. The solutions are expected to be as singular as the initial data, e.g. in the case of the Dirac initial data they would be generalized functions. Furthermore the product $\dot{W}_1 \dot{W}_2$ of the white noises can be given only in the sense of random nonlinear generalized functions, i.e. random Colombeau distributions.

Thus expecting the solutions of (3) to be a generalized solutions in $\mathcal{G}_\Omega(\mathbb{R}^2)$, we indicate them by capital letters and also without losing the generality we take $c_1 = 1 = -c_2;\ \lambda_1 = 1 = -\lambda_2$. Hence the system of equations is reformulated as:

$$D_1 U_1 = (\partial_t + \partial_x)U_1 = (U_1 + \dot{W}_1)(U_2 + \dot{W}_2)$$

$$D_2 U_2 = \partial_t - \partial_x)U_2 = -(U_1 + \dot{W}_1)(U_2 + \dot{W}_2)$$

$$U_j(x, 0) = \Gamma_j,\ (j = 1, 2) \quad (4)$$

where U_1, U_2 are generalized solutions in $\mathcal{G}_\Omega(\mathbb{R}^2)$, $\dot{W}_1$ and $\dot{W}_2$ are space-time white noises and $\Gamma_1, \Gamma_2 \in \mathcal{G}(\mathbb{R}^2)$.

Now $\dot{W}_1, \dot{W}_2 \in (S^*)$ (Hida distributions (c.f. Sects. 5.7 and 5.8), their product is not defined in (S^*), but meaningful in the stochastic Colombeau distribution space $\mathcal{G}_\Omega$ along the lines in [Çap-Ak 1] and [Çap-Ak 2].

We try to solve (4) at the representative platform $u_1, u_2;\ (U_1 = u_1 + \mathcal{N},\ U_2 = u_2 + \mathcal{N})$ and show that u_1 and u_2 are moderate. Then their classes will constitute a generalized solution. More explicitly:

(I) Find the classical solutions $u_1(\phi, x, t, \omega), u_2(\phi, x, t, \omega)$ for fixed ϕ and ω)

$$D_1 u_1 = (u_1 + \alpha)(u_2 + \beta)$$

$$D_2 u_2 = -(u_1 + \alpha)(u_2 + \beta)$$

$$u_1(x, 0) = \gamma_1(x),\ u_2(x, 0) = \gamma_2(x) \quad (5)$$

where $\alpha = \alpha(\phi, t, x, \omega), \beta = \beta(\phi, t, x, \omega)$ are representatives of white noises $\dot{W}_1$ and $\dot{W}_2$ respectively as random Colombeau distributions and $\gamma_1(\phi, x)$ and $\gamma_2(\phi, x)$ are representatives of Γ_2 and Γ_2.

(II) Show that $u_1(\phi, x, .)$ and $u_2(\phi, x, .)$ are in $\mathcal{E}_M$ i.e. are moderate and also that for fixed $\phi \in \mathcal{A}(\mathbb{R}^2)$, the mappings $(x, t, \omega) \to u_i(\phi, x, t, \omega),\ (i = 1, 2)$ are jointly measurable on $\mathbb{R}^2 \times \Omega$. Then their classes will form a random Colombeau solution of (4).

Since $D_1(-u_1) = D_2 u_2$, $\exists X$ such that $D_1 X = u_2$, $D_2 X = -u_1$. To find the form of X

$$D_1 X = \frac{\partial X}{\partial t} + \frac{\partial X}{\partial x} = u_2 \quad (\dagger)$$

$$D_2 X = \frac{\partial X}{\partial t} - \frac{\partial X}{\partial x} = -u_1 \quad (\ddagger)$$

By adding the two equations and substracting ($\ddagger$) from ($\dagger$) and using the initial conditions we easily obtain

$$X(x.t) = \frac{1}{2}\{\int_0^t [-u_1(x,s) + u_2(x,s)^x]ds + \int_0^x [\gamma_1(r) + \gamma_2(r)]dr\}$$

Following Hasimoto's method (c.f. [Has]) we let $Y = e^{-X}$ and obtain the differential equation of Y as

$$Y_{tt} - Y_{xx} + (\alpha + \beta)Y_x(\alpha - \beta)Y_t - \alpha\beta = 0 \quad (6)$$

with initial conditions

$$Y(x,0) \doteq Y_1(x) = e^{-X(x,0)} = e^{-\frac{1}{2}\int_0^x (\gamma_1(r)+\gamma_2(r))dr}$$

$$Y_t(x,0) \doteq Y_2(x) = -X_t(x,0)e^{-X(x,0)} = \frac{1}{2}[\gamma_1(x) - \gamma_2(x)]e^{-\frac{1}{2}\int_0^x (\gamma_1(r)+\gamma_2(r))dr} \quad (7)$$

It is difficult to find a solution to this hyperbolic system in its generality. However we make the natural assumption that the noises propagate along the characteristics allowing us to obtain a solution in a nice closed form. That means suppressing the other variables ϕ and ω we assume $\alpha = \alpha(x - t)$, $\beta = \beta(x + t)$. In this way we obtain the following solution to the initial value problem (6), (7) (for details see [Çap-2]).

$$Y(x,t) = \exp[\frac{1}{2}\int_{x-t}^{x+t} \beta(v)dv]Y_1(x - t) + I(x - t)J(x + t) \times \int_{x-t}^{x+t} \frac{Y_2(u) + Y_1'(u) - \beta(u)Y_1(u)}{2I(u)J(u)}du$$

where $I(u) = e^{\frac{1}{2}\int_0^u \alpha(s)ds}$ $J(u) = e^{\frac{1}{2}\int_0^u \beta(r)dr}$.

Then $u_1 = -D_2(-\ln Y)$, $u_2 = D_1(-\ln Y)$ is the solution at the representative platform.

<u>Note.</u> In the no noise deterministic case $\alpha = \beta \equiv 0$, we retrieve the determistic solution

$$Y_{det}(x,t) = \frac{1}{2}[Y_1(x - t) + Y_1(x + t) + \int_{x-t}^{x+t} Y_2(u)du.$$

For the stochastic predator-prey model the initial data are usually multiples of delta functions:

$\Gamma_1(x) = \Delta_1 \delta(x - \xi_1)$, $\Gamma_2(x) = \delta(x - \xi_2)$; $\Delta_1 \geq 0$, $\Delta_2 \geq 0$.

Suitable Colombeau representations can be taken as $\gamma_1(\phi, x) = \Delta_1 \phi(\xi_1 - x)$, $\Delta_2 \phi(\xi_2 - x))$, $\phi \in \mathcal{A}$.

Remark Because of the noise terms α and β, Y may take on negative values for certain ω and ϕ. To remedy this situation we should have bounded perturbations. For this purpose in [Çap-3] the Brownian populations involved with predators and preys are considered as doubly reflected ones. This also confirms well to the cyclic nature of the predator-prey problems in which the predator and prey populations fluctuate and swing between two bounds.

Another Approach to Random Colombeau Generalized Functions and Processes

Çapar & Aktuğlu in [Çap-Ak 2] randomize a particular bilinear $\bar{C}$-form in order to define a Colombeau random generalized function (C.r.g.f.). Let $O \subset \mathbb{R}^d$ be an open subset of $\mathbb{R}^d$ and $(\Omega, \mathcal{F}, P)$ be a complete probability space. Let also $T \in \mathcal{G}(O)$ be represented by $T = f_T + \mathcal{N}$, $f_T \in \mathcal{E}_M(O)$. Then for $\psi \in \mathcal{D}(O)$, $< T, \psi, > \in \bar{C}$ is the bilinear $\bar{C}$-form

$$< T, \psi, > = \int_O (\psi.T)(x)dx = h_T + \mathcal{N}_0 \in \bar{C}$$

where $h_T(\psi, \phi) = \int_O f_T(\phi, x)\psi(x)dx$.

Definition 2 (Generalized constant-valued random variables) A mapping $Y : \Omega \to \bar{C}$ is called

(i) strongly measurable if there exists a sequence of $\bar{C}$-valued simple functions $\{Y_m\}$ such that $\lim_{n\to\infty} \|Y_m - Y\|^- = 0$, P a.s.
(ii) representative measurable if a.s. in $\omega \in \Omega$ we have a representation $Y(\omega) = z(\omega) + \mathcal{N}_0$, where $z(\omega, \phi)$ is for every $\phi \in \mathcal{A}$ measurable as a usual complex-valued function,
(iii) Borel measurable if it is $\mathcal{F} \leftrightarrow \mathcal{B}_{\bar{C}}$ measurable.

For our purposes we define a $\bar{C}$-valued r.v. as the one which is representative measurable.

Definition 3 A C.r.g.f is a measurable map $T : \Omega \to \mathcal{G}(O)$. The measurability is in the weak sense, i.e. for every $\psi \in \mathcal{D}(O)$, $< T(\omega), \psi >$ is representative measurable.

Definition 4 A generalized random process generated by a C.r.f.g. $T(\omega)$ is the collection of $\bar{C}$-valued r.v.'s $\{< T(\omega), \psi >\}_{\psi \in \mathcal{D}(O)}$.

Note. The mapping $\psi \to\; <T(\omega), \psi>$ is linear and also continuous. Thus Definition 4 is in accordance with 4.3, Definition 2 in view of $\mathcal{D}(O) \subset S(O)$, the inclusion being dense. The continuity is clear from $<T(\omega), \psi> = \int_O (\psi . T(\omega)(x) dx = \int_O f_T(\phi, x, \omega) \psi(x) dx + \mathcal{N}_0$.

Also if $g_T(\phi, x, \omega)$ is another representative of $T(\omega)$, then for fixed ω, $|f_T(\phi_\epsilon, x, \omega) - g_T(\phi_\epsilon, x, \omega)| = O(\epsilon^{-N})$ for some $N \in \mathbb{N}$, therefore $<\psi, T(\omega)>$ yields the same generalized complex number when g_T is used.

APPLICATIONS. (Processes defined as nonlinear functions of the white noise.) Consider a canonical probability space $(\Omega, \mathcal{F}, \mu)$, where $\Omega = S^*(\mathbb{R})$, tempered distributions, $\mathcal{F}$ is the σ-algebra in S^* generated by the weak-star topology and μ is the probability measure provided by the Minlos theorem satisfying

$$\int_{S^*} e^{i<\xi, \omega>} d\mu(\omega) = e^{-\frac{1}{2}\|\xi\|^2_{L^2}}, \quad \xi \in S(\mathbb{R})$$

The white-noise is the random linear generalized function given by

$$\dot{W} : \Omega \to \mathcal{D}'(\mathbb{R}); \quad <\dot{W}(\omega), \psi> = <\omega, \psi> = - <W(\omega), \psi'>, \; \psi \in \mathcal{D}(\mathbb{R})$$

where W is the one parameter Wiener process (Brownian motion) (see also 4.4, Example 1 and Sects. 4.3 and 5.7).

The imbedding of $\dot{W}(\omega)$ into $\mathcal{G}_\Omega(\omega)$ will be the C.r.g.f $T_{\dot{W}}(\omega)$ with the representative $f_{\dot{W}}(\omega, \phi, x) = - <W_t(\omega), \phi'(t - x)>, \; \phi \in \mathcal{A}(\mathbb{R})$.

To view the white noise process as the generalized process generated by a C.r.g.f $\dot{W}(\omega)$ let $<T_{\dot{W}}(\omega), \psi> = g_\psi(\omega) + \mathcal{N}_0, \; \psi \in \mathcal{D}(\mathbb{R})$. suppressing ω

$$g_\psi(\phi) = \int_{\mathbb{R}} f_{\dot{W}}(\phi, x) \, \psi(x) dx = - \int_{\mathbb{R}^2} W(t) \phi'(x)(t - x)\, \psi(x) dt dx, \; \phi \in \mathcal{A}(\mathbb{R}), \; \psi \in \mathcal{D}(\mathbb{R}).$$

The nonlinear functions of the white noise (e.g. $\dot{W}^2$) are no more random linear distributions. However they have simple representations as random processes generated by C.r.g.f. s. For $\dot{W}^2$ we have =

$$f_{\dot{W}^2}(\phi . x) = \left[\int_{\mathbb{R}} W(x) \phi'(x)(t - x)\right]^2 = \int_{\mathbb{R}^2} W(t_1) W(t_2)\, \phi'(t_1 - x)\phi'(t_2 - x) dt_1 dt_2,$$

and the representative of the random Colombeau stochastic process $<\dot{W}^2, \psi>$ will be

$$h_\psi(\phi) = \int_{\mathbb{R}^2} W(t_1) W(t_2) \phi'(t_1 - x)\phi'(t_2 - x)\psi(x) dt_1 dt_2 dx; \; \phi \in \mathcal{A}(\mathbb{R}), \; \psi \in \mathcal{D}(\mathbb{R}).$$

Generalized constant valued expected value and covariance functionals can be calculated as

$\bar{m}_{\dot{W}^2}(\psi) = m_{\dot{W}^2}(\psi) + \mathcal{N}_0, \; m_{\dot{W}^2} \in \mathbb{C},$ where
$m_{\dot{W}^2}(\psi) = \int_{\mathbb{R}^2} \min(t_1, t_2)\phi'(t_1 - x)\phi'(t_2 - x)\psi(x)dt_1 dt_2 dx, \; \psi \in \mathcal{D}(\mathbb{R}).$
$c_{\dot{W}^2}(\psi_1, \psi_2) = k_{\dot{W}^2}(\psi_1, \psi_2) + \mathcal{N}_0,$

$$k_{\dot{W}^2}(\psi_1, \psi_2) = \int_{\mathbb{R}^6} E\left[\prod_{i=1}^{4} W(t_i)\right] \phi'(t_1 - x)\phi'(t_2 - x)\phi'(t_3 - x)\phi'(t_4 - 3 - x)\psi_1(x)\psi_2(y) \prod_{i=1}^{4} dt_i dx dy$$

$\psi_1, \psi_2 \in \mathcal{D}(\mathbb{R})$. Using the joint density of $\prod_{i=1}^{4} W(t_i)$, $E[\prod_{i=1}^{4} W(t_i)]$ is obtained, following a tedious calculation as $t_{(1)}(2t_{(2)} + t_{(3)})$; $(t_{(.)}$ denotes the ordered t_i, (c.f. [Çap-Ak 2]).

Likewise for any slowly increasing function $F \in \mathcal{C}^{\infty}(\mathbb{R})$, the functional $F(\dot{W})$ of the white noise is well defined as a generalized process generated by a C.r.f.g.

Chapter 5
Infinite Dimensional Distributions

5.1 Introduction

Firstly about the terminology of 'infinite dimensional distributions'. The reason why the distributions to be treated in this chapter are called infinite dimensional is that the underlying spaces will be some Banach space, a topological vector space or the dual of a countably Hilbertian nuclear space unlike the distributions like $\mathcal{D}'$ or $\mathcal{S}'$ where the underlying space is usually the finite dimensional $\mathbb{R}^n$.

Probability measures, especially the Gaussian measures will be of prime importance in these spaces. We had one case of these, namely the Wiener measure in Sect. 4.2.

Let Ω_0 denote the space of all continuous functions on $\mathbb{R}_+$, null at 0 equipped with the topology of uniform convergence on bounded sets. Then it is a Fréchet space. Let $\mathcal{B}(\Omega_0)$ denote the Borel σ-algebra on Ω_0. It is proved by N. Wiener that there exists a unique probability measure μ on $(\Omega_0, \mathcal{B}(\Omega_0))$ such that the coordinate evaluation process defined by $W_t(\omega) \equiv \omega(t)$, $t \in \mathbb{R}_+$, $\omega \in \Omega$ is a Brownian motion and $W_0 = 0$ a.s. (By the self-similarity of the Brownian motion we could as well consider the space $(\Omega_0([0, 1])$ with the sup norm and its Borel σ-algebra. Then the Wiener measure still exists on this measurable space).

We shall further elaborate on Wiener measure and Wiener space in Sect. 5.7 because of their intimate connection with the white noise, thus with the construction of one of the two major infinite dimensional distributions, namely the Hida distributions, the other one being the Meyer-Watanabe distributions.

© The Author(s), under exclusive license to Springer Nature Switzerland AG 2026
U. Çapar, *A Guide to Generalized Functions*,
https://doi.org/10.1007/978-3-032-09184-0_5

5.2 Probability Measures on Infinite Dimensioanal Spaces

5.2.1 Gaussian Measures on Hilbert Spaces

The characteristic function of the standard normal (Gaussian) measure in $\mathbb{R}^n$ is given by $\Phi(x) = e^{-\frac{1}{2}|x|^2}$ where $|x|$ is the norm in $\mathbb{R}^n$. Let now H be a separable Hilbert space with inner product $(\,,\,)$ and norm $\|\,.\,\|$. One may wonder whether

$$\Phi(\xi) = e^{-\frac{1}{2}\|\xi\|^2}, \ \xi \in H \quad (1)$$

could be the characteristic functional of a zero mean vector Gaussian probability measure on the σ-algebraσ-algebra of H. It satisfies the conditions of the Bochner's theorem, i.e. it is positive definite, continuous in ξ and $\Phi(0) = 1$. The answer to this question is negative. Suppose it is the case and $\Phi(\xi)$ is the characteristic functional of such a measure μ on H. Then we should have

$$\Phi(\xi) = \int_H e^{i(x,\xi)} d\mu(x)$$

If $\{\xi_n\}$ is an orthonormal basis in H, for $\xi = t\xi_n$, $t \in \mathbb{R}$ (1) yields

$$\int_H e^{it(x,\xi_n)} d\mu(x) = e^{-\frac{t^2}{2}} \quad (2)$$

But this creates a contradiction as $\sum_n |(x, \xi_n)|^2 = \|x\|^2 < \infty$, we have $\lim_{n\to\infty}(x, \xi_n) = 0$. Then (2) yields $1 = e^{-\frac{t^2}{2}}$.

An intuitive explanation of this contradiction could be as follows:

Equation (2) shows that (x, ξ_n) is a standard Gaussian random variable. On the other hand by (1)

$\Phi(\sum_{i=1}^n t_k \xi_k) = e^{-\frac{1}{2}\sum_1^n t_k^2}, \ t_k \in \mathbb{R}$.

This means that $\{(x, \xi_n), \ n \geq 1\}$ is an independent system of random variables. By the Law of Large Numbers then we have $\lim_{N\to\infty} \frac{\sum_1^N (x, \xi_n)^2}{N} = 1$ a.e.(μ). As $\{(x, \xi_n)\}$ are the coordinates of x, this implies

$$\sum_1^N (x, \xi_n)^2 \to \|x\|^2 = \infty$$

Namely the support of μ is much larger than H.

In fact the probability measure determined by (1) is the 'white noise measure'. We will see in Subsection 5.2.3 by the Minlos Theorem (Thm 2, Sect. 4.3) that the support of the white noise measure is the dual of a countably Hilbertian nuclear space which contains a separable Hilbert space H.

A general result in this direction is as follows:

A Borel probability measure μ on H is a Gaussian measure if and only if its characteristic functional Φ (equivalently its Fourier transform $\hat{\mu}$) can be expressed as

$$\Phi(x) = \exp i(m, x) - \frac{1}{2}(Sx, x),$$

where $m, x \in H$, S is a positive, symmetric trace class operator on H, (c.f. Sect. 5.2.3).

(m is the mean vector and S is the covariance operator of μ respectively. (c.f. [Vak] 3.1.4, [Hu-Yan] Theorem 4.11).

However our interest in this chapter is more on zero mean Gaussian measures on Banach spaces containing some separable Hilbert space H.

5.2.2 Gaussian Measures on Banach Spaces

Let B be a separable Banach space with norm $\| . \|_B$ and the dual norm $\| . \|_{B^*}$. Let $< . >$ be the canonical bilinear form on $B \times B^*$.

For $y_i \in B^*$, $(i = 1, 2, \cdots, n)$ and F a Borel subset of $\mathbb{R}^n$, the subset of B of the form

$$\{x \in B : (< x, y_1 >, \cdots, < x, y_n >\in F\} \quad (1)$$

is called a <u>cylinder set</u> with base $(y_1, \cdots, y_n)$. The collection $\mathcal{C}(B)$ of all cylinder sets is only an algebra. It is well-known that the σ-algebra generated by this algebra coincides with the Borel σ-algebra $\mathcal{B}(B)$. Let μ be a nonnegative set function on $\mathcal{C}(B)$, with $\mu(B) = 1$. If μ is a probability measure when restricted to the σ-algebra generated by cylinder sets with a fixed base, then μ is called a <u>cylindrical probability</u> on B.

A complex-valued function f on B is called a <u>cylindrical function</u> if it is measurable with respect to the σ-algebra generated by cylinder sets with some fixed base. If the cylindrical function is bounded, its integral $\int_B f(x) d\mu(x)$ can be defined. As a particular cylindrical function consider for fixed $y \in B^*$, $f(x) = i < x, y >$. Then the integral

$$\hat{\mu}(y) = \int_B e^{i<x,y>} d\mu(x), \quad y \in B^* \quad (2)$$

is called the <u>characteristic functional</u> (or the Fourier transform) of the cylindrical probability measure μ.

Clearly the characteristic functional of any cylindrical measure is positive definite and continuous on B^* and $\hat{\mu}(0) = 1$. Conversely if Φ is a positive definite continuous functional on B^* and $\Phi(0) = 1$, then there exists a unique cylindrical measure μ such that its characteristic functional is Φ.

Remark Counterpart of cylindrical measures on Hilbert spaces can be constructed by identifying $H^* \cong H$.

Question: What kind of cylindrical measures can be extended to Borel measures on B?

In a particular case when the Banach space B is the completion of some Hilbert space H with respect to a weaker norm, the answer to this question is affirmative according to the well-known Gross theorem.

Before stating this theorem we have to introduce the concept of a weaker norm and a measurable norm.

Definition 1 Let H be a separable Hilbert space with inner product $(\,,\,)$ and norm $\|\,.\,\|$. Let $|\,.\,|$ be another norm satisfying $|x| \leq c\|x\|$ for some constant c and $x \in H$. Then $|\,.\,|$ is said to be <u>weaker</u> than $\|\,.\,\|$.

Let B be the completion of H with respect to the weaker norm $|\,.\,|$. Then B is a separable Banach space and H is a linear subspace of B.

Denote by $\mathcal{P}$ the set of all finite dimensional orthogonal projections on H. For $P \in \mathcal{P}$ define $f(x) = |Px|$, $x \in H$; f is a cylindrical function on H.

Let μ be a cylindrical measure on H.

Definition 2 A weaker norm $|\,.\,|$ is said to be measurable with respect to μ (or simply μ-measurable), if for any $\epsilon > 0$ there exists $P_\epsilon \in \mathcal{P}$ such that for any $P \in \mathcal{P}$ orthogonal to P_ϵ, $\mu\{x \in H : |Px| > \epsilon\} < \epsilon$.

Going back to Eq. (1), Φ there, could not be the characteristic functional of a Gaussian measure if H is infinite dimensional as explained. But if μ is a cylindrical measure on H with the characteristic functional $\hat{\mu}(x) = e^{-\frac{1}{2}\|x\|^2}$, then μ should be called a standard Gaussian cylindrical measure on H. Now the famous Gross theorem.

Theorem 1 (Gross) *If μ is a Gaussian cylindrical measure on H and the norm $|\,.\,|$ is μ-measurable, then μ can be extended (or lifted) to the σ-algebra of the Banach space B which is the completion of H, with respect to $|\,.\,|$. The extension of μ (again denoted by μ) is called the Gaussian measure on B.*

Proof ([Kal, Hu-Yan], Theorem 4.16)

Remark The extended measure is in fact a Radon measure; i.e. $\mu(F) = \sup\{\mu(K) : K(\text{compact}) \subset F, F \in \mathcal{B}(B)\}$.

<u>Note</u>. The canonical bilinear form $<\,,\,>$ on $B \times B^*$ coincides with the inner product $(\,,\,)$ when restricted to $H \times B^*$, i.e. $<x, y> = (x, y)$, $x \in H$, $y \in B^*$.

Definition 3 (H, B, μ) is called an <u>abstract Wiener space</u>, μ is the Wiener measure and H is called the <u>Cameron-Martin (C-M) space</u>.

In the converse direction let B be a separable Banach space and let μ be the zero mean Gaussian measure on B, i.e. for any n and $y_i \in B^*$, $(<y_i, x>, i = 1, \cdots, n)$

is a standard Gaussian vector on B. Then there exists a Hilbert space H densely imbedded in B such that (B, H, μ) is an abstract Wiener space.

The most well-known example of an abstract Wiener space is the classical Wiener space based on $\Omega_0 = C_0([0, 1])$ and the Wiener measure μ on $\mathcal{B}(C_0([0, 1])$ introduced in Sect. 5.1. For the C-M space H we consider the Hilbert subspace of $C_0([0, 1])$ consisting of absolutely continuous functions with square-integrable derivatives. For $h \in H$, the norm is defined by $\|h\|^2 = \int_0^1 |\frac{dh}{ds}|^2 ds$; H is clearly dense in $C_0([0, 1])$. The triplet $(C_0([0, 1]), H, \mu))$ is a special case of an abstract Wiener space.

<u>Note.</u> For a more general cylindrical measure theory see [Gel-Vil, Bad] where the extension of cylindrical measures to the duals of locally convex topological vector spaces are considered. However for our purpose the Minlos theorem (c.f. Sect. 4.3, Thm. 2), also in the next section will suffice.

5.2.3 *Countably Hilbertian Nuclear Spaces and the Minlos Theorem*

Gaussian measures on the duals of countably Hilbertian nuclear spaces play an important role in the white noise analysis, consequently in the construction of one of the two major infinite dimensional distributions, namely the Hida distributions. This is due to the fact that these spaces have richer norm structures in comparison to Banach spaces and the functional analytic properties of the duality theory of topological vector spaces can be utilized. Also these nuclear spaces have properties similar to those in finite dimensional spaces.

We recall some terminology and definitions from the theory of operators in Hilbert spaces. Let E and H be two separable Hilbert spaces, $L(E, H)$ denote the space of linear operators from E to H and $\mathcal{L}(E, H)$ is the subspace of $L(E, H)$ consisting of bounded linear operators. Inner products and the norms are denoted by $(\, , \,)$ and $\| \, \|$.

Let $A \in \mathcal{L}(E, H)$ and $\{e_n\}\{f_n\}$ be bases in E and H respectively. Then $\sum_n \|Ae_n\|^2 = \sum_n \|A^* f_n\|^2$ (A^* is the adjoint of A). If $\sum_n \|Ae_n\|^2 < \infty$ for some basis $\{e_n\}$, then A is called a <u>Hilbert-Schmidt operator</u> (in short a H-S operator). Put $\|A\|_2 \equiv (\sum_n \|Ae_n\|^2)^{\frac{1}{2}}$, and call it a <u>H-S norm</u> also denoted by $\|A\|_{HS}$.

The set of all H-S operators from E to H is denoted by $\mathcal{L}_{(2)}(E, H)$. For $A, B \in \mathcal{L}_{(2)}(E, H)$, $(A.B)_2 = \sum_n (Ae_n, Be_n)$ becomes an inner product and $\mathcal{L}_{(2)}(E, H)$ is a separable Hilbert space under this inner product.

If A is a densely defined closed operator in $L(E, H)$, then $T = (A^*, A)^{\frac{1}{2}}$ is a self-adjoint operator on E. There is also a unique linear isometry U from the range of T to H such that $A = UT$. This is called the <u>polar decomposition</u> of A and $T = |A|$ the absolute value of A, (see also the GLOSSARY).

If A is a self-adjoint compact operator on E, there exists an orthonormal basis $\{e_n\}$ and a sequence of non-zero numbers (eigenvalues) $\{\lambda_n\}$ with $\lambda_n \to 0$ as $n \to \infty$ such that

$$Ae_n = \lambda_n e_n, \quad Ax = \sum_n \lambda_n(x, e_n)e_n, \quad x \in E \quad (1)$$

If A is a compact operator (c.f. GLOSSARY) from E into H, then $T = A^*A$ is a compact operator on E, therefore it has polar decomposition $A = UT$, if T has representation $Tx = \lambda_n(x, e_n)e_n, \quad x \in E$, then

$$A \in \mathcal{L}_{(2)}(E, H) \iff \sum_n \lambda^2 < \infty \text{ and } \|A\|^2 = \sum_n \lambda_n^2 \quad (2)$$

If on the other hand A is a compact operator from E into H with polar decomposition as above and $\sum_n \lambda_n < \infty$, , then A is called a <u>trace class</u> (or nuclear) operator. We put

$$\|A\|_1 = \sum \lambda_n \quad (3)$$

$\|A\|_1$ is called the <u>trace norm</u> of A. All trace class operators are denoted by $\mathcal{L}_{(1)}(E, H)$. We obviously have the following inclusions $\mathcal{L}_{(1)}(E, H) \subset \mathcal{L}_{(2)}(E, F) \subset \mathcal{K}(E, H)$ where $\mathcal{K}(E, H)$ denotes the set of all compact operators from E into H.

$\mathcal{L}_{(1)}(E, H)$ is a Banach space under the norm $\| . \|_1$

Let E be a topological vector space (tvs) over $\mathbb{C}$ with the topology given by a family $\{| . |_n \, n = 1, 2, \cdots\}$ of inner product norms. For $e, f \in E$ define

$$d(e, f) = \sum_{n=1}^{\infty} 2^{-n} \frac{|e - f|_n}{1 + |e - f|_n} \quad (4)$$

Then d is a metric on E and a net is Cauchy in this metric if and only if it is Cauchy in each of $| . |_n$. In particular d generates the same topology on E.

A topological vector space E with a family $\{| . |_n\}, n \geq 1$ of inner product norms is called a <u>countably Hilbert space</u> if it is complete with respect to its topology.

By defining new norms $\| . \|_n = (\sum_{k=1}^{n} |x|_k^2)^{\frac{1}{2}}$ if necessary, we may assume that the family $\{| . |_n; n \geq 1\}$ of norms is non decreasing, i.e. $|x|_1 \leq |x|_2 \leq \cdots \leq |x|_n \leq \cdots \cdots \forall x \in E$.

Let E_n denote the completion of E with respect to the norm $| . |$. Then E_n is a Hilbert space and the following inclusions are continuous

$$E \subset \cdots\cdots E_{n+1} \subset E_n \subset \cdots\cdots \subset E_1$$

$E = \bigcap_{n=1}^{\infty} E_n$, E in its projective limit topology is complete, also being a metrizable locally convex space it is a Fréchet space. If E is further a nuclear space, i.e. for every n, there exists $m \geq n$ such that the inclusion map from E_m into E_n is a Hilbert-

Schmidt (H-S) operator, i.e. there is an orthonormal basis $\{e_k\}$ for E_m such that $\sum_{k=1}^{\infty} |e_k|_n^2 < \infty$, then it is called a <u>countably Hilbertian nuclear space</u>.

As a trace class operator is also a H-S operator and the product of two H-S operators is a trace class operator, E is a nuclear space if and only if for every n, there exists , $m \geq n$ such that the inclusion map from E_m into E_n is trace class operator.

Nuclear spaces have many properties similar to those of finite dimensional spaces $\mathbb{R}^n$. For instance a subset of such a space is compact if and only if it is closed and bounded. This implies that infinite dimensional Banach spaces can not be nuclear. If E is a nuclear space, then its continuous dual E^* is the inductive limit of the duals E_n^*. It is also known that the σ-algebras on E^* generated by the weak, strong and the inductive limit topologies are all the same. (c.f. [Oba, Kuo-1]).

Now we can recall the important Minlos theorem, (c.f. Sect. 4.3, Theorem 2).

(For the proof see [Gel-Vil] and [Hida-1]). We need it for the following: It is often useful to imbed a nuclear space E into some Hilbert space H continuously and densely. As H and its dual H^* are identified by the Riesz representation theorem. H is also continuously and densely imbedded into the dual space E^*. Then we have a triplet

$$E \subset H \subset E^* \quad (5)$$

Such a triplet was called a <u>Gelfand triplet</u>, (c.f. Sect. 4.3). A particular example of a Gelfand triple is $S(\mathbb{R}) \subset L^2(\mathbb{R}) \subset S'(\mathbb{R})$, where $S(\mathbb{R})$ is the Schwartz space of rapidly decreasing functions and its continuous dual $S'(\mathbb{R})$ is the space of tempered distributions.

$S(\mathbb{R})$ is known to be a nuclear space (c.f. [Kuo-1], 3.2). The functional $C(x) = e^{-\frac{1}{2}\|x\|_{L^2}^2}$, $x \in S(\mathbb{R})$ satisfies the properties (1)–(3) of the Minlos theorem (Sect. 4.3, Thm. 2), then the uniquely defined (zero mean) Gaussian measure μ on $(S'(\mathbb{R}), \mathcal{B}(S'))$ is called the <u>white noise measure</u> and $(S'(\mathbb{R}), \mathcal{B}(S'(\mathbb{R}), \mu))$ is called the <u>classical white noise space</u>.

The white noise space will be investigated in more detail in Sect. 5.7.

5.2.4   *A General Framework: Gaussian Probability Spaces*

A framework called Gaussian probability space by Paul Malliavin requires only a separable Hilbert space and no additional structures. But it contains both the white noise space and the abstract Wiener spaces as special cases.

Let H be e separable Hilbert space with inner product (,) and norm $\| . \|_H$. By applying the Kolmogorov extension theorem (c.f. [Bau]) there is a probability space $(\Omega, \mathcal{F}, \mu)$ and a family of Gaussian random variables $\{\mathcal{H} : W_h, h \in H\}$ on it such that $E(W_h) = 0$, $E(W_h W_g) = (h, g)$, $\forall f, g \in H$

If $(\Omega, \mathcal{F}, \mu)$ is a complete probability space, then $(\Omega, \mathcal{F}, \mu; H)$ is called a <u>Gaussian probability space</u>. $h \rightarrow W_h$ is a linear isometry from H into $L^2(\Omega, \mathcal{F}, \mu)$, so that H is isometric to the closed subspace $\mathcal{H}$ of $L^2(\Omega, \mathcal{F}, \mu)$.

If $\mathcal{F}$ is generated by the random variables $\{W_h\}\, h \in H$ and contains all μ-null sets, then the Gaussian probability space is called <u>irreducible</u>.

For finite dimensional Gaussian space consider the completion $\mathcal{F}$ of the Borel σ-algebra of $\mathbb{R}^n$ and the standard Gaussian measure $\mu = \gamma^n$, $W_h(x) = h.x$ so that $(\mathbb{R}^n, \mathcal{F}, \mu;\ \mathbb{R}^n)$ is a Gaussian space.

Example 1 (*Classical Wiener space*) Take as in the Introduction Sect. 5.1 $\Omega = C_0([0, 1])$ with the sup norm. $\omega(0) = 0$ for $\omega \in \Omega$. μ is the Wiener measure and $\mathcal{F}$ is the μ-completion of $\mathcal{B}(\Omega)$. Take $H = L^2([0, 1])$. For $h \in H$ define $W_h(\omega) = \int_0^1 h(t)d\omega(t)$ where the integral is the stochastic integral (the Wiener integral, GLOSSARY). Then $(\Omega, \mathcal{F}, \mu;\ H)$ is a Gaussian probability space. For any $h \in H$ denote by $\tilde{h}(t) = \int_0^t h(s)ds$ $(0 \le t \le 1)$. Then $\tilde{h}$ has square integrable derivatives. Consequently the map $J : h \to \tilde{h}$ is a continuous linear injection from H into Ω since $\|\tilde{h}\|_{C_0} = \sup_{0 \le t \le 1} | \int_0^t |h(s)ds \le \sup_{0 \le t \le 1}(t \int_0^t |h(s)|^2)^{1/2} \le (\int_0^2 |h(s)|^2)^{1/2} = \|h\|_H$. $\tilde{H} = J(H)$ whih is dense in Ω is the Cameron-Martin space consisting of functions with square integrable derivatives.

Example 2 (*Abstract Wiener space*) Let B be a separable Banach space, H be a separable Hilbert space densely imbedded into B. Denote the imbedding map by $J : H \longrightarrow B$. Then B^* is continuously imbedded into $H^* \cong H$ by the dual map J^*. Therefore $B^* \hookrightarrow H^* \cong H \hookrightarrow B$. Let μ be the Gaussian measure on B satisfying

$$\int_B e^{i<h,x>}d\mu(x) = e^{-\frac{1}{2}\|J^*h\|_H^2}, \ \text{ for all } h \in B^* \quad (\dagger)$$

where $<,\ >$ is the canonical bilinear form on $B^* \times B$. (H, B, μ) is then an abstract Wiener space. Let $\mathcal{F}$ be the μ-completion of $\mathcal{B}(B)$. For $h \in B^*$ define W_h by $W_h(x) =<h, x>$. By $(\dagger)$ $\{W_h : h \in B^* \subset H\}$ is a family of Gaussian variables on $(B, \mathcal{F}, \mu)$ and $E(W_h) = 0$; $E(W_h W_g) =< J^*h, J^*g >$, $\forall h, g \in B^*$ which also follows from $(\dagger)$.

Consequently the map $J^*h \to W_h$ is a linear isometry from $J^*(B^*)$ into $L^2(B, \mathcal{F}, \mu)$. As $J^*(B^*)$ is dense in H, it can be extended to an isometry $H \to L^2(B, \mathcal{F}, \mu)$. Therefore $(B, \mathcal{F}, \mu;\ H)$ is a Gaussian probability space, (c.f. [Hu-Yan]), (H is the C-M space)

Example 3 (*White noise space*) Let $H = L^2(\mathbb{R}^d)$, $S(\mathbb{R}^d)$ and $S'(\mathbb{R}^d)$ be the Schwartz space and the tempered distributions respectively. We then have

$$S(\mathbb{R}^d) \hookrightarrow L^2(\mathbb{R}^d) \hookrightarrow S'(\mathbb{R}^d)$$

$S(\mathbb{R}^d)$ being a Fréchet nuclear space, by the Minlos theorem there exists a unique probability measure μ on $\mathcal{B}(S'(\mathbb{R}^d))$ such that for all $x \in S(\mathbb{R})$

$$\int_S' e^{i<y,x>}d\mu(y) = e^{-\frac{1}{2}\|x\|_H^2}$$

where $< y, x >$ is the canonical bilinear form on $S'(\mathbb{R}^d) \times S(\mathbb{R}^d)$. Denoting again $\mathcal{F}$ the μ-completion of $\mathcal{B}(S'(\mathbb{R}^d))$, define $W_x(y) =< y, x >$. As in the previous example the map $x \to W_x$ can be extended to a linear isometry $L^2(\mathbb{R}^d) \longrightarrow L^2(S'(\mathbb{R}^d), \mathcal{F}, \mu)$ rendering $(S'(\mathbb{R}), \mathcal{F}, \mu; L^2(\mathbb{R}))$ a Gaussian probability space.

White noise spaces constructed on general Gelfand triplets are discussed in Sect. 5.7.

5.3 Elements of Malliavin Calculus, Gradient and Divergence Operators

In an abstract Wiener space (H, B, μ), (respectively in a Gaussian probability space $(\Omega, \mathcal{F}, \mu; H)$) a function $F : B \longrightarrow \mathbb{R}$ (or $\mathbb{C}$)(resp. $F : \Omega \longrightarrow \mathbb{R}$(or $\mathbb{C}$)) is called a <u>Wiener functional</u> if it is measurable with respect to $\mathcal{B}(B)$(resp. to $\mathcal{F}$).

Some examples of such functionals are Ito stochastic integrals, solutions of Ito stochastic differential equations. But these functionals need not to be continuous. In fact all the interesting functionals are defined up to equivalence classes. Therefore an attempt to differentiate a functional in the sense of Fréchet as

$$\lim_{\|h\| \to 0} \frac{\|F(\omega + h) - F(\omega) - A(\omega)\|_B}{\|h\|_H} = 0, \ \omega \in B \quad (1)$$

will not even be well-defined since $F(\omega + h)$ and $F(\omega)$ are equivalence classes of random variables.

As a remedy Paul Mallliavin introduced in 1976, by virtue of the quasi-invariance of the Wiener measure, a kind of weak differential calculus for Wiener functionals such that the above mentioned important functionals became smooth under his sense of differentiation. His weak differentiation was obtained by the disturbance of Brownian trajectories in the direction of the Cameron-Martin space elements.. This approach justified the terminology <u>stochastic calculus of variation</u> or <u>Malliavin calculus</u>. In the model of an abstract Wiener space (or in the framework of a Gaussian probability space) W_h, $h \in H$ is a zero mean normal random variable in $\mathcal{N}(0, \|h\|^2)$, (in some literature it is denoted by δh, $I(h)$ or $, [h](\omega)$). For f being a polynomial of n variables, a functional of the form

$$F(\omega) = f(W_{h_1}(\omega), \cdots \cdots W_{h_n}(\omega)), \ n \in \mathbb{N}, \ h_1, \cdots, h_n \in H$$

is called a <u>polynomial functional</u>. We denote by $\mathcal{P}$ the totality of all polynomial functionals. If f is a tempered C^∞ function, then F is called a <u>smooth functional</u>. The set of all smooth functionals is denoted by S_M. Occasionally we consider E valued functionals where E is a separable Hilbert space. Then the notations will be $\mathcal{P}(E)$ and $S_M(E)$:

$L^p(E)$, $p \in [1, \infty)$ will stand for the Banach space of equivalence classes of E-valued functionals with finite $\|F\|_{L^p}$ norm. We have

$$\mathcal{P}(E) \subset S_M(E) \subset L^p(E), \quad p \in [1, \infty)$$

and $\mathcal{P}(E)$ is dense in $L^p(E)$. For $E = \mathbb{R} : \mathcal{P} \subset S_M \subset L^p$ and $\mathcal{P}$ is dense in L^p. In particular the linear span of the exponential functionals $\{\exp W_h; h \in H\}$ is also dense in L^p This implies the following theorem and the C-M formula.

Theorem 1 (Cameron- Martin) *Consider an abstract Wiener space. Let F be a L^p functional for some $p > 1$. Then*

$$E[F(\omega + h)] = E[F(\omega)e^{(W_h - \frac{1}{2}\|h\|^2)}], \quad \omega \in B, \ h \in H \quad (2)$$

(H is the C-M space).

Proof It will be sufficient to prove the equality for the exponential functionals. Let such a functional be $F(\omega) = e^{W_k(\omega)}, k \in H$. As B^* is dense in H there exists a sequence $\alpha_n \in B^*, (n = 1, 2, \cdots)$ such that $J^*\alpha_n \to k$,where J^* is the dual map $B^* \hookrightarrow H$, As $\alpha_n(\omega + h) = <\alpha_n, \omega> +(J^*\alpha_n, h)_H$ we have in the limit $W_k(\omega + h) = W_k(\omega) + (k, h)_H$, a.s.

The left-hand side of (2) is then $E[e^{W_k(\omega+h)}] = E[e^{W_k(\omega)+(k,h)_H}]$. As $W_k(\omega)$ is a $\mathcal{N}(0, \|k\|_H^2)$ random variable we have l.h.s $= e^{\frac{1}{2}\|k\|_H^2+(k,h)_H}$ (by the series expansion of $e^{W_k(\omega)}$).

But this is equal to the right-hand-side of (2) since $E[e^{W_k}e^{(W_h-\frac{1}{2}h\|_H^2)}] = E[e^{W_{k+h}}e^{-\frac{1}{2}\|h\|_H^2}] = e^{\frac{1}{2}\|k+h\|_H^2}e^{-\frac{1}{2}\|h\|_H^2} = e^{\frac{1}{2}\|k\|_H^2+(k,h)_H}$ (by the linearity of the map $h \to W_h$) showing the theorem for exponential functionals. $\qquad\square$

This theorem removes the difficulty in defining the derivatives of Wiener funcionals in the Fréchet sense since $F(\omega)$, and $F(\omega + h)$ are equivalence classes. Suppose $F_1 = F_2 - \mu$-almost surely. Then

$$\mu\{\omega : \ F_1(\omega + h) \neq F_2(\omega + h)\} = E[I_{\{\omega F_1(\omega+h)\neq F_2(\omega+h)\}}] = E[I_{\{\omega:F_1(\omega)\neq F_2(\omega)\}}e^{W_h-\frac{1}{2}\|h\|^2}]$$

last step folloing from the Cameron-Martin theorem. But the last expectation is zero since $\mu\{I_{\{\omega:F_1(\omega)\neq F_2(\omega)\}} = 0$.

Important Note. The Cameron-Martin theorem is also valid for an irreducible Gaussian probability space $(\Omega, \mathcal{F}, \mu; H)$. (For a proof see [Hu-Yan], Theorem 2.5 using the L^p martingale convergence theorem, c.f. GLOSSARY). Therefore the quasi-invariance of the Wiener measure along the members of the C-M space is also valid for the translations along the members of H in the Gaussian probabilty space. For this reason we will not make any distinction between H of the two models in the following.

Definition 1 For any smooth functional $F \in S_M$, the <u>weak derivative</u> (or Sobolev derivative) D_h in the direction of $h \in H$ is defined by

$$D_h F(\omega) = \lim_{\lambda \to 0} \frac{F(\omega + \lambda h) - F(\omega)}{\lambda}, \quad \omega \in \Omega, h \in H \quad (3)$$

The Gradient operator D is defined by

$$(DF, h)_H = D_h F \, h \in H, \, DF \in S_M(H) \quad (4)$$

Thus D is a linear operator from S_M into $S_M(H)$. Because of the denseness of these spaces in the corresponding L^p spaces, D is also a closable operator from L^p to $L^p(H)$.

For E-valued functionals the counterpart of (3) and (4) will be

$$(DF, h \otimes e)_{H \otimes E} = \lim_{\lambda \to 0} \frac{1}{\lambda}(F(\omega + \lambda h) - F(\omega), e), \, e \in E, \, DF \in S_M(H \otimes E) \quad (5)$$

For f in the form $F = f(W_{h_1}, \cdots, W_{h_n})$ we have
$(DF, h)_H = \frac{d}{d\lambda} F(\omega + \lambda h)|_{\lambda=0}, \, h \in H, \, (*)$
$F(\omega + \lambda h) = f(W_{h_1}(\omega + \lambda h), \cdots, W_{h_n}(\omega + \lambda h)) = f(W_{h_1}(\omega) +$
$\lambda(h_1, h)_H, \cdots, W_{h_n}(\omega) + \lambda(h_n, h)_H)$, (see the proof of the C-M theorem).
Substituting in (*) and for $\lambda = 0$ we have then

$$D_h F = (DF, h)_H = \sum_{j=1}^{n} \partial_j f(W_{h_1}, \cdots, W_{h_n})(h_j, h)_H \quad (6)$$

where ∂_j denotes the partial derivative of f with respect to the j-th variable.

On the other hand if $F = f(\phi_1, \cdots, \phi_n)$, $\phi_j \in S_M$, $(j = 1, \cdots, n)$ and f is a smooth functional on $\mathbb{R}^n$, then the following chain rule is valid

$$DF = \sum_{j=1}^{n} \partial_j f(\phi_1, \cdots, \phi_n) D\phi_j \quad (7)$$

If $F \in S_M(E)$ has the form $F(\omega) = \sum_{k=1}^{n} F_k(\omega) e_k$, $m \in \mathbb{N}$, $e_j \in E$ $(j = 1, \cdots m)$ then

$$DF = \sum_{k=1}^{m} DF_k \otimes e_k, \text{ where } F_k \in \mathcal{P} \text{ (or } S_M) \quad (8)$$

Higher order gradients are defined recursively as

$$D^j F = D(D^{j-1} F) \in S_M(H^{\otimes j}) \text{ (tensor products are symmetrized).} \quad (9)$$

In the case of E-valued functionals

$$DF \in S_M(H \otimes E), \, D^2 F = D(DF) \in S_M(H \otimes H \otimes E), \, D^j F \in H^{\otimes j} \otimes E. \quad (10)$$

As the gradient DF is an H-valued functional, H may be considered as the tangent space.

This means H-valued functionals are vector fields. The adjoint operator δ is then the divergence of the vector fields.

Definition 2 For any smooth vector field $V \in S_M(H)$, its divergence $\delta V \in S_M$ is determined by

$$E[G\delta V] = E[(DG, V)_H], \text{ for all } G \in S_M. \quad (11)$$

More generally for $V \in S_M(H \otimes E)$, its divergence $\delta V \in S_M(E)$ is defined by

$$E[(G, \delta V)_E] = E[(DG, V)_{H \otimes E}] \text{ for all } V \in S_M(H \otimes E). \quad (12)$$

If $V \in S_M(H)$ has the form $\sum_{k=1}^{m} F_k h_k$, $F_k \in S_M$ and $G = \sum_{j=1}^{n} f(W_{h_1}, \cdots, W_{h_n})$, then

$$\delta V = \sum_{k=1}^{m} (F_k W_{h_k} - (DF_k, h_k)). \quad (13)$$

It is not difficult to show that this expression satisfies the defining equation of the divergence given by (11), (c.f. [Hu-Yan], Proposition 2.9).

For $F \in S_M$, the <u>Ornstein-Uhlenbeck operator</u> (O-U operator) $\mathcal{L}$ is defined as

$$\mathcal{L}F = -\delta DF. \quad (14)$$

(δ should not be confused with the Dirac delta function).

By the adjointness of δ and D, $\mathcal{L}$ is essentially self-adjoint whose closure is a self-adjoint operator in $L^2(\Omega, \mathcal{F}, \mu)$ of the Gaussian probability space.

When we have a one-dimensional Gaussian space $(\mathbb{R}, \mathcal{B}(\mathbb{R}), \gamma)$, define two operations independently of Definitions 1 and 2 on the space of real polynomials $\mathcal{P}$ by $D = \frac{d \cdot}{dx}, \delta = -\frac{d}{dx} + x$. $\mathcal{P}$ is dense in $L^2(\mathbb{R}, \gamma)$. By the integration by parts we have for all $\phi, \psi \in \mathcal{P}$

$$(D\phi, \psi)_{L^2(\mathbb{R}, \gamma)} = (\phi, \delta\psi)_{L^2(\mathbb{R}, \gamma)},$$

Hence D, δ are mutually adjoint and can be extended to closed operators in $L^2(\mathbb{R}, \gamma)$. Also

$$\mathcal{L} = -\delta D = \frac{d^2 \cdot}{dx^2} - x\frac{d \cdot}{dx}. \quad (15)$$

<u>Hermite Polynomials</u> We recall that the Hermite polynomials H_n are defined as

$$H_n(x) = (-1)^n e^{\frac{x^2}{2}} \frac{d^n}{dx_n} e^{-\frac{x^2}{2}}, x \in \mathbb{R}, n \in \mathbb{N}_0 \quad (16)$$

In fact they were the coefficients in the expansion of $e^{tx-t^2/2}$ into a Taylor series as funtion of t, i.e.

$$e^{tx-\frac{t^2}{2}} = \sum_{n=0}^{\infty} \frac{t^n}{n!} H_n(x), \quad (17)$$

Equation (16) yields $H_0 = 1$, $H_1 = x$, $H_2 = x^2 - 1$, $H_3 = x^3 - 3x$. We notice that $H_1 = \delta 1$, $H_2 = \delta\delta 1 = x^2 - 1$, $H_3 = \delta^3 1$.

Properties of Hermite polynomials

(1) $H_n = \delta^n 1$

(2) $D H_n = n H_{n-1}$

(3) $H_{n+1}(x) = x H_n(x) - n H_{n-1}(x)$, $n \geq 1$

(4) $\delta D H_n = -\mathcal{L} H_n = n H_n$.

(5) $\int_{\mathbb{R}} H_m(x) H_n(x) d\gamma(x) = n! \delta_{mn}$, (i.e. Hermite polynomials constitute an orthogonal system in $L^2(\mathbb{R}, \gamma)$, $m \neq n \Rightarrow \delta_{mn} = 0$).

Proof (1) Using induction accept $H_n = \delta^n 1$. By direct calculation on (15) we find $\delta H_n(x) = -\dfrac{d H_n(x)}{dx} + x H_n(x) = H_{n+1}(x)$. Thus by the inductive hypothesis $H_{n+1} = \delta H_n = \delta\delta^n 1 = \delta^{n+1} 1$.

(2) Follows from the direct differentiation of both sides of (16).

(3) The right-hand side is equal to δH_n by (2). Then the result follows from (1).

(4) By (2) and (1) $\delta D H_n = \delta n H_{n-1} = n\delta\delta^{n-1} 1 = n\delta^n 1 = n H_n$.

(5) It follows from (16) that

$$\sum_{m,n=0}^{\infty} \frac{s^m t^n}{m!n!} \int_{\mathbb{R}} H_m(x) H_n(x) d\gamma(x) = \int_{\mathbb{R}} \exp\{(s+t)x - \frac{s^2+t^2}{2}\} d\gamma(x) = \exp\{-\frac{s^2+t^2}{2} + \frac{(s+t)^2}{2}\} = e^{st}.$$

Comparison of the coefficients of $s^m t^n$ yields (5).

Notes. (i) The reason why $-\mathcal{L} = \delta D$ is called a number operator is apparent from (4).

(ii) (5) shows that Hermite polynomials constitute an orthogonal system in $L^2(\mathbb{R}, \mathcal{F}, \gamma)$.

5.4 Wiener Chaos Decomposition, OU Semigroup

Firstly some notation: For any sequence of non-negative integers $\alpha = \{\alpha_j\}_{j \in \mathbb{N}}$ we denote by $|\alpha| = \sum_j \alpha_j \alpha! = \prod_j (\alpha_j!)$ and by Λ, the set of sequences in which there are only a finite number of non-zero elements. For $\alpha \in \Lambda$ and $x = \{x_j\} \in \mathbb{R}^{\infty}$ define

$$H_\alpha(x) = \prod_j H_{\alpha_j}(x_j). \quad (1)$$

This is a finite product. If $(\mathbb{R}^\infty, \mathcal{B}^\infty, \gamma^\infty)$ is the infinite product space of $(\mathbb{R}, \mathcal{B}(\mathbb{R}), \gamma)$. The counterpart of the property (5) of Hermite polynomials given in the previous section is

$$\int_{\mathbb{R}^\infty} H_\alpha(x)\, H_\beta(x)\, d\gamma^\infty(x) = \alpha!\, \delta_{\alpha\beta},\, \alpha, \beta \in \Lambda. \quad (2)$$

This follows from the property (5) of Hermite polynomials due to the independence of components. For the infinite product space define the O-U operator as

$$\mathcal{L} = -\sum_{j=1}^{\infty} \delta_j D_j \quad (3)$$

where D_j denotes the differentiation with respect to the j-th component and δ_j is its adjoint operator.

H_α are eigen functions of $\mathcal{L}$ and the counterpart of property (4) of the previous section will be

$$\mathcal{L}\, H_\alpha = -|\alpha|\, H_\alpha, \alpha \in \Lambda. \quad (4)$$

Let $\Lambda_n = \{\alpha \in \Lambda : |\alpha| = n\}$ and let $\mathcal{H}_n$ be the closed subspace in $L^2(\mathbb{R}^\infty, \mathcal{B}^\infty, \gamma^\infty)$ generated by $\{H_\alpha : \alpha \in \Lambda_n\}$. Then we have the following orthogonal decomposition

$$L^2(\mathbb{R}^\infty, \mathcal{B}^\infty, \gamma^\infty) = \bigoplus_{n=0}^{\infty} \mathcal{H}_n \quad (5)$$

<u>Numerical model</u> Let $(\Omega, \mathcal{F}, \mu; H)$ be a Gaussian probability space and $\{h_j\}$ be an orthonormal basis in H. Then for $\omega \in \Omega$, let $T(\omega) = \{W_{h_j}(\omega)\}_{j \in \mathbb{N}}$. Then $T : \Omega \to \mathbb{R}^\infty$ is $\mathcal{F} \leftrightarrow \mathcal{B}^\infty$ measurable and measure preserving, i.e. $\gamma^\infty = \mu \circ T^{-1}$. For $p \in [1, \infty]$ and $\phi \in L^p(\mathbb{R}^\infty, \mathcal{B}^\infty, \gamma^\infty)$ the dual map of T yields an isomorphism $T^* : L^p(\mathbb{R}^\infty, \mathcal{B}^\infty, \gamma^\infty) \to L^p(\Omega, \mathcal{F}, \mu)$. This is also a homomorphism when restricted to polynomial functionals $\mathcal{P}$.

The Gaussian probabilty space $(\mathbb{R}^\infty, \mathcal{B}^\infty, \gamma^\infty; l^2)$ is regarded as a numerical model of $(\Omega, \mathcal{F}, \mu; H)$ and of its special cases like abstract Wiener space.

The numerical model is very helpful in the proofs of many important results.

Theorem 1 *Wiener Chaos Decomposition, also called the wiener-Ito chaos decomposition. Let $(\Omega, \mathcal{F}, \mu; H)$ be a Gaussian probability space as above. Define*

$$H_\alpha(\omega) = \prod_j H_{\alpha_j}(W_{h_j})(\omega)\, \alpha \in \Lambda, \quad (6)$$

so that $\{(\alpha!)^{-1/2} H_\alpha; \alpha \in \Lambda\}$ constitute a basis of $L^2(\Omega, \mathcal{F}, \mu)$ and

$$L^2(\Omega, \mathcal{F}, \mu) = \bigoplus_{n=0}^{\infty} \mathcal{H}_n \cong \Gamma(H) \quad (7)$$

where $\Gamma(H)$ is the symmetric Fock space over H.
Denote by J_n the orthogonal projection onto subspace $\mathcal{H}_n$.
Outline of the proof. Consider a numerical model $(\mathbb{R}^\infty, \mathcal{B}^\infty, \gamma^\infty; l^2)$.
Since , $L^2(\mathbb{R}^\infty, \mathcal{B}^\infty, \gamma^\infty)$ is isomorphic to $L^2(\Omega, \mathcal{F}, \mu)$ the decomposition (7) follows from (5). (For the symmetric Fock space see the GLOSSARY).

For the space of square integrable E valued functionals the chaos decomposition is still valid:

$$L^2(\Omega, \mathcal{F}, \mu;\ E) \cong L^2(\Omega, \mathcal{F}, \mu) \otimes E = \left(\oplus_{n=0}^{\infty}\right) \otimes E.$$

Consider again the O-U operator $\mathcal{L}F = -\delta DF$, $F \in S_M$ given by (Sect. 5.3 (14)). From Sect. 5.3 (6) we have

$$DF = \sum_{j=1}^{n} \partial_j f(W_{h_1}, \cdots, W_{h_n}) h_j.$$

On the other hand in the expression of δ, (Sect. 5.3, (13)) take $V = \sum_{j=1}^{n} F_j h_j$, where $F_j = \partial_j f(W_{h_1}, \cdots, W_{h_n}) \in S_M$, substituting these for $\mathcal{L}$ we obtain

$$\mathcal{L}F = \sum_{j,k=1}^{n} \partial_k \partial_j f(W_{h_1}, \cdots, W_{h_n})\,(h_k, h_j) - \sum_{j=1}^{n} \partial_j f(W_{h_1}, \cdots, W_{h_n}) W_{h_j}. \quad (8)$$

Let $F \in L^2(\Omega, \mathcal{F}, \mu)$. Then using (7), it can be expressed as the direct sum $F = \sum_{n=0}^{\infty} J_n F$. We note that $J_n F \in \mathcal{H}_n$ and $\mathcal{H}_n$ is generated by $\{H_\alpha : \alpha \in \Lambda, |\alpha| = n\}$. Since for such H_α, $\mathcal{L}H_\alpha = -n H_\alpha$ by (5), we have the following alternative form of the O-U operator

$$\mathcal{L} = -\sum_{n=0}^{\infty} n J_n. \quad (9)$$

with the domain $\mathcal{D}(\mathcal{L}) = \{F \in L^2 : \sum_n n^2 \|J_n F\|^2 < \infty\}$. ($-\mathcal{L} = \sum_n n J_n$ is the number operator).

Definition 1 The   semigroup $T_t = \sum_{n=0}^{\infty} e^{-nt} J_n$ is   called   the Ornstein-Uhlenbeck (O-U) semigroup.

Definition 2 In the framework of a Gaussian probability space

$$\mathcal{E}(h) = e^{W_h - \|h\|_H^2} \quad (10)$$

is the exponential functional. In the numerical model $(\mathbb{R}^\infty, \mathcal{B}^\infty, \gamma^\infty; l^2)$ we define

$$\mathcal{E}_n(h) = \exp\left\{\sum_{j=1}^n h_j x_j - \frac{1}{2}\sum_{j=1}^n h_j^2\right\}, h = \{h_j\} \in l^2, x = \{x_j\} \in \mathbb{R}^\infty \quad (11)$$

It is shown by the L^p martingale convergence theorem that $\mathcal{E}_n(h) \xrightarrow{L^p} \mathcal{E}(h)$, (c.f. [Hu-Yan] Theorem 2.5).

Furthermore in terms of Hermite polynomials Sect. 5.3, (17) and Sect. 5.4, (1)

$$\mathcal{E}_n(h)(x) = \prod_{j=1}^n \sum_{k=0}^\infty \frac{h_j^k}{k!} H_k(x_j) = \sum_{\alpha \in \mathbb{N}_0^n} \prod_{j=1}^n \frac{h_j^{\alpha_j}}{\alpha_j!} H_{\alpha_j}(x_j) = \sum_{\alpha \in \Lambda_n} \frac{h^\alpha}{\alpha!} H_\alpha(x) \quad (12)$$

Thus

$$\mathcal{E}(h)(x) \overset{L^p}{=} \lim_{n\to\infty} \mathcal{E}_n(h)(x) = \sum_{\alpha \in \Lambda} \frac{h^\alpha}{\alpha!} H_\alpha(x) \quad (13)$$

where $h^\alpha = \prod_j h_j^{\alpha_j}$, $\alpha = \{\alpha_j\} \in \Lambda$. Then its projection onto $\mathcal{H}_n$ in the chaos decomposition is

$$J_n \mathcal{E}(h) = \sum_{\alpha \in \Lambda_n} \frac{h^\alpha}{\alpha!} H_\alpha, n \in \mathbb{N}_0 \quad (14)$$

Theorem 2 *In any numerical model* $(\mathbb{R}^\infty, \mathcal{B}^\infty, \gamma^\infty; l^2)$, *for* $t > 0$ *and* $F \in L^2$, *the action of the O-U semigroup is given by*

$$(T_t F)(x) = \sum_{n=0}^\infty e^{-nt} J_n F(x) = \int_{\mathbb{R}^\infty} F(e^{-t}x + \sqrt{1 - e^{-2t}}\, y)\, d\gamma^\infty(y) \quad (15)$$

Proof As the linear span of exponential functionals is dense in L^2, it is sufficient to show (15) for $F = \mathcal{E}(h)$, $h \in H$. For this purpose we start by $F = \mathcal{E}_n(h)$. Calling $X = e^{-tx} + \sqrt{1 - e^{-2t}}\, y$,

$$\mathcal{E}_n(h)(X) = \exp\{\sum_{j=1}^n h_j(e^{-tx_j} + \sqrt{1 - e^{-2t}}y) - \frac{1}{2}\sum_{j=1}^n h_j^2\}.$$

Integrating $\mathcal{E}_n(h)(X)$ with respect to γ^n we use the identity $\int_{\mathbb{R}} e^{cy} \frac{1}{\sqrt{2\pi}} e^{-y^2/2}\, d\gamma(y) = e^{c^2/2}$ to find

$$\int_{\mathbb{R}^n} \mathcal{E}_n(h)(x)\, d\gamma^n(y) = \exp\{\sum_1^n h_j e^{-t} x_j - 1/2 \sum_1^n h_j^2\} \int_{\mathbb{R}^n} e^{\sum_1^n h_j \sqrt{1-e^{-2t}}\, y}\, d\gamma^n(y)$$

$$= \exp\{\sum_1^n e^{-t} h_j x_j - 1/2 \sum_1^n h_j^2\} \prod_{j=1}^n e^{\frac{1}{2} h_j^2 (1-e^{-2t})} = \exp\{\sum_1^n e^{-t} h_j x_j - \frac{1}{2} \sum_1^n (e^{-t} h_j)^2 = \mathcal{E}_n(e^{-t} h)(x)$$

when n tends to infinity the right hand side of (15) becomes

$$\mathcal{E}(e^{-t}h)(x) = \sum_{\alpha \in \Lambda} \frac{(e^{-t}h)^\alpha}{\alpha!} H_\alpha(x) = \sum_{n=0}^\infty \sum_{\alpha \in \Lambda_n} \frac{(e^{-t}h)^\alpha}{\alpha!} = \sum_{n=0}^\infty e^{-nt} \sum_{\alpha \in \Lambda_n} \frac{h^\alpha}{\alpha!} H_\alpha(x) = \sum_{n=0}^\infty e^{-nt} J_n \mathcal{E}(h)(x)$$

$$= T_t \mathcal{E}(h)(x) \text{ verifying (15) for exponential functionals.} \qquad \square$$

For finite dimensional Gaussian spaces O-U semigroup (15) has an interesting interpretation in terms of stochastic differential equations. A special form of Ito linear stochastic differential equation is

$$dX_t = cX_t dt + \sigma dW_t, \ X = X_0 \text{for} t = 0; \ c, \sigma \text{ are real constants} \qquad (16)$$

(Here X_t is an $\mathbb{R}^n$-valued stochastic process and W_t is an n-dimensioanal standard Gaussian process). Equation (16) is also called a Langevin equation. Its solution is the stochastic process

$$X_t = e^{ct} X_0 + \sigma \int_0^t e^{-(s-t)} dW_s \qquad (17)$$

where the initial condition X_0 is a constant. Equation (17) is also called an Ornstein-Uhlenbeck (O-U) process which is also a Gaussian process. Then if $X_0 = x \in \mathbb{R}^n$, $E(X_t) = e^{ct} x$, $V(X_t) = \frac{\sigma^2}{2c}(e^{2ct} - 1)$, (c.f. [Miko]).

Theorem 3 *If the process X_t is given by (17), then for $L^2(\mu)$ where $\mu = \gamma^n$, (the standard Gaussian measure on $\mathbb{R}^n$,) the relation*

$$T_t \phi(x) = E^x(\phi(X_t))$$

defines a semigroup of continuous linear operators on $Ł^2$.
(X^x denotes the expectation when the process starts at $X_0 = x$.

Proof Utilizing the strong Markov property of the Brownian motion (c.f. [Weiz-Wink] Corollary 9.1.11 also the GLOSSARY), for $s, t \geq 0$ which states that

$$E^x(\phi(X_{s+t})| \mathcal{F}_s) = E^{X_s}(\phi(X_t))$$

where $\phi \in C_b(\mathbb{R}^n)$, $\mathcal{F}_s$ is the σ-field generated by $\{X_t :, t \leq s\}$, and the left-hand side is the conditional expectation given the events in $\mathcal{F}_s$ (events up to time s. Calling $h(x) = T_t \phi(x) = E^x(\phi(X_t))$, we have

$$T_{s+t}\phi(x) = E^x(\phi(X_{s+t})) = E^x(E^s \phi(X_t)) = E^x(h(X_s)) = T_s(h(x)) = T_s T_t \phi(x)$$

showing the semigroup property. The result follows then from the denseness of $C_b(\mathbb{R}^n) \in L^2(\mathbb{R}^n, \mu)$. $\qquad\qquad\square$

For $c = -1$, $\sigma = \sqrt{2}$, the solution of the Langevin differential equation $dX_t = -X_t + \sqrt{2}dW_t$ with X_0 will be

$$X_t = e^{-t}x + \sqrt{2}\int_0^t e^{-(s-t)}dW_s; \quad EX_t = e^{-t}x, \quad V(X_t) = 1 - e^{-2t}.$$

Hence $Y = \dfrac{X_t - e^{-t}x}{\sqrt{1 - e^2 t}}$ is governed by the n-dimensional standard Gaussian distribution. Thus $X_t = e^{-t}x + Y\sqrt{1 - e^{-2t}}$ and Theorem 3 yields

$$(T_t\phi)(x) = E^x(\phi(X_t)) = \int_{\mathbb{R}^n} \phi(e^{-t}x + \sqrt{1 - e^{-2t}}\, y)\, d\gamma^n(y) \quad (18)$$

is the transition semigroup generated by the O-U process.

More properties of the O-U semigroup:

(I) The unique infinitesimal generator of the O-U semigroup is the operator $\mathcal{L}$. To verify this, considering the topology of L^2

$$E\{|t^{-1}(T_t F - F) - \mathcal{L}F|^2\} \overset{(10)}{=} E\{|\frac{\sum_0^\infty e^{-nt}J_n F}{t} + \sum_0^\infty nJ_n F|^2\} = \sum_0^\infty \{(\frac{e^{-nt} - 1}{t} + n)^2 E(|J_n F|^2)\}.$$

As $\frac{e^{-nt}-1}{t} + n \to 0$ as $t \downarrow 0$ the expression goes to zero. Thus the L^2 limit of $t^{-1}(T_t F - \mathcal{L}F)$ vanishes.

Conversely if $t^{-1}(T_t F - F) \overset{L^2}{\to} G$, then for every n, $J_n G = \lim_{t\downarrow 0} t^{-1}(T_t J_n F - J_n F) = -nJ_n F$. Thus by (9) $G = \sum_n J_n G = -\sum_n nJ_n F = \mathcal{L}F$.

(II) T_t is positive, i.e. $F \geq 0 \Rightarrow T_t F \geq 0$ by (15).

(III) Contractivity: For the n dimensional Gaussian space consider the transformation $\mathbb{R}^n \times \mathbb{R}^n \to \mathbb{R}^n$ given by $Q_t : (x, y) \to e^{-t} + \sqrt{1 - e^{-2t}}\, y$. By the rotational invariance of $\mu = \gamma^n = (\mu \times \mu) \circ Q_t^{-1}$. For any $p \geq 1$ and $\phi \in L^p$, $(\|T_t\phi\|_p)^p = \int_{\mathbb{R}^n} |\int_{\mathbb{R}^n} \phi(e^{-t}x + \sqrt{1 - e^{-2t}}\, y\, d\mu(y)|^p d\mu(x) \leq \int_{\mathbb{R}^n \times \mathbb{R}^n} |\phi(e^{-t}x + \sqrt{1 - e^{-2t}}\, y|^p\, d\mu(y)d\mu(x) = \int_{\mathbb{R}^n} |\phi(x)|^p\, d\mu(x)$. Thus $\|T_t\phi\|_p \leq \|\phi\|_p$.

(IV) A more general contractivity property (hyper-contractivity) is stated as follows: (Nelson) For $p > 1, t > 0$ let $q(t) = e^{2t}(p - 1) + 1 > p$. Then for all $\phi \in L^p(\mathbb{R}^n)$ we have $\|T_t(\phi)\|_{q(t)} \leq \|\phi\|_p$, [Nel], for proof see [Wat], Theorem 8.6.

By a similar invariance property of the product measure in (15), the contractivity of the O-U semigroups is established, (c.f. [Üst-Zak] B.5).

5.5 Sobolev Norms and Sobolev Spaces of Functionals

For a real sequence $\beta = \{\beta_n\}_{n=0}^{\infty}$ we can define an operator $T_\beta = \sum_{n=0}^{\infty} \beta_n J_n$, T_β can be extended to a self-adjoint operator in L^2, (denoted also by T_β) with the domain $\mathcal{D}(T_\beta) = \{F \in L^2 : \sum_{n=0}^{\infty} \beta_n^2 \|J_n F\|^2 < \infty\}$ and

$$T_\beta F = \sum_{n=0}^{\infty} \beta_n J_n F; \quad F \in \mathcal{D}(T_\beta). \quad (1)$$

(Recall that J_n is the orthogonal projection onto $\mathcal{H}_n$ (Sect. 5.4)). We note that for $\beta_n = n$ we have the number operator and for $\beta_n = e^{-nt}$, $t \geq 0$ we retrieve the semigroup of O-U operators.

If $\beta_n = (1+n)^{s/2}$, $s \in \mathbb{R}$ we can define the following operator by the Wiener chaos decomposition and for E valued polynomial functionals

$$(I - \mathcal{L})^{s/2} F = \sum_{n=0}^{\infty} (1+n)^{s/2} J_n F, \quad F \in \mathcal{P}(E).$$

Using the denseness of polynomias functionals in $L^p(E)$ we define the following norm

$$\|F\|_{s,p} = \|(I - \mathcal{L}^{s/2} F\|_{L^p(\mu;E)}, \quad s \in \mathbb{R}, \ p \in (1, \infty), \ F \in S_M(E). \quad (2)$$

Also following norms are defined

Definition 1 For $k \in \mathbb{N}$, $[1, \infty)$, $F \in S_M(E)$ denote

$$\|F\|_{k,p} = \sum_{j=0}^{k} \|D^j F\|_{L^p(\mu;H^{\otimes j}\otimes E)} = \|F\|_{L^p(\mu;E)} + \sum_{j=1}^{k} \|D^j F\|_{L^p(\mu;H^{\otimes j}\otimes E)}, \quad (3)$$

(For $D^j F$ see Sect. 5.3, (9)).
Consider also

$$\|F\|_{k,p}^{-} = \left(\sum_{j=0}^{k} \|D^j F\|_{L^p(\mu;H^{\otimes j}\otimes E)}^p \right)^{1/p}.$$

(These are related to Sobolev norms given in Sect. 2.3). Using the inequalities $a^p + b^p \leq (a+b)^p \leq 2^p(a^p + b^p)$, $a \geq 0, b \geq 0$, we see that the norms $\|\cdot\|_{k,p}$ and $\|\cdot\|_{k,p}^-$ are equivalent, i.e. they create the same topology.

The norms $\|\cdot\|_{k,p}$ have the following properties:

(I) Monotonicity: If $1 \leq p \leq q < \infty, k \leq l$, then $\|\cdot|_{k,p} \leq \|\cdot\|_{l,q}$ By (4) this is obvious.

(II) Consistence: For $E = \mathbb{R}$, $p, q \in [1, \infty)$, $k, l \in \mathbb{N}$, let $\{F_n\} \subset S_M$ be a Cauchy sequence with respect to both norms $\|\cdot\|_{k,p}$ and $\|\cdot\|_{l,q}$. Then

$$\lim_{n\to\infty} \|F_n\|_{k,p} = 0 \Rightarrow \lim_{\to\infty} \|F_n\|_{l,q} = 0.$$

To show this result we note that for $j \in \mathbb{N}$, $j \leq l$, $\{D^j F_n\}$ is a Cauchy sequence in $L^q(H^{\otimes j})$. By the completeness of $L^q(H^{\otimes j})$, this sequence has a limit G_j. Obviously $G_0 = 0$. By induction it is not difficult to show that $G_j = 0$.

The following important inequalities are due to P.A. Meyer.

Theorem 1 (Meyer Inequalities) *For any $p \in (1, \infty)$, $k \in \mathbb{N}$ and $F \in S_M(E)$ there are constants $c_{k,p}$, $C_{k,p}$ such that*

$$c_{k,p}\|(I - \mathcal{L})^{k/2}F\|_{L^p(\mu;E)} \leq \sum_{j=1}^{k} \|D^j F\|_{L^p(\mu;H^{\otimes j}\otimes E)} \leq C_{k,p}\|(I - \mathcal{L}^{k/2}F\|_{L^p(\mu E)}. \quad (4)$$

(For the lengthy proof see [Mey-1, Sug, Wat-1]).

<u>Note.</u> These inequalities also show that when $s = k \in \mathbb{N}$ the norms $\|F\|_{s,p}$ and $\||F_{s,p}\|| = \|(I - \mathcal{L}^{s/2}F\|_{L^p(\mu E)}$ are equivalent.

Referring to the operator $T_\beta = \sum_{n=0}^{\infty} \beta_n J_n$, in (1), it may be asked for which real sequences $\{\beta_n\}$, T_β can be extended to a bounded linear operator in L^p., The following important theorem provides sufficient condition for such an extension.

Theorem 2 (Meyer's L^p Multiplier Theorem) *If there exist $n_0 \in \mathbb{N}$ and $\Theta > 0$ such that $\beta_n = \sum_{k=0}^{\infty} a_k (n^{-\Theta})^k, n \geq n_0$, where $\{a_k\}$ is a sequence of real numbers satisfying $\sum_{k=0}^{\infty} |a_k|(n_0^{-\Theta})^k < \infty$, then for all $p \in (1, \infty)$, T_β extends uniquely to a bounded linear operator in $Ł^p$.*

(The extensive proof is given in [Shi].)

<u>Application.</u> The condition of the theorem means that there exist a function ϕ which is analytic in some neighborhood of the origin and $\beta > 0$ such that $\beta_n = \phi(n^{-\Theta})$. If ϕ is defined as $\phi(x) = (\frac{x}{1+x})^{-s/2}$, $s \geq 0$ then it is analytic in the neighborhood of 0 and $(1+n)^{s/2} = \phi(n^{-1})$, $(\Theta = 1)$. Therefore for $s \leq 0$ $(I - \mathcal{L})^{s/2} = \sum_n(1 + n)^{s/2} J_n$ can be extended to a bounded linear operator in $L^p(1 < p < \infty)$. By the duality given in the next section the result is true for all $s \in \mathbb{R}$.

5.6 Meyer-Watanabe Generalized Functions

Let $\mathbb{D}_s^p(E)$ denote the Banach space obtained by the completion of $\mathcal{P}(E)$, (or of $S_M(E)$) with respect to the norm $\|\|F\|\|_{s,p} = \|(I - \mathcal{L})^{s/2}F\|_{L^p(\mu;E)}$; $p \in (1,\infty)$, $s \in \mathbb{R}$. By the note following Theorem 1 when $s = k \in \mathbb{N}$, the norms $\|\|F\|\|_{s,p}$ and $\|F\|_{s,p}$ are equivalent, thus $\mathbb{D}_k^p$ is also obtained by the completion of $\mathcal{P}(E)$ with resoect to the norm $\|F\|_{k,p\cdot}$,

The norms $\|\|.\|\|_{s,p}$ also satisfy monotonicity, i.e. for $s, s' \in \mathbb{R}$ $s \le s'$ and $1 < p \le p' < \infty$ we have $\|\|F\|\|_{s,p} \le \|\|F\|\|_{s',p'}$, $F \in \mathcal{P}(E)$. To see this we first show that for $s > 0$

$$(I - \mathcal{L}^{-s/2}) = \frac{1}{\Gamma(s/2)} \int_0^\infty e^{-t} t^{s/2-1} T_t \, dt, \ T_t = \sum_n e^{-nt} J_n \quad (\dagger)$$

When projected onto $\mathcal{H}_n$ of the Wiener chaos decomposition this gives

$$(1+n)^{-s/2} = \frac{1}{\Gamma(s/2)} \int_0^\infty e^{-t} t^{s/2-1} e^{-nt} \, dt \text{ or}$$

$$\Gamma(s/2) = (1+n)^{s/2} \int_0^\infty t^{s/2-1} e^{-(n+1)t} dt$$

which is an identity in view of the one of the equivalent definitions of the gamma function $\Gamma(x) = c^x \int_0^\infty t^{x-1} e^{-ct} \, dt$, (take $c = 1 + n$, $x = s/2$) This verifies $(\dagger)$ by the uniqueness of the direct sum decomposition.

Since T_t is a contraction semigroup on $L^p(E)$ $p \in (1,\infty)$, $(I - \mathcal{L})^{-s/2}$ is also a contraction operator on $L^p(E)$ by $(\dagger)$. Thus for $s > 0$

$$\|(I - \mathcal{L})^{-s/2})F\|_{L^p(E)} \le \|F\|_{L^p(E)}. \quad (1)$$

This suffices to show the monotonicity of $\|\|.\|\|_{s,p}$. (start by $F \longleftrightarrow (I - \mathcal{L})^{-s'/2}F$, $s' \ge s$). By this monotonicity we have

$$\mathbb{D}_{s'}^{p'}(E) \subset \mathbb{D}_s^p(E) \, ; \, 1 < p \le p' < \infty, s \le s'. \quad (2)$$

If T belongs to the dual space $(\mathbb{D}_s^p(E))'$, then for $F \in \mathbb{D}_s^p(E)$, $< T, F >=< (I - \mathcal{L})^{-s/2} T, (I - \mathcal{L})^{s/2} F >$ where $< , >$ is the duality bracket. By the definition $(I - \mathcal{L})^{s/2} F$ belongs to $L^p(\mu; E)$. Hence $(I - \mathcal{L})^{-s/2}T$ belongs to $L^q(\mu; E)$ where $q^{-1} = 1 - p^{-1}$. Consequently T belongs to $(\mathbb{D}_s^p(E))'$ iff $(I - \mathcal{L})^{-s/2} T$ belongs to $L^q(\mu; E)$, i.e. iff T belongs to $\mathbb{D}_{-s}^q(E)$. Thus $\|\|.\|\|_{-s,q}$ is the dual norm of $\|\|.\|\|_{s,p}$ and

$$(\mathbb{D}_s^p(E))' = \mathbb{D}_{-s}^q \quad (3)$$

Using (2) and (3) we have for $1 < p \le q < \infty, 0 \le r \le s < \infty$

$$\mathbb{D}_s^p(E) \subset \mathbb{D}_r^p(E) \subset L^p(\mu; E) \subset \mathbb{D}_{-r}^p(E) \subset \mathbb{D}_{-s}^p(E). \quad (4)$$

<u>Note.</u> By (1) for $r > 0$, $\|\|F\|\|_{-r,p} \le \|F\|_{L^p(\mu;E)}$ implying $L^p(\mu; E) \subset \mathbb{D}_{-r}^p(E)$, $(\mathbb{D}_0^p \equiv L^p(\mu; E))$.
Also

$$\mathbb{D}_s^q(E) \subset \mathbb{D}_r^q(E), \ \mathbb{D}_{-r}^q(E) \subset \mathbb{D}_{-s}^q(E), \ \text{and} \ L^q(\mu; E) \subset L^p(\mu; E). \quad (5)$$

We define

$$\mathbb{D}^\infty(E) = \bigcap_{s>0} \bigcap_{1<p<\infty} \mathbb{D}_s^p(E) \quad (6)$$

equipped with the projective limit topology and

$$\mathbb{D}^{-\infty}(E) = \bigcup_{s>0} \bigcup_{1<p<\infty} \mathbb{D}_{-s}^p(E) \quad (7)$$

equipped with the inductive limit topology.

Then $\mathbb{D}^\infty(E)$ is a complete countably normed space (a Fréchet space) and $\mathbb{D}^{-\infty}(E)$ is its continuous dual.

We have a clear analogy with the Schwartz distribution theory and it is natural to call the elements of $\mathbb{D}^{-\infty}(E)$ the generalized Wiener functionals or <u>Meyer-Watanabe distributions.</u>

<u>Remarks:</u> (1) For $E = \mathbb{R}$ the corresponding symbols are $\mathbb{D}^\infty$ and $\mathbb{D}^{-\infty}$.
2. By the monotonicity of $\|\| \cdot \|\|_{s,p}$, $\mathbb{D}^\infty(E)$ can also be given by $\mathbb{D}^\infty = \bigcap_{k\in\mathbb{N}} \bigcap_{1<p<\infty} \mathbb{D}_k^p(E)$.

Theorem 1 ([Mey-2, Sug, Kre-Kre]) *The operator $D : S_M(E) \to S_M(H \otimes E)$ can be extended uniquely to an operator $D : \mathbb{D}^{-\infty}(E) \to \mathbb{D}^{-\infty}(H \otimes E)$, so that its restriction $D : \mathbb{D}_{s+1}^p(E) \to \mathbb{D}_s^p(H \otimes E)$ is continuous for every $p \in (1, \infty)$ and $s \in \mathbb{R}$. (long proof depending on an auxiliary lemma and proposition can be found in the above references).*
    In particular $D : \mathbb{D}^\infty(E) \to \mathbb{D}^\infty(H \otimes E)$ is continuous.

<u>Note</u> As in the Schwartz theory of distributions $\mathbb{D}^\infty(E) \subset \mathbb{D}^{-\infty}(E)$ which is easily seen from (5), (6) and (7). The duality between the operators D and δ implies immediately the following Corollary:

Corollary 1 *The operator $\delta : S_M(H \otimes E) \to S_M(E)$ can be extended uniquely to an operator $\delta : \mathbb{D}^{-\infty}(H \otimes E) \to \mathbb{D}^{-\infty}(E)$ so that its restriction $\delta : \mathbb{D}_{s+1}^p(H \otimes E) \to \mathbb{D}_s^p(E)$ is continuous for all $p \in (1, \infty)$ and $s \in \mathbb{R}$.*
    (In particular $\delta : \mathbb{D}^\infty(H \otimes E) \to \mathbb{D}^\infty(E)$ is continuous).
    In these extended senses $\mathcal{L} = -\delta D$ is still valid and we have

Corollary 2 *The operator $\mathcal{L}$ uniquely extends to an operator from $\mathbb{D}^{-\infty}$ into $\mathbb{D}^{-\infty}$ such that $\mathcal{L} : \mathbb{D}^p_{s+2}(E) \to \mathbb{D}^p_s(E)$ is continuous for all $p \in (1, \infty)$ and $s \in \mathbb{R}$.*
(In particular $\mathcal{L} : \mathbb{D}^\infty \to \mathbb{D}^\infty$ is continuous).
These results imply, in particular, that the operators D, δ and $\mathcal{L}$ on the smooth functionals can be extended to Sobolev spaces $\mathbb{D}^p_s$.

Differential calculus involving the tensor products of these operators on $\mathcal{P}(E)$ or $S_M(E)$ follows basically the rules of finite dimensional calculus. For instance if $F, G \in \mathcal{P}(E)$ we have $D(F \otimes G) = (DF) \otimes G + F \otimes (DG)$ and more generally

$$D^l(F \otimes G) = \sum_{l=1}^{j} \binom{j}{l} (D^l F) \otimes (D^{j-l} G). \quad (8)$$

Theorem 2 *Let E_1, E_2 be two separable Hilbert spaces. If $p, q \in (1, \infty), k \in \mathbb{N}$ and $p^{-1} + q^{-1} = r^{-1} < 1$, then*

(i) There exists a constant $C = C(p, q, k) > 0$ such that for $F \in \mathcal{P}(E)$, $G \in \mathcal{P}(E) \||F \otimes G\||_{k,r} \le C \||F\||_{k,p} \||G\||_{k,q}$,
(ii) If $E_1 = E_2 = \mathbb{R}$ and $F, G \in \mathbb{D}^\infty$, then $FG \in \mathbb{D}^\infty$, thus $\mathbb{D}^\infty$ is a topological algebra.

(We use short-hand notation for the L^p norms, e.g. $\||F \otimes G\||_{k,r} \equiv \|(I - \mathcal{L}^{k/2})F \otimes G\|_{L^r(\mu; E_1 \otimes E_2)}$ and $\|D^j(F \otimes G)\|_r \equiv \|D^j(F \otimes G)\|_{L^r(H^{\otimes j} \otimes E_1 \otimes E_2)}$.)

Proof (i) By Meyer's inequalities (Sect. 5.5 (4)) and (8)

$$\||F \otimes G\||_{k,r} \preceq \sum_{j=0}^{k} \|D^j(F \otimes G)\|_r \preceq \left(\sum_{j=1}^{k} \|D^j F\|_p \right) \left(\sum_{j=1}^{k} \|D^j G\|_q \right) \preceq \||F_{k,p}\|| \||G\||_{k,q}$$

showing (i). (ii) For $E_1 = E_2 = \mathbb{R}$ the tensor product is replaced by the ordinary product. If $F, G \in \mathbb{D}^\infty$, then by part (i) and by Remark 2 preceding Theorem 1 for all $r \in (1, \infty)$ and $k \in \mathbb{N}$ we have $FG \in \mathbb{D}^r_k$, hence $FG \in \mathbb{D}^\infty$.

5.7   An Introduction to White Noise Analysis

So far we have considered white noise in different contexts. Firstly in Sect. 4.2 we have defined the white noise measure in relation to random tempered distributions. In Sect. 4.4 we have considered the solution of some nonlinear, white noise driven stochastic differential equations. Finally in the Subsection 5.2.2, as an application of Minlos' theorem, we have constructed the white noise measure on the space $S'(\mathbb{R})$ of tempered distributions.

What is white noise? For time series analysts and statisticians it is a stochastic process Z_t which is the continuous analogue of a sequence of independent, identically distributed (iid) or uncorrelated random variables. $\{Z_t'\}s$ have zero means and infinite fluctiations (variances). For $s \neq t$, Z_s and Z_t are independent or uncorrelated. Such a process can not be conceived as a genuine, conventional stochastic process.

For physicists and probabilists a white noise is the formal derivative $\dot{W}_t$ of a Brownian motion process W_t. But it is well-known that Brownian paths are everywhere continuous but nowhere differentiable. To circumvent this difficulty we should view a stochastic process not as a random function, but rather as a random generalized function so that the paths will be differentiable of any order in the distribution sense. This is exactly what we are going to do by a slight modification of the Subsection 5.2.3 approach to a white noise space.

In parallel to the role played by the Gaussian probability space and O-U operators in the construction of the infinite dimensional Meyer-Watanabe distributions, white noise space and the second quantization operators are essential in the definition of the Hida distributions.

Consider once more the Fréchet space Ω_0 introduced in Sect. 5.1, of all continuous functions on $\mathbb{R}_+$, null at zero, the Wiener measure ν on $(\Omega_0, \mathcal{B}(\Omega_0))$ and the Brownian motion $W_t(\omega) \equiv \omega(t)$, $t \in \mathbb{R}_+$, $\omega \in \Omega_0$ with $W_0 = 0$, $a.s.$
On the other hand consider $\Omega_0^- \subset \Omega_0$ defined in Sect. 4.2, Eq. (1) by

$$\Omega_0^- = \{\omega \in \Omega_0 : \lim_{t \to \infty} (1 + t^2)^{-\frac{1}{2}} |\omega(t)| = 0. \quad (1)$$

We observe that $\Omega_0^- \subset S'(\mathbb{R})$ is a subset of Schwartz tempered distributions. Thus the Brownian motion restricted to Ω_0^- has sample paths which are tempered distributions. Then $\dot{W}_t$ will be well defined. In other words, as explained in Sect. 4.2, both the sample paths and their generalized derivatives $\dot{W}_t$ are $S'(\mathbb{R})$-valued random elements.

In this way we may call $\dot{W}_t$ as the <u>white noise</u> and its probability distribution on the sample path space $\Omega_0^- \cap S'(\mathbb{R}))$, the <u>white noise measure</u>. $(S'(\mathbb{R}), \mu)$ is called the white noise space. A measurable function on $(\Omega_0^-, \mathcal{F}, \mu)$ is a <u>white noise functional</u>.

The advantage of this approach is justified by the fact that it allows us to utilize linear topological structure of $S'(\mathbb{R})$ as the dual of a nuclear space.

A Wiener functional $f(\omega)$ can be regarded also a white noise functional F by the relation $F = f \circ J^{-1}$, where $J(\omega) = \dot{\omega}$ is a measurable map $\Omega_0^- \to S'(\mathbb{R})$.

From the Gelfand triplet $S(\mathbb{R}) \hookrightarrow L^2(\mathbb{R}, dx) \hookrightarrow S'(\mathbb{R})$ we produce another Gelfand triplet $(S) \hookrightarrow L^2(S'(\mathbb{R}), \mu) \hookrightarrow (S^*)$ by the use of the second quantization operator. (S) is the <u>Hida testing functional space</u> and it will be shown to be smaller than the Meyer-Watanabe testing functional space $\mathbb{D}^\infty$, however continuously and densely imbedded in it. In turn the <u>Hida distribution space</u> (S^*) is larger than the Meyer-Watanabe distribution space $\mathbb{D}^{-\infty}$. The space $L^2(S'(\mathbb{R}), \mu)$ will usually be denoted by (L^2). The details of the construction of the Hida distribution via the second quantization operators will be investigated in the next Section. There are also several other constructions like those of Kubo and Takanake, Kondratiev and Streit and construction by a Hilbert space and an operator.

5.8 Construction of Hida Testing Functionals and Distribution Spaces

We start by the construction via the standard second quantization operator. For this purpose we first introduce the multiple Wiener-Ito integrals and the decomposition of L^2 functions in terms of them.

In the general framework of a Gaussian probability space $(\Omega, \mathcal{F}, \mu; H)$, consider a special case where the separable Hilbert space is given by $H = L^2(\Theta, \mathcal{B}, \nu)$ where ν is a σ-finite non-atomic measure on $(\Theta, \mathcal{B})$. Let $\mathcal{B}_0 = \{A \in \mathcal{B} : \nu(\mathcal{A}) < \infty\}$ and define on $\mathcal{B}_0$ the following random set function

$$W(A) \equiv W(I_A), \quad A \in \mathcal{B}_0. \quad (1)$$

(I_A is the indicator function of A).

Referring to Sect. 5.2.4 $W(.)$ is a $L^2(\Omega, \mathcal{F}, \mu)$-valued random set function with the properties

(i) $W(A) \sim N(0, \nu(A))$,
(ii) $E[W(A) W(B)] = \nu(A \cap B)$.

Hence if $A_1, A_2, \cdots, A_n \cdots$ are disjoint, then $W(A_1), \cdots, W(A_n)$ are mutually independent and if $\bigcup_n A_n \in \mathcal{B}_0$, then $W(\cup_n A_n) = \sum_n W(A_n)$(in L^2-convergence) which follows from the linear isometry $h \to W_h$ as explained in Sect. 5.2.4. This σ-additive set function $W(.)$ is called the Gaussian orthogonal random measure on $(\Theta, \mathcal{B})$ with constructive measure ν.

When $\Theta = [0, 1]$ and ν is the Lebesgue measure, then as any $h \in H$ can be approximated by the linear combinations of $\{I_{A_n} \, A_n \in \mathcal{B}_0\}$, W_h is just the stochastic integral of h with respect to this random measure

$$W_h = \int_\Theta h(t) dW(t). \quad (2)$$

This is the well-known Wiener stochastic integral, (c.f. GLOSSARY). As for the construction of the multiple Wiener-Ito integral, consider $f \in L^2(\Theta^n, \mathcal{B}^n, \nu^n)$ of the following form

$$f = \sum_{j_1,\cdots,j_n=1}^{N} a_{j_1,\cdots,j_n} I_{A_{j_1} \times \cdots \times A_{j_n}}$$

where $A_1, \cdots, A_N \in \mathcal{B}_0$ are disjoint and if any two of indices $j_1, \cdots, j_n$ are equal, then $a_{j_1,\cdots,j_n} = 0$. Define the n-fold stochastic integral of f as

$$I_n(f) = \sum_{j_1,\cdots,j_n=1}^{N} a_{j_1,\cdots,j_n} W(A_{j_1}) \cdots W(A_{j_n}) \quad (3)$$

This definition is independent of representation of f and I_n is linear in f.
For Σ_n being the permutation group of $\{1, 2, \cdots, n\}$ we can consider the symmetrization $\tilde{f}$ of f

$$\tilde{f}(t_1, \cdots, t_n) \equiv \frac{1}{n!} \sum_{\sigma \in \Sigma_n} f(t_{\sigma(1)}, \cdots, t_{\sigma(n)}).$$

By definition $I_n(f) = I_n(\tilde{f})$, thus we can assume that $f \in H^{\hat{\otimes}n}$, i.e. f is symmetric with respect to $t_1, \cdots, t_n$.

Under these conditions for any $n \in \mathbb{N}$, I_n uniquely extends to a linear isometry from $H^{\hat{\otimes}n}$ into $L^2(\Omega, \mathcal{F}, \mu)$. I_n is referred as the n-fold Wiener-Ito stochastic integral of f with respect to the random measure W and denoted by

$$I_n(f) = \int_{\Theta^n} f(t_1, \cdots, t_n)\, dW(t_1) \cdots dW(t_n). \quad (4)$$

(for proof c.f. [Hu-Yan] Ch. II, Proposition 1.6)).

When $\Theta = [0, 1]$ and ν is the Lebesgue measure we obtain a classical Wiener space. Then $W_t = W([0, t])$, $t \geq 0$ is a Brownian motion and we retrieve the n-fold iterated Ito stochastic integral

$$I_n(f) = n! \int_0^1 \int_0^{t_n} \cdots \int_0^{t_2} dW_{t_1} \cdots dW_{t_n}. \quad (5)$$

Theorem 1 (Wiener-Ito) *For any $f \in L^2(S', \mu) \equiv (L^2)$, there exists a unique sequence $\{f_n\}_{n \in \mathbb{N}_0} \in \Gamma(H)$ (the symmetric Fock space in H) such that*

$$f = \sum_{n=0}^{\infty} I_n(f_n). \quad (6)$$

The proof depends on the Wiener-Ito decomposition of (L^2) as $\bigoplus_{n=0}^{\infty} \mathcal{H}_n$, given in Sect. 5.4, Theorem 1. (c.f. [Hu-Yan] Ch. IV, Theorem 1.3). In an equivalent expression of the theorem f_n can be assumed to exist in $\hat{L}^2(\mathbb{R}^n)$, (the symmetrized $L^2(\mathbb{R}^n)$). The correspondence between (L^2) and $H^{\otimes n}$ will be denoted by $f \sim \{f_n\}$.

Also consider the recursion relation : If $f \in H^{\hat{\otimes}n}$, $g \in H$

$$I_n(f)I_1(g) = I_{n+1}(f \otimes g) + n I_{n-1}(f \otimes_1 g), \quad (7)$$

where $f \otimes_1 g \in H^{\otimes n-1}$, ([Hu-Yan] Ch. II, Proposition 1.7).
The (L^2) norm $\|f\|_0$ is given by

$$\|f\|_0 = \left(\sum_{n=0}^{N} n! |f_n|_0^2\right)^{1/2} \quad (8)$$

where $|f_n|_0$ denotes the $L^2(\mathbb{R}^n)$ norm for any n.

We start by the differential operator

$$A = -\frac{d^2}{dx^2} + x^2 + 1. \quad (9)$$

For $f = \sum_{n=0}^{\infty} I_n(f_n)$ satisfying $\sum_{n=0}^{\infty} n!|A^{\otimes n} f_n|_0^2 < \infty$ we define

$$\Gamma(A)f = \sum_{n=0}^{\infty} I_n(A^{\otimes n} f_n), \quad (10)$$

(For the tensor product of operators see the GLOSSARY).

The operator $\Gamma(A)$ is called the second quantization operator of A. It is densely defined on (L^2) and has the following properties, the detailed proofs of which can be found in [Kuo-3], Theorem 1.5.1, also see some proofs in Sect. 5.10 for the more general case, i.e. construction from a Hilbert space and an operator.

 (i) $\Gamma(A)$, has a set of eigenfunctions which form an orthonormal basis for (L^2),
(ii) $\Gamma(A)^{-1}$ is a bounded operator on (L^2) with $\|\Gamma(A)^{-1}\| = 1$,
(iii) For any $p > 1$, the operator $\Gamma(A)^{-p}$ is a Hilbert-Schmidt operator of (L^2).

For each $p \geq 0$, define

$$\|f\|_p = \|\Gamma(A)^p f\|_0, \text{ let}(S_p) \equiv \{f \in (L^2) : \|f\|_p < \infty\}$$

and (S) is the projective limit of $\{(S_p) : p \geq 0\}$.

Then (S) is a nuclear space by the property iii) of $\Gamma(A)$. It is the Hida testing functionals space. Its dual $(S)^*$ in the inductive limit topology is the space of generalized functions or Hida distributions. Thus we have produced another Gelfand triplet

$$(S) \subset (L^2) \subset (S)^*$$

and also the following continuous and dense inclusions

$$(S) \subset (S_p) \subset (L^2) \subset (S_p)^* \subset (S)^*.$$

The dual of (S) is the inductive limit of $(S_p)^*$, i.e. $(S)^* = \bigcup_{p \geq 0}(S_p)^*$ and the norm on the dual space $(S_p)^*$ is given by

$$\|\Phi\|_{-p}^2 = \|\Gamma(A)^{-p}\Phi\|_0^2 = \sum_{n=0}^{\infty} n!|(A^{-p})^{\otimes n} F_n|_0^2, \; F_n \in S^*(\mathbb{R}^n). \quad (11)$$

where $(S_p)^* \ni \Phi = \sum_{n=0}^{\infty} I_n(F_n), \; F_n \in S^*(\mathbb{R}^n) \equiv S'(\mathbb{R}^n)$.

Remark This whole construction can be refined starting by $\hat{L}^2_c$, i.e. L^2 is complexified and symmetrized.

The bilinear pairing between $(S)^*$ and (S), is denoted by $<< \,.\,,\,.\,>>$ and related to the inner product of (L^2) by $<< \phi, \psi >>=< \phi, \bar{\psi} >_{(L^2)}$, $\phi \in (L^2)$, $\psi \in (S)$.

Next we see the close relation between multiple Wiener-Ito stochastic integrals and the Hermite polynomials.

Theorem 2 *(a) If $h \in H = L^2(\Theta, \mathcal{B}, v)$ and $, \|h\| = 1$, then $H_n(W_h) = I_n(h^{\otimes n})$.*
(b) For any $\alpha = \{\alpha_j\}_{j\in\mathbb{N}} \in \Lambda, |\alpha| = \sum_j \alpha_j = n$, and $\{h_j\}_{j\in\mathbb{N}}$ is an orthonormal system in H we have

$$\prod_j H_{\alpha_j}(W_{h_j}) = I_n(\hat{\otimes} h_j^{\otimes \alpha_j}) \quad (12)$$

Proof (a) We prove by induction. For $n = 1$ it reduces to (2). Accept it for $n = m$, then by (7) and property (3) of Hermite polynomials (Sect. 5.3)

$$I_{m+1}(h^{\otimes m+1}) = I_m(h^{\otimes m}) I_1(h) - m I_{m-1}(h^{\otimes m-1}) = H_m(W_h)W_h - m H_{m-1}(W_h) = H_{m+1}(W_h).$$

(b) By part (a) $\prod H_{\alpha_j}(W_{h_j}) = \prod_j I_{\alpha_j}(h_j^{\otimes \alpha_j}$ (†).
On the other hand by $I_n(f)I_m(g) = \sum_{r=0}^{n\wedge m} r!\binom{n}{r}\binom{m}{r}I_{n+m-2r}(f \hat{\otimes}_r g)$ ($\sqrt{}$), (c.f. [Hu-Yan] Ch. II. (1.44)). As $\{h_j\}$ is an orthonormal system, for $j \neq k$, $r > 0$, we have $, h_j^{\otimes \alpha_j} \otimes_r h_k^{\otimes \alpha_k} = 0$. Then the only non-zero term in ($\sqrt{}$) corresponds to $r = 0$. Thus in view of (†) we obtain $I_{\alpha_j}(h_j^{\otimes \alpha_j})I_{\alpha_k}(h_k^{\otimes \alpha_k}) = I_{\alpha_j+\alpha_k}(h_j^{\otimes \alpha_j} \hat{\otimes} h_k^{\otimes \alpha_k})$. As this is generalized to any $\{\alpha_j\}$ the proof of (b) is completed.

5.9 Some Examples of Infinite Dimensional Generalized Functions

Hermite polynomials and functions are indispensable tools of stochastic analysis. Hermite polynomials were eigen-functions of the Ornstein-Uhlnbeck operator $\mathcal{L}$. Likewise Hermite functions are the eigen-functions of the differential operator A given by Sect. 5.8, (9).

Hermite function is defined by

$$e_n(x) = \pi^{-1/4} 2^{-n/2} (n!)^{-1/2} H_n(x) e^{-x^2/2} \quad (1)$$

where H_n is the Hermite polynomial of degree n.

$\{e_n : n \geq 0\}$ is an orthonormal basis for $L^2(\mathbb{R})$. Moreover these Hermite functions satisfy

$$A e_n = (2n + 2) e_n; \, n = 0, 1, \cdots . \quad (2)$$

A^{-1} is a bounded operator and by (2) $\|A^{-p}\|_{HS}^2 = \sum_{n=0}^{\infty} (2n + 2)^{-2p}$, therefore for $p > 1/2$, A^{-p} is a Hilbert-Schmidt operator of $L^2(\mathbb{R})$.

For each $p \geq 0$ and $f \in L^2(\mathbb{R})$ define $|f|_p = |A^p f|_0$ where $|\,.\,|_0$ is the $L^2(\mathbb{R})$ norm. Consequently $|f|_p$ is given by

$$|f|_p = \sum_{n=0}^{\infty} (2n + 2)^{2p} (f, e_n)^2)^{1/2}, \quad (3)$$

$(\,,\,)$ is the inner product of $L^2(\mathbb{R})$ and

$$|f|_{-p} = |A^{-p} f|_0, \, p > 0, \quad (4)$$

where $|\,.\,|_0$ is the $L^2(\mathbb{R}$ norm.

(I) As pointed out in the introduction to Sect. 5.7, the sample paths of the white noise process are the generalized derivatives $\dot{W}(\omega); \, \omega \in \Omega_0^-$, which are in the space of tempered distributions. But $\dot{W}_t$ as a mathematical object is a Hida distribution, i.e. $\dot{W}_t \in (S)^*$. This follows from $\dot{W}_t = < \,.\, \delta_t >\, = I_1(\delta_t) = \int_{\Omega_0^-} \delta_t dW_s$, (where $, < , >$ is the canonical bilinear form on $S' \times S$). Also by Sect. 5.8 (11) we have $\|\dot{W}_t\|_{-p} = |A^{-p}\delta_t|_0 = |\delta_t|_{-p}$. The Dirac delta function can be expanded in terms of the Hermite functions as $\delta_t = \sum_{n=0}^{\infty} e_n(t) e_n.$, Regarding δ_t as the L^2 limit of L^2 functions and using (2) we find

$$|\delta_t|_{-p} = \left(\sum_{n=0}^{\infty} (2n + 2)^{-2p} e_n(t) \right)^{1/2} .$$

But [Sze] gives the estimate $|e_n(t)| = O(n^{-1/4})$. Thus for p sufficiently large $|\delta|_{-p} < \infty$, so that $\dot{W}_t \in (S_p)^* \subset (S)^*$.

(II) By Theorem 2 of Sect. 5.8

$$H_2 \left(\frac{< \,.\,, f >}{< f, f >^{1/2}} \right) = I_2 \left(\frac{f^{\otimes 2}}{< f, f >} \right)$$

By the properties of Hermite polynomials given in Sect. 5.3 $H_2(x) = x^2 - 1$, thus

$$< \,.\,, f >^2 = I_2(f^{\otimes 2}) + < f, f > \quad (5)$$

If $f \in S(\mathbb{R})$, the function $< , f >$ is defined everywhere on $S'(\mathbb{R})$. If $f \in S'(\mathbb{R})$, it is a generalized function, which can be seen by regarding the continuous and dense imbeddings of the Gelfand triplet $S(\mathbb{R}) \subset L^2(\mathbb{R}) \subset S'(\mathbb{R})$.

If $\xi \in S(\mathbb{R})$, then $< ., \xi)^2$ is defined everywhere on $S'(\mathbb{R})$ and in fact belongs to $L^2(S'(\mathbb{R}), \mu) \equiv (L^2)$. But it also belongs to (S) since its $\| . \|_p$ norm is finite for some p. If $f \in L^2(\mathbb{R})$ then $< ., f >^2 \in (L^2)$, this is because we can construct a Cauchy sequence $\{< ., \xi_n >, \xi_n \in S(\mathbb{R})\}$: converging to $< ., f >$ in (L^2).

But if $\eta \in S'(\mathbb{R})$, then $< , , \eta >^2$ has no meaning. We need to consider its <u>renormalization</u>. This renormalization is motivated by (5), ([Kuo-1], 3.4) and is given by

$$:< ., \eta >^2: \equiv I_2(\eta^{\otimes 2}). \quad (6)$$

Then by Sect. 5.8 (11)

$$\| :< ., \eta >^2: \|_{-p} = \sqrt{2} |(A^{-p})^{\otimes 2} \eta^{\otimes 2}|_0 \overset{(3)}{=} \sqrt{2} |\eta|^2_{-p}$$

and by (2) $|\eta|^2_{-2} < \infty$ for some $p \geq 0$. This shows that $:< ., \eta >:^2 \in (S)^*$.
In particular for $\eta = \delta_t$, we have the generalized function $: \dot{W}^2_t :$. In general for $\eta \in S'(\mathbb{R})$, we define the renormalization $:< ., \eta >^n:$ of $< ., \eta >^n$ by $I_n(\eta^{\otimes n})$.

(III) Exponential Functional: Its definition was given in Sect. 5.4, Definition 2. Then in white noise set up it should be

$$\mathcal{E}(f) = e^{< .,f >-\frac{1}{2}|f|^2_0}, f \in L^2(\mathbb{R}), |f|_0 \equiv \|f\|_{L^2}. \quad (7)$$

To find its Wiener-Ito decomposition we again resort to Sect. 5.8, Theorem 2:

$$I_n(f^{\otimes n}) = |f|^n_0 H_n(|f|^{-1}_0 < ., f >)(\dagger)$$

Substituting $t = |f|_0$, $x = |f|^{-1}_0 < ., f >$ in ($\dagger$) we obtain
$\sum_{n=0}^{\infty} \frac{I_n(f^{\otimes n})}{n!} = \sum_{n=0}^{\infty} \frac{t^n}{n!} H_n(x) \overset{5.3(17)}{=} e^{tx-t^2/2} = e^{< .,f >-|f|^2_0/2}$. Thus we obtain

$$e^{< .,f >} = e^{\frac{1}{2}|f|^2_0} \sum_{n=0}^{\infty} \frac{1}{n!} I_n(f^{\otimes n}) \quad (8)$$

We easily see that $e^{< .,\xi >} \in (S)$ if $\xi \in S(\mathbb{R})$ and $e^{< .,f >} \in (L^2)$ if $f \in L^2(\mathbb{R})$, however $e^{< .,\eta >}$ has no meaning if $\eta \in S'(\mathbb{R})$. Again we need a renormalization. This is suggested by (8). For $\eta \in S'(\mathbb{R})$ define the renormalization as

$$: e^{< .,\eta >} : \equiv \sum_{n=0}^{\infty} I_n(\eta^{\otimes n}). \quad (9)$$

We have by Sect. 5.8 (11), Sect. 5.9 (4) and the linearity of the multiple Wiener Ito integral

$$\| : e^{<\cdot,\eta>} : \|_{-p} = \sum_{n=0}^{\infty} n! \, |\frac{1}{(n!)^2}(A^{-p})^{\otimes n}\eta^{\otimes n}|_0^2 = \sum_{n=0}^{\infty} \frac{1}{n!} \, |\eta|_{-p}^{2n} = e^{|\eta|_{-p}^2}.$$

For some $p \geq 0$, $|\eta|_{-p} < \infty$, thus $: e^{<\cdot,\eta>} :\, \in (S)^*$.
For $f \in L^2(\mathbb{R})$, according to (8), the renormalized exponential function and exponential functional coincide, i.e.

$$: e^{<\cdot,f>} := e^{<\cdot,f>} - \frac{1}{2} < f, f > .$$

Remarks. 1. Complexified versions of examples (II) and (III) for $L_c^2(\mathbb{R})$ and $S_c(\mathbb{R})$ can be given, ([Kuo-1]).
2. $< .\eta >^2$, $\eta \in (S)'(\mathbb{R})$ is given a meaning of a generalized function in a different sense in [Çap-1] (c.f. Sect. 5.12), where Meyer-Watanabe and Hida distributions are imbedded into Colombeau type algebras, so that the product operation is defined in a natural way without resorting to the renormalization and/or via somewhat indirect and artificial Wick product, (See next section).

5.10 More General Constructions of the Hida Distributions

Let $E \hookrightarrow H \hookrightarrow E^*$ be a Gelfand triplet where H is a real separable Hilbert space, E is a countably Hilbertian nuclear space which is continuously and densely imbedded into H and E^* is the continuous dual of E, $(H^* \cong H)$. By Minlos' theorem there exists a unique Gaussian measure μ on $(E^*, \mathcal{B}(E^*))$ such that $\int_{E^*} e^{i<x,f>}\mu(dx) = e^{-\frac{1}{2}|f|^2}$, $f \in E$ where $|.|$ denotes the norm in H and $<,>$ the canonical bilinear form on $E^* \times E$. Let $f \in E$, the counterpart of W_h in a Gaussian probability space is W_f with $W_f(x) =< x, f >$, $x \in E^*$. W_f is a Gaussian random variable on (E^*, μ) and

$$E(W_f) = 0, \; E(W_f^2) = |f|^2.$$

Thus the linear map $f \to W_f$ from E to $L^2(E^*, \mu) \equiv (L^2)$ can be extended from H to (L^2). Consequently associated with any Gelfand triplet $E \hookrightarrow H \hookrightarrow E^*$, there is a Gaussian probability space $(E^*, \mathcal{B}(E^*), \mu; H)$ that is called a Canonical Gaussian Probability space. The classical white noise space $(S'(\mathbb{R}^d), \mathcal{B}(S'(\mathbb{R}^d)), \mu; L^2(\mathbb{R}^d))$ is an example of a canonical Gaussian probability space.

In the following we construct particular Gelfand triplets such that the nuclear space E will become the space of testing functionals and its dual E^* will be the space of distributions (generalized functions).

Construction From a Hilbert Space and an Operator

Let H be real, separable Hilbert space with norm $|\,.\,|$. Let A be an operator on H such that there exists an orthonormal basis $\{h_j = 1, 2, \cdots\}$ for H satisfying

(i) $Ah_j = \lambda_j h_j$; $j = 1, 2, \cdots$
(ii) $1 < \lambda_1 \le \lambda_2 \le \cdots \le \lambda_n \le \cdots$
(iii) $\sum_{j=1}^{\infty} \lambda_j^{-\alpha} < \infty$ for some positive constant α.

By (ii) the operator A^{-1} is bounded with the operator norm $\|A^{-1}\| = \sup_j |A^{-1} h_j| = \sup_j \frac{1}{\lambda_j}|h_j| = \frac{1}{\lambda_1}$.
By (iii) the operator $(A^{-1})^{\frac{\alpha}{2}}$ is a Hilbert-Schmidt (H-S) operator since:

$$\|(A^{-1})^{\frac{\alpha}{2}}\|_{HS}^2 = \sum_{j=1}^{\infty} |(A^{-1})^{\frac{\alpha}{2}} h_j|^2 = \sum_{j=1}^{\infty} |(\frac{h_j}{\lambda_j})^{\frac{\alpha}{2}}|^2 = \sum_{j=1}^{\infty} \frac{1}{\lambda_j^{\alpha}} < \infty$$

We introduce a family of the Hilbertian norms and also the spaces related to these norms

$$\|\xi\|_p = |A^p \xi|, \xi \in Dom(A^p), \ p \in \mathbb{R}; \ \mathcal{E}_p = \{\xi \in H : \|\xi\|_p < \infty\} \quad (1)$$

For $q \ge p \ge 0$ we have $Dom(A^q) \subset Dom(A^p)$ which is seen by $\|\xi\|_p = |A^{-(q-p)} A^q \xi| \le C\|\xi\|_q$.
   Thus $\mathcal{E}_q \subset \mathcal{E}_p$.
   If $f_{p,q} : \mathcal{E}_q \to \mathcal{E}_p$ is the continuous injection, then $\{\mathcal{E}_p, f_{p,q}\}$ becomes a projective system of Hilbert spaces. Thus

$$\cdots\cdots \subset \mathcal{E}_q \subset \cdots \subset \mathcal{E}_p \subset \mathcal{E}_0 = H; \ , \ q \ge p \ge 0.$$

Let $E = \bigcap_{p \ge 0} \mathcal{E}_p$. Then E is a countably Hilbertian (nuclear) space. This is seen by observing that $\mathcal{E}_{q+\frac{\alpha}{2}} \hookrightarrow \mathcal{E}_q$ and if α is the one provided by (iii), then the inclusion map is a Hilbert-Schmidt operator. In fact an orthonormal system in $\mathcal{E}_{q+\frac{\alpha}{2}}$ is $\left\{\dfrac{h_j}{\lambda_j^{q+\frac{\alpha}{2}}}\right\}$. We have $\sum_{j=1}^{\infty} \dfrac{\|h_j\|_q^2}{\lambda_j^{2q+\alpha}} = \sum_{j=1}^{\infty} \dfrac{1}{\lambda_j^{\alpha}} < \infty$. E is isomorphic to the following projective limit

$$E \cong \operatorname*{proj\,lim}_{p \to \infty} \mathcal{E}_p. \quad (2)$$

As for the dual space of E we have

$$E^* = \bigcup_{p \ge 0} \mathcal{E}_p^* \text{ and } E^* \cong \operatorname*{ind\,lim}_{p \to \infty} \mathcal{E}_p^*. \quad (3)$$

i.e. it is isomorphic to the inductive limit of duals $\mathcal{E}_p^*$ to $\mathcal{E}_p$. $E \subset H \subset E^*$ is a Gelfand triplet.

Next we establish the association between the Hilbertian norms given by (1) with negative indices and the dual spaces. Let $\mathcal{E}_{-p}$ be the completion of H with respect to $\|\cdot\|_{-p}$. Then the identity map of H onto itself extends to a continuous injection $f_{-q,-p} : \mathcal{E}_{-p} \to \mathcal{E}_{-q}$, $(q \geq p \geq 0)$, therefore $\{\mathcal{E}_{-p}, f_{-q,-p}\}$ becomes an inductive system of Hilbert spaces, i.e.:

$$H = \mathcal{E}_0 \subset \cdots \subset \mathcal{E}_{-p} \subset \cdots \subset \mathcal{E}_{-q} \subset \cdots \ (q \geq p \geq 0). \quad (4)$$

To connect the two inductive systems (3) and (4):

$A^{-p} : H \to \mathcal{E}_{-p}$ is a bounded operator satisfying $|A^{-p}\xi| = \|\xi\|_{-p}$, $\xi \in H$. Therefore A^{-p} extends to an isometric isomorphism I_A^{-p} of $\mathcal{E}_{-p}$ onto H. The inner product $(\ ,\ )_p$ of $\mathcal{E}_p$ is then given by

$$(\xi, \eta)_p = (A^p \xi, A^p \eta)_H, \xi, \eta \in \mathcal{E}_p. \quad (5)$$

By the Riesz's representation theorem for the dual of a Hilbert space there is an isometric isomorphism $R_p : \mathcal{E}_p^* \to \mathcal{E}_p$ such that $(x^*, \xi)_H = (R_p(x^*), \xi)_p$, $x* \in \mathcal{E}_p^*$, $\xi \in \mathcal{E}_p$.

Then by (5):

$$(R_p(x^*), \xi)_p = (A^p R_p(x^*), A^p \xi)_H = ((I_A^{-p} \circ (I_A^{-p})^{-1}) A^p R_p(x^*), \ A^p \xi)_H$$

Thus for $h_p = (I_A^{-p})^{-1} \circ A^p \circ R_p : \mathcal{E}_p^* \to \mathcal{E}_{-p}$ becomes an isometric isomorphism.

Hence the dual norm on $\mathcal{E}_p^*$ is given by $\|\cdot\|_{-p}$ and the two inductive systems $(\mathcal{E}_p^*, f_{p,q}^*)$ and $(\mathcal{E}_{-p}, f_{-p,-q})$ are isometrically isomorphic under h_p. We have

$$E \subset \mathcal{E}_p \subset H \subset \mathcal{E}_p^* \subset E^*, \ p \geq 0. \quad (6)$$

Now we apply the same procedure to the second quantization operator to produce another Gelfand triplet related to the testing functionals and distributions.

Let (L^2) denote the complex Hilbert space $L^2(E^*, \mu)$, where μ is the Minlos measure. By the Wiener-Ito theorem $\phi \in (L^2)$ can be expressed uniquely as $\phi = \sum_{n=0}^{\infty} I_n(f_n)$, $f_n \in H_c^{\hat{\otimes}n}$ where $H_c^{\hat{\otimes}n}$ is the complexification of the symmetric tensor product $H^{\hat{\otimes}n}$.

Moreover for the $(L^2)-$norm $\|\phi\|_0$ of ϕ we have

$$\|\phi\|_0^2 = \sum_n n! |f_n|^2 \quad (7)$$

where $|\cdot|$ denotes the $H_c^{\hat{\otimes}n}-$norm.

Recall Theorem 2(b) of the Sect. 5.8

$$I_n(h_1^{\hat{\otimes}n_1} h_2^{\hat{\otimes}n_2} \hat{\otimes} \cdots) = H_{n_1}(W_{h_1}) H_{n_2}(W_{h_2}) \cdots \qquad (8)$$

The second quantization operator (Sect. 5.8, (10)) $\Gamma(A)$ of A is a densely defined operator on (L^2) given by

$$\Gamma(A) = \sum_{n=0}^{\infty} I_n(A^{\otimes n} f_n), \ f_n \in H_c^{\hat{\otimes}n}. \qquad (9)$$

$\Gamma(A)$ has properties comparable to those of A. For $\{n_j\} \in \Lambda, \ |n_j| = n$

$$\phi_{\{n_j\}} = \frac{1}{(n_1! n_2! \cdots)^{1/2}} I_n(h_1^{\otimes n_1} \hat{\otimes} h_2^{\otimes n_2} \hat{\otimes} \cdots).$$

By (8) and the orthogonality of Hermite polynomials the family $\{\phi_{\{n_j\}} : |n_j| = n; \ n = 0, 1, 2, \cdots\}$ forms an orthonormal basis for (L^2).

$$\Gamma(A)\phi_{\{n_j\}} = \frac{1}{\sqrt{n_1! n_2! \cdots}}(\lambda_1^{n_1} \lambda_2^{n_2} \cdots) I_n(h_1^{\otimes n_1} \hat{\otimes} h_2^{\otimes n_2} \hat{\otimes} \cdots) = (\lambda_1^{n_1} \lambda_2^{n_2} \cdots)\phi_{\{n_j\}}.$$

It is also seen that $(\Gamma(A)^{-1})^{\alpha/2}$ is a Hilbert-Schmidt operator. In fact

$$(\Gamma(A)^{-1})^{\alpha/2}\phi_{\{n_j\}} = (\frac{1}{\lambda_1^{n_1}} \cdot \frac{1}{\lambda_2^{n_2}} \cdots \phi_{\{n_j\}})^{\frac{\alpha}{2}}$$

thus

$$\|(\Gamma(A)^{-1})^{\alpha/2}\|_{HS}^2 = \sum_{n_1} \sum_{n_2} \cdots [(\frac{1}{\lambda_1})^{\alpha}]^{n_1} [(\frac{1}{\lambda_2})^{\alpha}]^{n_2} \cdots = \frac{1}{\prod_{j=1}^{\infty}(1 - \lambda_j^{-\alpha})} < \infty, \ \text{as} \ 0 < (\frac{1}{\lambda_j})^{\alpha} < 1.$$

Following the steps leading to the inclusions in (6), for each $p \geq 0$ define

$$\|\phi\|_p = \|\Gamma(A)^p \phi\|_0 \text{and}(\mathcal{E}_p) = \{\phi \in (L^2); \|\phi\|_p < \infty\}. \qquad (10)$$

For $q \geq p \geq 0$

$$\|\phi\|_p = \|\Gamma(A)^{-(q-p)}\Gamma(A)^q \phi\|_0 \leq \|\Gamma(A)^{-(q-p)}\|_0 \|\phi\|_q$$

implying that $(\mathcal{E})_q \subset (\mathcal{E})_p$.

The inclusion map $(\mathcal{E}_{p+\alpha/2} \hookrightarrow (\mathcal{E})_p$ is still a Hilbert-Schmidt operator:
By an analysis similar to the one before (2) we find that the families

$$\left\{\frac{\phi_{\{n-j\}}}{(\lambda_1^{n_1} \lambda_2^{n_2} \cdots)^{p+\alpha/2}}; |n_j| = n; n = 0, 1, 2, \cdots\right\} \quad \text{form an orthonormal basis for}$$

$\mathcal{E}_{p+\alpha/2}$. Then the Hilbert-Schmidt norm of the inclusion map is

$$\sum_{n_1}\sum_{n_2}\cdots\left[\frac{(\lambda_1^{n_1}\lambda_2^{n_2}\cdots)^p}{(\lambda_1^{n_1}\lambda_2^{n_2}\cdots)^{p+\alpha/2}}\right]^2 = \frac{1}{\prod_{j=1}^{\infty}(1-\lambda_j^{-\alpha})} < \infty.$$

Let (E) = the projective limit of $\{(\mathcal{E}_p)\}$; $p \geq 0$, and $(E)^*$ the dual space of (E). Then (E) is a nuclear space and we have the Gelfand triplet $(E) \subset (L^2) \subset (E)^*$ and the continuous inclusions

$$(E) \subset (\mathcal{E}_p) \subset (L^2) \subset (\mathcal{E}_p)^* \subset (E)^*.$$

Let $(\mathcal{E})_{-p}$ be the completion of (L^2) with respect to the norm $\|\phi\|_{-p} = \|\Gamma(A)^{-p}\phi\|_0$. Acting as in the case of single operator A, we find an isometric isomorphism between $(\mathcal{E})_{-p}$ and $(\mathcal{E}_p)^*$. Thus the dual norm on $(\mathcal{E}_p)^*$ is found to be $\|\,.\,\|_{-p}$.

<u>Construction of Kondratiev and Streit</u>. This is a generalization of the construction via the second quantization operator.

The basic idea is to multiply the terms of the n-th chaos by a factor $(n!)^{\beta}$ in order to obtain a smaller testing functional space, in return a larger distribution space.

As it goes in parallel to the construction method by the Hilbert space and an operator, the detailed proofs will not be given. As before each $\phi \in (L^2)$ is represented as $\phi = \sum_{n=0}^{\infty} I_n(f_n)$; $f_n \in H_c^{\hat\otimes n}$ and $\|\phi\|_0 = (\sum_{n=0}^{\infty} n!|f_n|_0^2)^{\frac{1}{2}}$.

Let $0 \leq \beta < 1$ be a fixed number. For each $p \geq 0$ define

$$\|\phi\|_{p,\beta} = (\sum_{n=0}^{\infty}(n!)^{1+\beta}|(A^p)^{\otimes n}f_n|_0^2)^{1/2}. \quad (11)$$

$(\mathcal{E}_p)_{\beta} = \{\phi \in (L^2) : \|\phi\|_{p,\beta} < \infty. \ (\mathcal{E}_0)_{\beta} \neq (L^2)$ unless $\beta = 0$.
Like in the previous section $(\mathcal{E}_q)_{\beta} \subset (\mathcal{E}_p)_{\beta}$ for any $q \geq p \geq 0$.
The inclusion map $(\mathcal{E}_{p+\alpha/2})_{\beta} \hookrightarrow (\mathcal{E}_p)_{\beta}$ is a Hilbert-Schmidt operator.
Let $(E)_{\beta}$: The projective limit of $\{(\mathcal{E}_q)_{\beta} : q \geq 0\}$.
 $(E)_{\beta}^*$: The dual space of $(E)_{\beta}$.
Then $(E)_{\beta}$ is a nuclear space and we have a Gelfand triplet $(E)_{\beta} \subset (L^2) \subset (E)_{\beta}^*$ with continuous inclusions

$$(E)_{\beta} \subset (\mathcal{E}_q)_{\beta} \subset (L^2) \subset (\mathcal{E}_q)_{\beta}^* \subset (E)_{\beta}^*, q \geq 0$$

where the norm on $(\mathcal{E}_q)_{\beta}^*$ is given by

$$\|\phi\|_{-q,-\beta} = (\sum_{n=0}^{\infty}(n!)^{1-\beta}|(A^{-q})^{\otimes n}f_n|_0^2)^{1/2}$$

i.e. $(\mathcal{E}_q)_{\beta}^*$ is the completion of (L^2) with respect to the norm $\|,\,.\,\|_{-q-\beta}$.
Compared with the previous test functional space (E) and the distribution space $(E)^*$, the new Gelfand triplet has the following relationships

$$(E)_{\beta} \subset (E) \subset (E)^* \subset (E)_{\beta}^*, 0 \leq \beta < 1,$$

i.e. smaller testing functional space and larger distribution space.

Also $(E)_0 = (E)$ and for $0 \leq \beta_2 \leq \beta_1 < 1$ $(E)_{\beta_1} \subset (E)_{\beta_2}$. Moreover for any $q \geq 0$

$$(\mathcal{E}_q)_\beta \subset (\mathcal{E}_q) \subset (L^2) \subset (\mathcal{E}_q)^* \subset (\mathcal{E}_q)_\beta^*.$$

Again the effect of multiplying by a factor, $(n!)^{1+\beta}$ is observed.

A variation of the Kondratiev & Streit construction does not use a Hilbert space operator A, but rather starts with the Gelfand triplet $E \subset H \subset E^*$, given at the start of the Sect. 5.10. Thus by the nuclearity of E there exists a sequence of increasing and compatible norms $\{\|.\|_p\}_{p \in \mathbb{N}}$ such that E is the projective limit of H_p which is the completion of H with respect to $\|.\|_p$; $H_0 = H$, $\|.\|_0 = \|.\|$.

Let H_{-p} be the dual of H. For any $p \in \mathbb{N}$, there exists $p' > p$ such that the imbedding from $H_{p'}$ into H_p, is a Hilbert-Schmidt operator. Moreover $|.|_p \geq |.|.$ Considering this standard sequence $\{\|.\|_p\}$ of norms on E and for $p, q \in \mathbb{N}$, $\beta \geq 0$, define

$$\|\phi\|_{p,q,\beta}^2 = \sum_{n=0}^{\infty} (n!)^{1+\beta} \, 2^{nq} \, |f_n|_p^2; \; f_n \in H_p^{\hat{\otimes}n} \quad (12)$$

with $\phi = \sum_n I_n(f_n)$ according to the Wiener - Ito decomposition (chaos decomposition) of (L^2). Let $H_{p,q,\beta} = \{\phi \in (L^2) : \|\phi\|_{p,q,\beta} < \infty\}$ and let $(E)_\beta$ be the projective limit of $\{(H_{p,q,\beta} \, p, q \in \mathbb{N}\}$, i.e. $(E_\beta) = \bigcap_{p,q \geq 1} H_{p,q,\beta}$ furnished with the projective limit topology.

Then $(E)_\beta$ is a countably Hilbertian nuclear space continuously and densely imbedded in (L^2).

Its dual $(E)_\beta^* \cong \bigcup_{p,q \geq 1} H_{-p,-q,-\beta}$ is the space of Hida distributions.

An alternative representation of $H_{-p,-q,-\beta}$ is $\{F : \|F\|_{-p,-q,\beta} < \infty\}$ where

$$\|F\|_{-p,-q,-\beta} \equiv \sum_{n=0}^{\infty} (n!)^{1-\beta} 2^{-nq} |g_n|_{-p}^2, \; F \sim \{g_n\}, \; g_n \in H_{-p}^{\hat{\otimes}n}. \quad (13)$$

The canonical bilinear form on $(H_{p,q,\beta}) \times (H_{-p,-q,\beta})$ is given by $<< \phi, F >> = \sum_{n=0}^{\infty} n! < f_n, g_n >$, where $\phi \sim \{f_n\}$, $F \sim \{g_n\}$ and $< ., . >$ is the canonical bilinear form on $H_p^{\hat{\otimes}n} \times H_{-p}^{\hat{\otimes}n}$. For $\beta = 0$, (E) and $(E)^*$ are the classical Hida testing functional and distributions respectively. As a particular case $E = S(\mathbb{R}^d)$, the space of C^∞ rapidly decreasing functions. The Meyer-Watanabe distributions can also be constructed on the canonical Gaussian probability space $(E^*, \mathcal{B}(E^*), \mu)$.

This allows us to compare them with the Hida distributions as in the statement of the next theorem. However for the proof of this theorem we make a change in (12) by allowing $p \in \mathbb{Z}$, $q, r \in \mathbb{R}$, accordingly $\|\phi\|_{p,q,r} = \sum_{n=0}^{\infty} (n!)^{1+r} \, 2^{nq} \, |f_n|_p^2$ and $H_{p,q,r} = \{\phi \in (L^2) : \|\phi\|_{p,q,r} < \infty\}$. As we will see this will not effect the proof.

Theorem $(E) \subset \mathbb{D}^\infty$; (E) *is dense in* $\mathbb{D}^\infty$ *and the imbedding map is continuous.*

Proof Denote by $\mathcal{P}$ the polynomial functionals on E^*, i.e. any element $\phi \in \mathcal{P}$ has the form

$\phi(x) = f(W_{\xi_1}(x), \cdots, W_{\xi_n}(x)); \ \xi_1, \cdots, \xi_n \in E, n \in \mathbb{N}$ where f is a polynomial on $\mathbb{R}^n$.

As the construction of $\mathbb{D}^\infty$ starts by completion of polynomial functionals by the Sobolev norms and also in view of the chaos distribution, the polynomials belong to every $H_{p,q,0}$. We have $\mathcal{P} \subset \mathbb{D}^\infty \cap (E)$ and $\mathcal{P}$ is dense in both $\mathbb{D}^\infty$ and (E). It will be sufficient to prove

$$\|(I - \mathcal{L})^k \phi\|_{L^2} \leq \|\phi\|_{0,q,0} \text{for} \phi \in \mathcal{P}, \ k \geq 1, r \geq 2 \text{and} q > 0$$

since this will clearly indicate the implication $\phi \in (E) \Rightarrow \phi \in \mathbb{D}^\infty$.

For $\mathcal{P} \ni \phi = \sum_n I_n(f_n)$ we apply the hypercontractivity property of the O-U semigroup $e^{t\mathcal{L}}$ acting on $(I - \mathcal{L}^k)\phi$. (c.f. Sect. 5.4, Property IV) of the O-U semigroup with the replacement $q(t) \leftrightarrow r$). Taking then $t = \frac{1}{2}\ln(r-1)$ and defining $r = e^{2t}(p-1) + 1$ we have

$$\|e^{t\mathcal{L}}(I - \mathcal{L})^k \phi\|_{L^r} \leq \|(I - \mathcal{L})^k \phi\|_{L^2},$$

Then as $-\mathcal{L} = n$

$$\|(I - \mathcal{L})^k\|_{L^r} \leq e^{-t\mathcal{L}}(I - \mathcal{L}^k)\phi\|_{L^2} = \|\sum_n e^{tn}(1+n)^k I_n(f_n)\|_{L^2} \leq \|\sum_n e^{(t+k)n} I_n(f_n)\|_{L^2}$$

By taking $2^{q/2} = e^{t+k}$

$$\|\sum_n 2^{\frac{nq}{2}} I_n(f_n)\|_{L^2} = \|I_n(2^{\frac{nq}{2}} f_n)\|_{L^2} \overset{(7)}{=} (\sum_n n! \, 2^{nq} |f_n|^2)^{\frac{1}{2}} = \|\phi\|_{0,q,0}.$$

As a consequence $\mathbb{D}^{-\infty}$ is densely and continuously imbedded into $(E)^*$.

5.11 Wick Products, *S*-Transform and U_β Functionals

For Wick tensor product we follow closely the approach of [Kuo-1].

Definition 1 (*Trace operator*) For a Gelfand triplet let $\tau \in E^* \hat{\otimes} E^*$ be defined as $< \tau, f \otimes g >=< f, g >, \ f, g \in E$.

If $\{h_j\}_j \in \mathbb{N}$ are the eigenvectors in the construction of Hida distributions via a Hilbert space and an operator then obviously $\tau = \sum_{j=1}^\infty h_j \otimes h_j$.

Definition 2 The Hermite polynomial of degree n with parameter σ^2 is defined by

$$H_{n,\sigma^2}(x) \equiv: x^n :_{\sigma^2} = (-\sigma^2) e^{\frac{x^2}{2\sigma^2}} \frac{d^n}{dx^n} e^{-\frac{x^2}{2\sigma^2}}$$

Notice that for $\sigma = 1$ we retrieve the formula (16) of the Sect. 5.3.

These Hermite polynomials are the coefficients in the expansion of power series of the generating function $e^{tx-\frac{\sigma^2 t^2}{2}}$

$$\sum_n \frac{t^n}{n!} : x^n :_{\sigma^2} = e^{tx-\frac{\sigma^2 t^2}{2}}$$

If we expand right side and compare the coefficients of t^n on both sides we find

$$: x^n :_{\sigma^2} = n! \sum_{k=0}^{[n/2]} \frac{(-\sigma^2)^k x^{n-2k}}{2^k k!(n-2k)!}$$

which can also be rewritten as

$$: x^n :_{\sigma^2} = \sum_{k=0}^{[n/2]} \binom{n}{2k} (2k-1)!!(-\sigma^2) x^{n-2k}. \quad (1)$$

where $(2k-1)!! = 1.3. \cdots (2k-3)(2k-1)$. Conversely we have

$$x^n = \sum_{k=0}^{[n/2]} \binom{n}{2k} (2k-1)!! \, \sigma^{2k} : x^{\otimes(n-2k)} :_{\sigma^2} .(\sqrt{}).$$

Being motivated by (1) we give the following definition

Definition 3 $: x^{\otimes n} := \sum_{k=0}^{[n/2]} \binom{n}{2k} (2k-1)!! \, x^{\otimes n-2k} \hat{\otimes} \, \tau^{\hat{\otimes}k}$ is called the n-fold Wick tensor product of x.
Clearly $: x^{\otimes n} :\in (E^*)^{\hat{\otimes}n}$ and $: x^{\otimes 0} := 1$, $: x^{\otimes 1} :\equiv x$.

Similar to the formula $(\sqrt{})$ we have

$$x^{\otimes n} = \sum_{k=0}^{[n/2]} \binom{n}{2k} (2k-1)!! : x^{(n-2k)} : \hat{\otimes} \, \tau^{\hat{\otimes}k}.$$

Lemma 1 *For any $x \in E^*$ and $\xi \in E$ following equalities hold*

(a) $<: x^{\otimes n} : \xi^{\otimes n} >=:< x, \xi >^n:_{|\xi|^2}$,
(b) $\| <: .^{\otimes n} :, \xi^{\otimes n} \|_0 = \sqrt{n!} \, |\xi|^n$.

Proof (a) $<: x^{\otimes n} :, \xi^{\otimes n} >= \left\langle \sum_{k=0}^{[n/2]} \binom{n}{2k}(-1)^k x^{\otimes(n-2k)} \hat{\otimes} \tau^{\hat{\otimes}k}, \xi^{\otimes n} \right\rangle$

$$= \sum_{k=0}^{[n/2]} \binom{n}{2k}(2k-1)!!(-1)^k < x^{\otimes(n-2k)}, \xi^{\otimes(n-2k)} >< \tau^{\hat{\otimes}k} \xi^{\otimes k} >= \sum_{k=0}^{[n/2]} (2k-1)!! \, (-|\xi|^2)^k < x, \xi >^{n-2k}$$

which in view of (1) implies (a).

(b) The proof of property (5) of Hermite polynomials in Sect. 5.3 can be repeated with minor alterations for the Hermite polynomials with parameter σ^2 as given in Definition 2 to yield $\int_{\mathbb{R}^n} : x^n :_{\sigma^2} d\mu_{\sigma^2}(x) = n!\sigma^{2n}$. By part a) $<: .^{\otimes n} :, \xi^{\otimes n} >=:< ., \xi >^n: |\xi|^2$, substituting in the integral and recalling that in our Gelfand triplet $< x, \xi >$ is $N(0, |\xi|^2)(x \in E^*)$ we get part (b).

Corollary 1 *By an obvious generalization for* $\xi_1, \xi_2, \cdots \in E$ *orthogonal in* H *and for all* $x \in E^*$

(a) $<: x^{\otimes n} :, \xi_1^{\otimes n_1}\hat{\otimes}\xi_2^{\otimes n_2}\hat{\otimes} \cdots >=:< x, \xi_1 >^{n_1}:_{|\xi_1|^2}:< x, \xi_2 >^{n_2}:_{|\xi_2|^2} \cdots$ *where*
$\qquad n_1 + n_2 + \cdots = n,$
(b) $\| \langle : .^{\otimes n} :, \xi_1^{\otimes n_1} \hat{\otimes}\xi_2^{\otimes n_2}\hat{\otimes} \cdots \rangle \|_0 = \sqrt{n_1!n_2!\cdots}|\xi_1|^{n_1}|\xi_2|^{n_2}\cdots .$

<u>Note.</u> If $h \in E^{\hat{\otimes}n}$ then considering the definition of the symmetrized tensor product we find by the corollary that
$\| <: .^{\otimes n} :, h > \|_0 = \sqrt{n!}\,|h|$. If $f \in H^{\hat{\otimes}n}$, We construct a sequence $\{h_j\} \in E^{\hat{\otimes}n}$ such that $h_j \to f$ in $H^{\hat{\otimes}n}$. Then the sequence $\{<: .^{\otimes n} : h_j >\}$ is Cauchy in (L^2). The (L^2)-limit of this sequence gives $<: .^{\otimes n} : f >$. Thus for $f \in H^{\hat{\otimes}n}$, $<:^{\otimes n}:, f >$ is defined for almost all $x \in E^*$.

If $f = f_1 + if_2 \in H_c^{\hat{\otimes}n}$ with $f_1, f_2 \in H^{\hat{\otimes}n}$, we define:
$<: x^{\otimes n} :, f >=<: x^{\otimes n} :, f_1 > +i <: x^{\otimes n} : if_2 >$ and $\| <: .^{\otimes n} :, f \|_0 = \sqrt{n!}|f|..$

Next we show that the multiple Wiener integral can be stated in terms of Wick tensors.

Theorem 1 *Let* $f \in H_c^{\hat{\otimes}n}$. *Then for any* $f \in H_c^{\hat{\otimes}n}$ *and for almost all* $x \in E^*$

$$I_n(f)(x) =<: x^{\otimes n} :, f > .$$

<u>Sketchy Proof.</u> Firstly Corollary 1 holds also for $h_1, h_2, \cdots \in H$ *and for almost* $x \in E^*$, $n_1 + n_2 + \cdots = n$. *To see this it suffices to note that for any* h *and for almost all* $x \in E^*$ *the equality* $<: x^{\otimes n} :,^{\otimes n} >=:< x^{\otimes n}, h >^n: |h|^2$ *holds. But this easily follows by constructing a sequence* $\{\xi_j\} \subset E$ *such that* $\xi_j \to h$ *in* H.

In Sect. 5.10 formula (8) $H_n(W_h(.))$ is $:< , h >:$ of Definition 2. Thus comparing (8) with the above modification of Corollary 1 we conclude that the assertion holds for $f = h_1^{\otimes n_1}\hat{\otimes}h_2^{\otimes n_2}\hat{\otimes} \cdots$ hence for all $f \in H_c^{\hat{\otimes}n}$.
As a consequence of this theorem an element $\phi \in (L^2)$ can be expressed in terms of Wick tensors by

$$\phi(x) = \sum_{n=0}^{\infty} <: x^{\otimes n} :, f_n >, \quad \mu - a.e., x \in E^* \text{and} f_n \in H_c^{\otimes n}.$$

A similar representation can be given for the dual members. Using the Kondratiev-Streit construction any member $\Phi \in (\mathcal{E}_p)_\beta^*$ has a Ito-Wiener decomposition indicated by $\Phi \sim F_n$. Accordingly we have $\Phi(x) = \sum_{n=0}^{\infty} <: x^{\otimes n} :, F_n >$ where $F_n \in$

$(\mathcal{E}_{p,c}^*)^{\hat{\otimes}n}$ and $x \in (\mathcal{E}_{p,c}) \subset (L_c^2)$, thus $: x^{\otimes n} :$ should be considered in $(L_c^2)^{\otimes n}$. The norm of $\Phi \in (\mathcal{E}_p)^*$ is given by

$$\|\Phi\|_{-p,-\beta}^2 = \sum_{n=0}^{\infty} (n!)^{1-\beta} |(A^{-p})^{\otimes n} F_n|^2.$$

Also the canonical bilinear form between $\phi = \sum_{n=0}^{\infty} <: x^{\otimes n} :, f_n >\in (E)_\beta$ and Φ is given by

$$<< \Phi, \phi >>= \sum_{n=0}^{\infty} n! < F_n, f_n > . \quad (2)$$

Definition 4 For $h \in H, : e^{<\cdot,h>} := e^{<\cdot,h>-1/2|h|^2}$.
Note that this is the counterpart of the exponential functional $\mathcal{E}(h)$ given in Sect. 5.4 (10), (also compare with Sect. 5.9 (7)).

Lemma 2 *Consider $h \in H$. Then the Ito-Wiener decomposition of $: e^{<\cdot,h>} :$ is given by*

$$:< e^{<x,h>} := \sum_{n=0}^{\infty} \frac{1}{n!} <: x^{\otimes n} :, h^{\otimes n} >= \sum_{n=0}^{\infty} \frac{1}{n!} I_n(h^{\otimes n}), (i.e. :< e^{<\cdot,h>} :\sim \{\frac{h^{\otimes n}}{n!}\}).$$

Proof By the generating function of the Hermite polynomials with parameter $\sigma^2 = |h|^2$ and $t = 1$ we obtain

$$: e^{<x,h>} := e^{<x,h>-\frac{1}{2}|h|^2} = \sum_{n=0}^{\infty} \frac{1}{n!} :< x, h >^n :_{|h|^2} .$$

This proves the Lemma in view of the Note following Corollary 1.

Definition 5 For $h \in H_c$, the renormalization $: e^{<\cdot,h>} :$ of $e^{<\cdot,h>}$, is defined by $: e^{<\cdot,h>} := \sum_{n=0}^{\infty} \frac{1}{n!} <: \cdot^{\otimes n} :, h^{\otimes n} > .$

<u>Note.</u> We note that also for any $h \in H_c$ we have $: e^{<\cdot,h>} := e^{<\cdot,h>-\frac{1}{2}<h,h>}$.

Lemma 3 *For $\xi \in E_c : e^{< \cdot, \xi >} :\in (E_c)_\beta$.*

Proof Using the variation of the Kondratiev-Streit construction discussed in Sect. 5.10 and Lemma 2 we have

$$\| : e^{<\cdot, xi>} : \|_{p,q,\beta} = \sum_{n=0}^{\infty} (n!)^{1+\beta} 2^{nq} \left(\frac{|\xi|_p^2}{n!} \right)^2 = \sum_{n=0}^{\infty} (n!)^{\beta-1} 2^{nq} |\xi|_p^{2n} < \infty.$$

Definition 6 The (extended) S-transform $S\Phi$ of a generalized function $\Phi \in (E_c)_\beta^*$ is defined to be the function

$$S\Phi(\xi) = << \Phi, \; : e^{<\cdot, f>} : >>, \; f \in E_c.$$

Its restriction to E is also called the S-transform.

Definition 6' (*Alternative forms of the S-transform*)
(i) Using Definition 4

$$S\Phi(\xi) = e^{-\frac{1}{2}<\xi,\xi>} \; << \Phi, \, e^{<\cdot, \xi>} >>, \xi \in E_c.$$

(ii) Also by Sect. 5.9, (7), the exponential functional $,\mathcal{E}(f) = e^{<\cdot, f> - \frac{1}{2}|f|^2} = e^{W_f(\cdot) - \frac{1}{2}|f|^2}$, (as $W_f(\cdot) = < \cdot, f >$, c.f. Sect. 5.10), hence

$$S(\Phi(f) = << \Phi, \mathcal{E}(f) >>, \; f \in E_c.$$

(iii) Besides if $\Phi(x) = \sum_{n=0}^{\infty} <: x^{\otimes n} : F_n >\in (E)_\beta^*, ,$ then its S-transform is given by $S\Phi(\xi) = \sum_{n=0}^{\infty} < F_n, \xi^{\otimes n} >, \xi \in E_c.$ This easily follows from Definition 5 and (2) since $: e^{<\cdot, \xi>} : \sim \frac{1}{n!} \xi^{\otimes n}$.

The S-transform is a fundamental tool in the white-noise distribution theory. Often we provide the S-transform of a distribution to specify it, (see for instance the definition of the Wick product of the distributions ahead). This is a consequence of 1-1 ness of the S-transform

Proposition 1 Let $\Phi, \Psi \in (E)_\beta^*$. If $S\Phi = S\Psi$ then $\Phi = \Psi$.

Proof It suffices to show that $S\Phi = 0$ implies $\Phi = 0$. Suppose $\Phi(x) = \sum_{n=0}^{\infty} <: x^{\otimes n} :, F_n > .$ Then by the Remark above $S\Phi(\xi) = \sum_{n=0}^{\infty} < F_n, \xi^{\otimes n} >, \xi \in E.$ For any $c \in \mathbb{R}$, $S\Phi(c\xi) = \sum_{n=0}^{\infty} c^n < F_n \xi^{\otimes n} >= 0$, which yields $F_0 = 0$. Also for $n \geq 1$, $< F_n, \xi^{\otimes n} >= 0$. The following polarization identity for $F \in (E^*)^{\otimes n}$ can be verified by direct calculation:

$$< F, \; \xi_1 \hat\otimes \cdots \hat\otimes \xi_n >= \frac{1}{n!} \sum_{k=1}^{n} (-1)^{n-k} \sum_{j_1 < \cdots < j_k} < F, (\xi_{j_1} + \cdots + \xi_{j_k})^{\otimes n} > .$$

Then by the above each term on the right hand side of the identity gives $F_n = 0$ for any $n \geq 1$. Hence $F_n = 0$ for all $n \geq 0$ and so $\Phi = 0$. $\qquad\square$

How to characterize $(H_c)_\beta^*$ is another important question. For this purpose one resorts to U_β functionals.

Definition 7 Let $F : H \to \mathbb{C}$ be a complex-valued function on H. Suppose that for $0 \leq \beta < 1$ it satisfies

(I) For any $f, g \in H$, the map $\lambda \to F(g + \lambda f)$ has an entire analytic continuation (ray-continuation) on $\mathbb{C}$;

(II) There exist constants C, $K > 0$ and , $p \in \mathbb{N}_0$, such that

$$|F(zf)| \leq \exp\{K|z|^{\frac{2}{1-\beta}}|f|_p^{\frac{2}{1-\beta}}\}, \forall f \in H,\ z \in \mathbb{C},$$

then F is called a U_β functional.

Theorem 2 *Let* $0 \leq \beta < 1$, *A map* $F : E \longrightarrow \mathbb{C}$ *is the S-transform of some element of* $(E_c)^*_\beta$ *if and only if it is a* U_β *functional.*

Proof We skip the sufficiency part. It results from a technical Lemma and the nuclear theorem (abstract Schwartz kernel theorem) relating the projective tensor products to the bounded linear operators on dual spaces, (c.f. [Hu-Yan], p. 175). As for the necessity let $\Phi \in< (E_c)^*_\beta$. There exists $p, q \in \mathbb{N}$, such that $\Phi \in (H_{-p,-q,\beta})_c$. Put $F(f) = S\Phi(f)$, $f \in E$. Then the extended S-transform of Φ (still denoted by F) is the entire analytic continuation of F onto E_c. Let $\Phi \sim \{g_n\}$. Then by the Remark to Definition 6 $S\Phi(f) = \sum_{n=0}^{\infty} < g_n, f^{\otimes n} > .\ |F(zf)| \leq \sum_{n=0}^{\infty} | < g_n, (zf)^{\otimes n} > | \leq \sum_{n=0}^{\infty} |z|^n |g_n|_{-p} |f|_p^n$.

Multiplying and dividing by $(n!)^{\frac{1-\beta}{2}} 2^{\frac{qn}{2}}$ and applying the Schwarz inequality we find

$$\leq \left(\sum_{n=0}^{\infty} (n!)^{1-\beta} 2^{-qn} |g_n|_{-p}^2\right)^{1/2} \left(\sum_{n=0}^{\infty} (\tfrac{1}{n!})^{1-\beta} 2^{qn} |z|^{2n} - f|_p^{2n}\right)^{1/2}$$

$$\overset{5.10,(13)}{=} \|\Phi\|_{-p,-q,-\beta} \left((\tfrac{1}{\beta})^{1-\beta} 2^{qn} |z|^{2n} |f|_p^{2n}\right)^{1/2}.$$

For some $0 < \rho < 1$ multiply and divide by ρ^n and apply the Hölder inequality for conjugate indices $p = \frac{1}{\beta}$, $p' = \frac{1}{1-p}$:

$$\leq \|\Phi\|_{-p,-q,-\beta} \left(\sum_{n=0}^{\infty} \rho^n\right)^{\beta} \left(\sum_{n=0}^{\infty} \tfrac{1}{n!} 2^{\frac{qn}{1-\beta}} |z|^{\frac{2}{1-\beta}} |f|_p^{\frac{2}{1-\beta}} \rho^{-\frac{n}{1-\beta}}\right)^{1-\beta} =$$

$$\|\Phi\|_{-p,-q,-\beta}(1-\rho)^{-\beta} \times \exp\left\{(1-\beta)2^{\frac{q}{1-\beta}} |z|^{\frac{2}{1-\beta}} |f|^{\frac{2}{1-\beta}} \rho^{-\frac{1}{1-\beta}}\right\}, \quad \text{which} \quad \text{yields}$$

that for some $C > 0$, $K > 0$, $|F(zf)| \leq C \exp\left\{K|z|^{\frac{2}{1-\beta}} |f|_p^{\frac{2}{1-\beta}}\right\}, \forall f \in E,\ z \in \mathbb{C}$. Hence comparing with Definition 7, (II) we find that F is a U_β functional.

Examples of Distributions in Relation to S-transforms

(i) Generalized Exponential

For $\zeta \in E_c^*$ as inspired by Definition 4, put $\mathcal{E}(\zeta) \sim \{\frac{1}{n!}\zeta^{\otimes n}\}$. Then $\mathcal{E}(\zeta) \in (E)_c^*$. If $\zeta \in H_{-p,c}$, then $\forall q \in \mathbb{N}$, we have $\|\mathcal{E}(\zeta)\|_{-p,-q,0} = \sum_{n=0}^{\infty} (n!)^{-1} 2^{-nq} |\zeta|_{-p}^{2n} = \exp\{\zeta^{-q}|\zeta|_{-p}^2\}$. It is a Hida distribution. Its S-transform is obtained by Definition 6', (iii) as $S[\mathcal{E}(\zeta)](f) = \sum_{n=0}^{\infty} < \frac{1}{n!}\zeta^{\otimes n}, f^{\otimes n} >= e^{<\zeta,f>}, \forall f \in E$.

(ii) Gaussian Measure with Complex Parameter.

First for real $\lambda > 0$ define $\mu^{(\lambda)}(B) = \mu(\frac{B}{\lambda})$, $B \in \mathcal{B}(E^*)$. For every $f \in E$ define $G(f) = \int_{E^*} [\mathcal{E}(f)](x)\,\mu^{(\lambda)}dx = \int_{E^*} e^{<x,f>-\frac{1}{2}|f|^2} d\mu^{(\lambda)}dx = e^{<x,\lambda f>-\frac{1}{2}|f|^2} d\mu(x) = \int_{E^*} [\mathcal{E}(\lambda f)](x)\,e^{\frac{\lambda^2-1}{2}} d\mu(x),$

Then in view of the Minlos theorem $\int_{E^*}[\mathcal{E}(\lambda f)](x)\,d\mu(x) = 1$, hence $G(f) = e^{\frac{\lambda^2-1}{2}}$. This is the S-transform of some element in $(E)^*$ by Theorem 2 since it is a U_β-functional with $z = 1$, $\beta = 0$.

Denote $S^{-1}(G)$ by $F(\lambda)$. Then $F(\lambda)$ can be regarded as the generalized Radon-Nikodym derivative of $\mu^{(\lambda)}$ with respect to μ and it is a Hida distribution.

Now for $\lambda \in \mathbb{C}$, $G(f) = e^{\frac{\lambda^2-1}{2}|f|^2}$, $f \in E$ is still the S-transform of some element in $(E)^*_c$. Denote $S^{-1}(G)$ by $F(\lambda)$. We can regard it as the Gaussian measure with complex parameter.

(iii) <u>Delta Functionals</u>

Based on the idea of the delta function in Schwartz theory, we want to define for every $y \in E^*$, a distribution δ_y with the property that for every $\phi \in (E)_c$, $<<\phi, \delta_y>> = \phi(y)$. Then $S\delta_y(f) = <<\mathcal{E}(f), \delta_y>> = \mathcal{E}(f)(y) = \exp\{<y, f> - \frac{1}{2}|f|^2\}$, $\forall f \in E$.

More generally for every $y \in E^*_c$, this is still the S-transform of some generalized Hida distribution which will still be denoted by δ_y.

<u>Wick Product of Distributions</u>

In general the direct product of distributions is not defined (c.f. also the next section). The so-called Wick-product partly remedies the situation.

Firstly consider two generalized exponentials $\mathcal{E}(\eta)$, $\mathcal{E}(\zeta)$; $\eta, \zeta \in E^*_c$ as in Example (i). Then $S[\mathcal{E}(\eta+\zeta)](f) = e^{<\eta+\zeta, f>} = S[\mathcal{E}(\eta)](f)S[\mathcal{E}(\zeta)](f)$. Thus the S-transform of $\mathcal{E}(\eta+\zeta)$ is the product of the S-transforms of $\mathcal{E}(\eta)$ and $\mathcal{E}(\zeta)$. We call $\mathcal{E}(\eta+\zeta)$ the Wick product of $\mathcal{E}(\eta)$ and $\mathcal{E}(\zeta)$ and denote by $\mathcal{E}(\eta) \diamond \mathcal{E}(\zeta) = \mathcal{E}(\eta+\zeta)$. More generally for $\Phi, \Psi \in (E_c)^*_\beta$, $0 \leq \beta < 1$ the product F of their S-transforms, $F(f) = S\Phi(f)S\Psi(f)$, $\forall f \in E$ is also an S-transform of some distribution since they are both U_β functionals. Thus their product is also a U_β functional by Definition 7, (I), (II) and according to Theorem 2 $S^{-1}(F)$ is a distribution called the Wick product of Φ and Ψ and denoted by $\Phi \diamond \Psi$.

We can determine the chaos decomposition of the Wick product.

Lemma 4 *Let $0 \leq \beta < 1$, $\Phi, \Psi \in (E_c)\beta^*$ and $\Phi \sim \{F_n\}$, $\Psi \sim \{G_n\}$. Then $\Phi \diamond \Psi \sim \{H_n\}$, where $H_n = \sum_{m+n=l} F_m \hat{\otimes} G_n$.*

Proof By the Definition 6', (iii) of the S-transform and the Wick product:

$$\sum_{l=0}^{\infty}(H_l, f^{\otimes l}) = S(\Phi \diamond \Psi) = S\Phi(f)S\Psi(f) = \sum_{m=0}^{\infty} <F_m, f^{\otimes m}> \sum_{n=0}^{\infty} <G_n, f^{\otimes n}> = \sum_{l=0}^{\infty}(\sum_{m+n=l} F_m \hat{\otimes} G_n, f^{\otimes l}), \forall f \in E,$$

showing the result.

Example By the examples (ii) and (iii) of distributions we have for $y \in E^*_c$, $[S(\mathcal{E}(y) \diamond F(0)](f) = e^{<y,f> - \frac{1}{2}|f|^2} = S\delta_y$. Thus $\delta_y = \mathcal{E}(y) \diamond F(0)$. This is because the complex linear space generated by the exponential functionals $\{\mathcal{E}(f)\}$ is dense in $(E_c)_\beta$. Thus by Definition 6', (ii) any element of $(E_c)^*_\beta$ is uniquely determined by its S-transform.

We also note that $\mathcal{E}(y) = \delta_y \diamond F(\sqrt{2})$.

In the next section we propose an alternative approach to the product of infinite dimensional distributions [Çap-1]. This article also establishes a connection to Chap. 3.

5.12 Imbedding of Infinite Dimensional Distributions into Colombeau Type Algebras

The Colombeau a-extension $\mathcal{G}_a(H) = \mathcal{X}_{M,a}/\mathcal{N}_a(H)$ introduced in Subsection 3.5.1 (Delcroix & Scarpalezos' construction) is an ifinite dimensional distribution since H was a locally convex, metrizable topological vector space and under identity mapping this is an example of the imbedding of an infinite dimensional distribution into a Colombeau type algebra.

The construction of the factor space Colombeauu a-extension was given in the Subsection 3.5.1. In this section we concentrate on the topological properties of Colombeau a-extension.

For $f \in \mathcal{X}_{M,a}$ and $n \in \mathbb{N}$, the $\underline{n\text{-valuation}}$ of $(f_\epsilon)_\epsilon \in \chi_{M,a}(H)$ denoted $v_n(f)$, is the supremum (possibly infinite) of $q \in \mathbb{Z}$ such that $\mu_n(f(\epsilon)) = O(a_q(\epsilon))$.

A family δ_n of ultrametric pseudo-distances on $\chi_{M,a}(H)$ is defined by $\forall n \in \mathbb{N}$, $\forall (f, g) \in (\chi_{M,a}(H))^2$, $\delta_n(f, g) = e^{-v_n(f-g)}$.

An element $(f_\epsilon)_\epsilon$ of $\chi_{M,a}(H)$ lies in $\mathcal{N}_a(H)$ iff $\forall n \in \mathbb{N}$, $v_n(f) = +\infty$, (or equivalently iff $\forall n \in \mathbb{N}$, $\delta_n(f, 0) = 0$).

The above definitions transfer naturally to the quotient space $\mathcal{G}_a(H)$. The ultrametric uniform structure and the topology constructed in this manner on $\mathcal{G}_a(H)$ is called respectively the $\underline{\text{sharp uniform structure}}$ of $\mathcal{G}_a(H)$ and the $\underline{\text{sharp topology}}$. This topology is Hausdorff and does not depend on the choice of seminorms $\{\mu_n\}$.

The valuations and pseudo-distances satisfy
(1) $\forall \lambda \in \mathbb{C}\backslash\{0\}$, $\forall f \in \mathcal{G}_a(H)$, $v_n(\lambda f) = v_n(f)$,
(2) $\forall (f, g) \in (\mathcal{G}_a(H))^2$, $v_n(f + g) \geq \inf(v_n(f) + v_n(g))$,
(3) $\forall (f, g, h) \in (\mathcal{G}_a(H))^3$, $\delta_n(f, h) \leq \sup(\delta_n(f, g)\delta_n(g, h))$.
(These properties of valuations and pseododistances are in parallel to those discussed in Subsection 3.4.10 (see Definition 3, (Subsection 3.4.10) and the linkage to generalized functions). The last item implies that the distance is ultrametric.

The space $\mathcal{G}_a(H)$ with the sharp topology is not a topological vector space on $\mathbb{C}$.

The space $\mathcal{G}_a(H)$ is an algebra on $\mathcal{G}_a(\mathbb{C})$. Moreover the product in $\mathcal{G}_a(H)$ is continuous for the sharp topology.This follows mainly from the fact that $\chi_{M,a}(H)$ is stable by product. Furthermore when any of the valuations of the terms of a product tends to infinity, so does the same valuation of the product. Thus the product is continuous in $\mathcal{G}_a(H)$. As a consequence $\mathcal{G}_a(H)$ is a topological ring.

Also for any asymptotic scale $\mathcal{G}_a(H)$ is a $\bar{\mathbb{C}}_a$ topological module for the corresponding sharp topologies on $\bar{\mathbb{C}}_a$ and $\mathcal{G}_a(H)$.

In Theorem 3.1 [Del-Scar1] it is stated that $\mathcal{G}_a(H)$ is complete for the sharp uniform structure. The proof of this theorem and also those of the above stated results are found in [Del-Scar2].

Imbedding of Infinite Dimensional Distributions into Simplified Colombeau Type Algebras

(This subsection is based on the article [Çap-1])

Let $\Omega_0 = \mathcal{C}_0([0, 1])$ be the set of all $\mathbb{R}^d$-valued functions on $[0, 1]$, null at zero with the sup-norm (c.f. Sect. 5.1),

μ : : the Wiener measure on $(\Omega_0, \mathcal{B}(\Omega_0))$,

For $t \in [0, 1]$, $\omega \in \Omega_o$, $W_t(\omega) \equiv \omega(t)$ is a $\mathbb{R}^d$-valued Wiener functional and $W(t, \omega)$ is a d-dimensional Wiener process, $\mathcal{F}_t$; $t \in [0, 1]$: the natural filtration generated by the process. $\mathcal{F} = \mathcal{F}_\infty$ is the μ-completion of the sigma-algebra $\mathcal{B}(\Omega_0)$, so that $(\Omega_0, \mathcal{F}, \mu)$ is our Wiener space.

H : the Cameron-Martin space (C-M space), is the Hilbert space formed by all $h \in \Omega_0$, such that each component of $h(t) = (h_1(t), \cdots, h_d(t))$ is absolutely continuous and has square integrable derivatives. The inner product in H : for $g, h \in H$

$$(g, h)_H = \int_0^1 \sum_{i=1}^d \dot{g}_i(t)\, \dot{h}_i(t)dt. \text{ and the Hilbertian norm is } |h|_H^2 = \int_0^1 |\dot{h}(t)|^2 dt. \quad (1)$$

A $\mathcal{F}$-measurable function $F : \Omega_0 \to \mathbb{R}$ is a Wiener polynomial $\mathcal{P}$ (resp. smooth S_M) functional and denoted like in Sect. 5.1 satisfying $\mathcal{P} \subset S_M \subset L^p \equiv L^p(\Omega_0, \mathcal{F}, \mu)$ and also $\mathcal{P}$ is dense in L^p ($1 \le p < \infty$).

The weak derivative (directional derivative in the directions of C-M elements), the gradient, the chain rule etc. of Sect. 5.3 (Malliavin calculus) are valid, (c.f. formulae (3)–(10)).

$\|F_{k,p}\|$ norms are given by (3)–(10). The Meyer-Watanabe testing functionals $\mathbb{D}^\infty(E)$, $\mathbb{D}^\infty$ and the Meyer-Watanabe generalized functions are given by Sect. 5.6, (6), (7) and by the Remark 1 prior to Theorem 1, Sect. 5.6.

Construction of the algebra $\mathcal{H}_s$

Let χ be the set of all functions from $(0, 1]$ into $\mathbb{D}^\infty$. A member of χ will then be denoted by $(f_\epsilon)_{\epsilon \in (0,1]}$, in an abbreviated notation by $(f_\epsilon)_\epsilon \in \chi \equiv (\mathbb{D}^\infty)^{(0,1]}$.

Definition 1 $(f_\epsilon)_\epsilon$ is called moderate if given (k, p), $k \in \mathbb{N}, 1 < p < \infty$, $\exists N \in \mathbb{N}$ such that $\| f_\epsilon \|_{k,p} = O(\epsilon^{-N})$ as $\epsilon \to 0$.

The set of all moderate elements in χ is denoted by $\mathcal{M}_s(\Omega_0, \mathcal{F}, \mu; H)$.

Definition 2 $(f_\epsilon)_\epsilon$ is called negligible if given (k, p), $k \in \mathbb{N}, 1 < p < \infty$, $\exists N \in \mathbb{N}$ such that $\| f_\epsilon \|_{k,p} = O(\epsilon^{m-N})$, $\forall m \ge N$ as $\epsilon \to 0$.

The set of all negligible elements of χ is denoted by $\mathcal{J}_s(\Omega_0, \mathcal{F}, \mu; H)$.

Definition 3 $\mathcal{H}_s = \mathcal{M}_s(\Omega_0, \mathcal{F}, \mu; H)/\mathcal{J}_s(\Omega_0, \mathcal{F}, \mu; H)$ will be called the Colombeau extension of $\mathbb{D}^\infty$, can also be named the space of Wiener-Colombeau distributions.

<u>Note.</u> The subindex 's' is reminiscent of the word 'simplified'. For $\mathbb{D}^\infty$ being replaced by $\mathcal{C}^\infty(\Omega)$, $\Omega \subset \mathbb{R}^d$ and $\| \cdot \|_{k,p}$ by single index semi-norms in terms of supremums of partial derivatives on increasing compacts we retrieve the simplified Colombeau distributions ([Del-Scar2] and 3.4.8).

Definition 4 Two members $(f_\epsilon)_\epsilon$, $(g_\epsilon)_\epsilon$ of χ are said to be ϵ-equivalent-if $\lim_{\epsilon \to 0}(f_\epsilon - g_\epsilon) = 0$.

Definition 5 Two members $(f_\epsilon)_\epsilon$, $(g_\epsilon)_\epsilon$ of χ are said to be law-equivalent if they have the same probability distribution.

<u>Conclusions</u>
(1) $\mathcal{M}_s$ is an algebra. Let $(f_\epsilon)_\epsilon \in \mathcal{M}_s$, define $(f_\epsilon)_\epsilon \, (g_\epsilon)_\epsilon \doteq (f_\epsilon \cdot g_\epsilon)_\epsilon$ since $\mathbb{D}^\infty \equiv \mathbb{D}^\infty(\mathbb{R})$ is an algebra.
Given (k, p), $\exists N_1, N_2 \in \mathbb{N}$, $\|f_\epsilon\|_{k,p} = O(\epsilon^{-N_1})$ and $\|g_\epsilon\|_{k,p} = O(\epsilon^{-N_2})$. Then $\|f_\epsilon \cdot g_\epsilon\|_{k,p} = O(\epsilon^{-N_1-N_2})$, hence $(f_\epsilon \cdot g_\epsilon) \in \mathcal{M}_s$.
(2) $\mathcal{J}_s$ is an ideal in $\mathcal{M}_s$. Let $(f_\epsilon)_\epsilon \in \mathcal{J}_s$ and $(g_\epsilon)_\epsilon \in \mathcal{M}_s$. Given (k, p), $\exists N_1, N_2 \in \mathbb{N}$ such that $\|f_\epsilon\|_{k,p} = O(\epsilon^{m-N_1}) \forall m \geq N_1$, $\|g_\epsilon\|_{k,p} = O(\epsilon^{-N_2})$. Then $\|f_\epsilon \cdot g_\epsilon\|_{k,p} = O(\epsilon^{m-N})$, $\forall m \geq N = N_1 + N_2$.
(3) $\mathcal{H}_s$ is a factor algebra. If $F \in \mathcal{H}_s$, it is of the form $(f_\epsilon)_\epsilon + \mathcal{J}_s$, $(f_\epsilon)_\epsilon \in \mathcal{M}_s$.
(4) By the monotonicity of the norms $\| \cdot \|_{k,p}$, if the conditions in Definitions 4 and 5 hold for a certain pair (k, p), then they also hold for all pairs in the cone $\{(k', p') : k' \leq k, \, p' \leq p\}$.
<u>Inclusions.</u>
(A) $\mathbb{D}^\infty \subset \mathcal{M}_s$. If $F \in \mathbb{D}^\infty$, then take $(f_\epsilon)_\epsilon \equiv F \ \forall \epsilon \in (0, 1]$. Then $\mathbb{D}^\infty$ is a faithful algebra of $\mathcal{H}_s$.
(B) $\mathbb{D}^{-\infty} \subset \mathcal{H}_s$, (Inclusion of Meyer-Watanabe distributions)
   Firstly consider the canonical $\mathbb{R}^d$-valued functional $W_1(\omega) = \omega(1)$, $(\omega \in \Omega_0)$. Then $DW_{1,i}(t) = te_i \in H$, $D_h W_{1,i} = h_i(1)((e_1, \cdots, e_d)$ is an orthonormal basis in $\mathbb{R}^d$) which satisfies Sect. 5.3, (4) in view of (1) of the inner product in C-M spaces.
   Given $T \in \mathbb{D}^{-\infty}$, we define the following functional in convolution sense :

$$f^\phi_{T,\epsilon}(\omega) = \left\langle T(\tilde{\omega}), \frac{1}{\epsilon^d}\phi\left(\frac{\omega(1) - \tilde{\omega}(1)}{\epsilon}\right)\right\rangle, \epsilon \in (0, 1], \, \omega, \tilde{\omega} \in \Omega_0, \, \phi \in \mathcal{D}(\mathbb{R}^d). \quad (2)$$

$(< ., . >$ denotes the bilinear form on $\mathbb{D}^\infty \times \mathbb{D}^{-\infty})$ and for fixed ϵ, $\phi(\frac{\omega(1)-\tilde{\omega}(1)}{\epsilon}) \in \mathbb{D}^\infty$.

   We show that $f^\phi_{T,\epsilon} \in \chi$ and also it is moderate.
   With repetitive use of Sect. 5.3, (4)

$$D_h f^\phi_{T,\epsilon}(\omega) = \lim_{h \to 0} \frac{1}{\epsilon^d}\left\langle T(\tilde{\omega}), \phi\left(\frac{\omega(1) + \lambda h(1) - \tilde{\omega}(1)}{\epsilon}\right) - \phi\left(\frac{\omega(1) - \tilde{\omega}(1)}{\epsilon}\right)\right\rangle$$

$$= \frac{1}{\epsilon^d}\left\langle T(\tilde{\omega}), \left(D\phi\left(\frac{\omega(1) - \tilde{\omega}(1)}{\epsilon}\right), h\right)_H\right\rangle \overset{5.3,(7)}{=} \frac{1}{\epsilon^{d+1}}\left\langle T(\tilde{\omega}), \left(\sum_{i=1}^d \partial_i\phi\left(\frac{\omega(1) - \tilde{\omega}(1)}{\epsilon}\right) te_i, h\right)\right\rangle$$

$$= \frac{1}{\epsilon^{d+1}} \left(\sum_{i=1}^{d} \left\langle T(\tilde{\omega}) \, \partial\phi \left(\frac{\omega(1) - \tilde{\omega}(1)}{\epsilon} \right) \right\rangle t e_i, \, h \right)_H$$

Thus

$$Df_{T,\epsilon}^{\phi}(\omega) = \frac{1}{\epsilon^{d+1}} \sum_{i=1}^{d} \left\langle T(\tilde{\omega}, \, \partial_i \phi \left(\frac{\omega(1) - \tilde{\omega}(1)}{\epsilon} \right) \right\rangle t e_i \in L^p(H), \, t \in (0, 1]$$

By a similar calculation

$$D^2 f_{T,\epsilon}^{\phi}(\omega) = \frac{1}{\epsilon^{d+2}} \sum_{i,j=1}^{d} \left\langle T(\tilde{\omega}), \, \partial_{i,j}^2 \phi \left(\frac{\omega(1) - \tilde{\omega}(1)}{\epsilon} \right) \right\rangle \tau e_i \otimes t e_j \in L^p(H \otimes H), \, \tau, t \in (0, 1].$$

By induction

$$D^k f_{T,\epsilon}^{\phi}(\omega) = \frac{1}{\epsilon^{d+k}} \sum_{i_1, i_2, \cdots, i_k} \left\langle T(\tilde{\omega}) \partial_{i_1, i_2, \cdots, i_k} \phi \left(\frac{\omega(1) - \tilde{\omega}(1)}{\epsilon} \right) \right\rangle \otimes_{r=1}^{k} t_r e_{i_r} \in L^p) H^{\otimes k}, \, t_r \in (0, 1].$$

The calculations involved and the verification of the inductive hypothesis are found in [Çap-1].

With a straightforward estimation using the Minkowski inequality we find $|D^k f_{T,\epsilon}^{\phi}|_{H^{\otimes k}}^{p} < \infty$ so that $D^k f_{T,\epsilon}^{\phi} \in L^p(H^{\otimes k})$, $k \in \mathbb{N}\backslash 0$. Then referring to Sect. 5.5, (2) we have $\|f_{T,k}^{\phi}\|_{k,p} < \infty$. and as (k, p) is arbitrary $(f_\epsilon)_\epsilon \in \chi$ by the definition of $\mathbb{D}^\infty$ in Sect. 5.6. Furthermore $\|f_{T,\epsilon}^{\phi}\|_{k,p} = O(\epsilon^{-k-p})$, hence $(f_{T,\epsilon})_\epsilon \in \mathcal{M}_s$. Its class in $\mathcal{H}_s$ is a representative of $T \in \mathbb{D}^{-\infty}$.

If $\phi^* \in \mathcal{D}(\mathbb{R}^d)$ is another Schwartz test function,then $\lim_{\epsilon \to 0}(f_{T,\epsilon}^{\phi}) - f_{T,\epsilon}^{\phi^*} = 0$ so that $f_{T,\epsilon}^{\phi}$ and $f_{T,\epsilon}^{\phi^*}$ are ϵ-equivalent.

(C) Inclusion of Hida Distributions
(I) Real Case
We consider the Gelfand triplet as introduced in the first paragraph of Sect. 5.10. $(E^*, \mathcal{B}(E^*), \mu; H)$ forms a canonical Gaussian probability space where μ is the Gaussian measure provided by the Minlos theorem. Denote the inner product and the norm in H (or $H^{\otimes n}$) by $(., .)$ and $|.|$ respectively.

Using the second quantization operator the Hida testing functional space (E) and its dual $(E)^*$ can be constructed, the details of which are covered in Sect. 5.10.

On the other hand the Meyer-Watanabe functional space $\mathbb{D}^\infty$ and the distribution space $\mathbb{D}^{-\infty}$ can also be constructed on $(E^*, \mathcal{B}(E^*), \mu)$ as well, the C-M space being replaced by H. The starting point of the weak differential calculus will be then

$$(Df, h)_H = \lim_{\lambda \to 0} \frac{F(x + \lambda h) - F(x)}{\lambda}, \, x \in E^*,, \, h \in H. \quad (3)$$

For $f \in E$, $W_f(x) = <x, f>$ as before. By the Minlos theorem $\int_{E^*} e^{i<x,f>} d\mu(x) = e^{-\frac{1}{2}|f|^2}$, so that W_f is a Gaussian random variable on E^*. The linear map $f \to W_f : H \to L^2(E^*, \mu)$ can be extended to a linear isometry from H into $L^2(E^*, \mu)$. Now

$$(DW_f, h) = \lim_{\lambda \to 0} \frac{< f, x + \lambda h > - < f, x >}{\lambda} = < f, h > = (f, h)_H$$

yielding $DW_f = f \in E \subset H$.

It is known from Sect. 5.10 that $(\mathbb{D}^\infty, \mathbb{D}^{-\infty})$ when constructed on $(E^*, \mathcal{B}(E^*), \mu)$ satisfies $(E) \subset \mathbb{D}^\infty$ and $\mathbb{D}^{-\infty} \subset (E)^*$ both inclusions being dense. For $T \in (E)^*$, the counterpart of (2) (inclusion B)) will be ($d = 1$):

$$g_{T,\epsilon}^{\phi,f}(x) = \frac{1}{\epsilon} << T(\tilde{x}), \left(\frac{W_f(\tilde{x}) - W_f(x)}{\epsilon} \right) >> ; x, \tilde{x} \in E^*, \ \phi \in \mathcal{D}(\mathbb{R}), \ |f| = 1 \quad (4)$$

where $<< ., . >>$ is the canonical bilinear form on $((E)^*, (E))$.

The Polynomial functionals in the variables $\{W_{f_j}\}$ ($j = 1, \cdots, n$) belong to $\mathbb{D}^\infty \cap (E)$ and are dense in both of them. The right member of the bilinear form in (4) can be regarded as the limit of the Taylor polynomials in W_f and is in (E) due to its completeness as a countably Hilbertian nuclear space (in fact a smooth functional).

It is shown that $(g_{T,\epsilon}^{\phi,f})_\epsilon \in \chi$ and it is also moderate:

$$(Dg_\epsilon, h)_H = \frac{1}{\epsilon^2} << T(\tilde{x}), \phi' \left(\frac{W_f(x) - W_f(\tilde{x}}{\epsilon} \right) >> (f, h)_H$$

$$= \frac{1}{\epsilon^2} \left(<< T(\tilde{x}), \phi' \left(\frac{W_f(x) - W_f(\tilde{x}}{\epsilon} \right) >> f, h \right)_H$$

which yields $Dg_\epsilon(x) = \frac{1}{\epsilon^2} << T(\tilde{x}), \phi' \left(\frac{W_f(x) - W_f(\tilde{x})}{\epsilon} \right) f >>$. Following a similar calculation of $(D_h Dg_\epsilon, \tilde{h})_H$ we find $D^2 g_{T,\epsilon}^{\phi,f}(x) = \frac{1}{\epsilon^3} << T(\tilde{x}), \phi'' \left(\frac{W_f(x) - W_f(\tilde{x})}{\epsilon} \right) >> (f \otimes f)$.

By induction

$$D^k g_{T,\epsilon}^{\phi,f}(x) = \frac{1}{\epsilon^{k+1}} << T(\tilde{x}), \phi^{(k)} \left(\frac{W_f(x) - W_f(\tilde{x})}{\epsilon} \right) >> f^{\otimes k} = \frac{1}{\epsilon^{k+1}} << T(\tilde{x}, \phi^{(k)} \left(\frac{< x, f > - < \tilde{x}, f >}{\epsilon} \right) >> f^{\otimes k}.$$

(For the calculation details and the verification of the inductive hypothesis c.f. [Çap-2]).

As clearly $|D^k g_{T,\epsilon}^{\phi,f}|_{H^{\otimes k}}^p < \infty$ we have $D^k g_{T,\epsilon}^{\phi,f} \in L^p(H^{\otimes k})$, so that referring to Sect. 5.3, (6) $(g_{T,\epsilon}^{\phi,f})_\epsilon \in \chi$ and furthermore $(g_{T,\epsilon}^{\phi,f})_\epsilon \in \mathcal{M}_s$. As $\|g_{T,\epsilon}^{\phi,f}\|_{k,p} = O(\epsilon^{-k-1})$, we then have $(E)^* \in \mathcal{H}_s$.

Different representatives obtained by $\tilde{\phi} \in \mathcal{D}(\mathbb{R})$ and/or by $\tilde{f} \in E$, $|\tilde{f}| = 1$ are either ϵ-equivalent or law equivalent or a combination of them according to Definitions 4 and 5.

(II) Complex Case

For any topological vector space V on $\mathbb{R}$, denote by $V_{\mathbb{C}}$ its complexification, i.e. $V_{\mathbb{C}} = V + iV$.

If V is a Hilbert space and $u_1 + iv_1$ and $u_2 + iv_2$ are in $V_{\mathbb{C}}$, then their inner product is $(u_1 + iv_1, u_2 + iv_2)_{\mathbb{C}} = (u_1, u_2) + (v_1, v_2) + i[(u_2, v_1) - (u_1, v_2)]$, thus $|u_1 + iv_1|_{\mathbb{C}}^2 = |u_1|_V^2 + |v_1|_V^2$.

Let $\chi_{\mathbb{C}}$ be the set of all functions $(0, 1] \to \mathbb{D}_{\mathbb{C}}^\infty = \mathbb{D}^\infty + i\mathbb{D}^\infty$. Previous definitions are modified accordingly to define $\mathcal{H}_{,\mathbb{C},s}$, $\epsilon-$and law equivalences. $\mathcal{H}_{\mathbb{C},s}$ is also an algebra.

Consider complex-valued functionals on the measure space $(E^*, \mathcal{B}(E^*), \mu)$. We let $E_{\mathbb{C}} \ni f$, $f = f_1 + if_2$, $(f_j \in E; j = 1, 2$. Then for $x \in E^*$, $W_f = <x, f_1 + if_2> = <x, f_1> + i <x, f_2>$. The real and the imaginary parts of W_f are independent Gaussian variables with characteristic functions $e^{-\frac{1}{2}|f_j|^2}$, $(j = 1, 2)$. If we take $|f_1|^2 = |f_2|^2 = \frac{1}{2}$. Then $W_f \in \mathbb{C}N(0, 1)$, i.e. a standard complex-valued Gaussian random variable.

The linear map $f \to W_f$ from $E_{\mathbb{C}}$ to $L_{\mathbb{C}}^2(E^*, \mu) \equiv (L^2)_{\mathbb{C}}$ can be extended to a linear isometry $H_{\mathbb{C}} \to (L)_{\mathbb{C}}^2$. For $T \in (E)_{\mathbb{C}}^*$ define

$$\Theta_{T,\epsilon}^{\phi,f}(x) = \frac{1}{\epsilon} << T(\tilde{x}), \phi\left(\frac{<x, f_1> - <\tilde{x}, f_1>}{\epsilon}\right) + i\phi\left(\frac{<x, f_2> - <\tilde{x}, f_2>}{\epsilon}\right) >>, f = f_1 + if_2; x, \tilde{x} \in E^*$$

(The use of two test functions ϕ_1, ϕ_2 will not bring in an essential difference). The real and imaginary parts of the right side of $<< ., . >>$ are both in (E) according to the real case. Therefore the right side belongs to $(E)_{\mathbb{C}}$. The calculation of the gradients is basically the same as in the previous section except modifications for the complex case. Thus we arrive at

$$D^k \Theta_{T,\epsilon}^{\phi,f}(x) \frac{1}{\epsilon^{k+1}} << T(\tilde{x}), \phi^{(k)}\left(\frac{<x, f_1> - <\tilde{x}, f_1>}{\epsilon}\right) >> f_1^{\otimes k}$$

$$+ i\frac{1}{\epsilon^{k+1}} << T(\tilde{x}), \phi^{(k)}\left(\frac{<x, f_2> - <\tilde{x}, f_2>}{\epsilon}\right) >> f_2^{\otimes k}.$$

In parallel to the evaluations of the real case, the $\|.\|_{k,p}$ norms of the real and imaginary parts of the right-hand sides in the last expression are finite. This then shows that for arbitrary $k \in \mathbb{N}$ and $1 < p < \infty$ we have $\|\Theta_{T,\epsilon}^{\phi,f}\|_{k,p} < \infty$. Thus $\Theta_{T,\epsilon}^{\phi,f} \in \chi_{\mathbb{C}}$ and it is also moderate and represents $T \in (E)_{\mathbb{C}}^*$ in $\mathcal{H}_{\mathbb{C},s}$. In this way $(E)_{\mathbb{C}}^* \subset \mathcal{H}_{\mathbb{C}}$.

Topology in $\mathcal{H}_s$

$\mathbb{D}^\infty$ is a countably normed space for a strictly increasing set $\{p_m\}$, $(1 < p_m < \infty)$ $(m = 1, 2, \dots)$ of numbers. In fact due to the monotonicity of $\|.\|_{k,p}$ norms, we have $\mathbb{D}^\infty = \bigcap_{k \in \mathbb{N}} \bigcap \mathbb{D}_k^{p_m}$. Arrange the set of countable norms as

$$\| \cdot \|_{1,p_1}; \ \| \cdot \|_{2,p_1}, \ \| \cdot \|_{1,p_2}; \ \| \cdot \|_{3,p_1}, \ \| \cdot \|_{2,p_2}, \ \| \cdot \|_{1,p_3}; \cdots; \ \| \cdot \|_{n,p_1}, \ \| \cdot \|_{n-1,p_2}, \cdots \| \cdot \|_{1,p_n}; \cdots$$

Then we have a new set of countable norms μ_n, $(n = 1, 2, \cdots)$ by

$$\mu_n = \left(\sum_{j=1}^{n} \| \cdot \|_{n+1-j,p_j} \right)^{1/2} \qquad (5)$$

which are increasing, also $\{\mu_n\}_{n \in \mathbb{N}}$ and $\{\| \cdot \|_{k,p}\}$ norm systems are equivalent and create the same topology. Furthermore $(f_\epsilon)_\epsilon \in \mathcal{H}_s$ is moderate in $\{\| \cdot \|_{k,p}$ norms $\Longleftrightarrow$ it is moderate in $\{\mu_n\}$ norms.

For $(f_\epsilon)_\epsilon \in \mathcal{M}_s$, the n-valuation of f, denoted by $v_n(f)$ is $\sup_{b \in \mathbb{Z}}\{\mu_n(f_\epsilon) = O(\epsilon^b)\}$. A family $\{\delta_n\}$ of ultrametric pseudo-distances on $\mathcal{M}_s$ is defined by $\delta_n(f, g) = e^{-v_n(f-g)}$, $f, g \in \mathcal{M}_s$.

$(f_\epsilon)_\epsilon$ lies in $\mathcal{J}_s \Longleftrightarrow \forall n \in \mathbb{N}, \ v_n(f) = +\infty$, (or equivalently $\forall n \in \mathbb{N}, \ \delta_n(f, 0) = 0$.

These definitions transfer naturally to the quotient space $\mathcal{H}_s$. The ultrametric uniform structure and the topologyconstructed are called the 'sharp uniform structure' and the 'sharp topology' respectively.

Theorem *The space $\mathcal{H}_s$ is complete for the sharp uniform structure.*

Proof Based on Proposition 1.31 of Garetto ([Gar-1]) and Theorem 3.1 [Del-Scar2], (see [Çap-2] for details.

Application to the Feynman Integrand
The Donsker's delta function in white noise theory set-up is given by

$$\delta_a(W_f) \equiv \delta(W_f - a), \ W_f = <x, f> \ x \in E^*, \ f \in E, \ a \in \mathbb{R}. \qquad (6)$$

It is shown to be a Hida distribution, therefore included in $\mathcal{H}_s$ with a representative

$$R(x) = \frac{1}{\epsilon} \phi \left(\frac{W_f(x) - a}{\epsilon} \right), \ \phi \in \mathcal{D}(\mathbb{R}).$$

As before we have

$$DR(x) = \frac{1}{\epsilon^2} \phi' \left(\frac{<f, x> - a}{\epsilon} \right) f, \cdots, D^k R(x) = \frac{1}{\epsilon^{k+1}} \phi^{(k)} \left(\frac{<f, x> - a}{\epsilon} \right) f^{\otimes k}.$$

Other representatives are ϵ-equivalent.

$\delta(B_t)$, (or $\delta(B_t - a)$ where B_t is a standard Brownian motion starting from 0 is also a Donsker's delta function. In the representative we may select $|f| = 1$, $f \in E$ and $W_{\sqrt{f}t}$

As discussed in 11.5, Example (ii) of generalized functions $\forall \lambda \in \mathbb{C}, \ G(f) = \int_{E^*} [\mathcal{E}(f)] \mu^{(\lambda)} dx = \exp \frac{\lambda^2 - 1}{2} |f|^2; \ f \in E$ and $\mathcal{E}(f)$ a generalized exponential functional, is the S-transform of some element in $(E)^*$. $F(\lambda) = S^{-1} G$ is a Hida distribution thus in $\mathcal{H}_{s,\mathbb{C}}$.

The following short review is from [Hu-Yan] and [Khan-Str], (the Gelfand triplet can be taken as $S(\mathbb{R}^d \subset L^2(\mathbb{R}^d) \subset S'(\mathbb{R}^d))$ where $S'(\mathbb{R}^d)$ is the set of tempered distributions.

Consider the Schrödinger equation

$$\frac{\partial \Psi}{\partial t} = i\left(\frac{\Delta}{2} - V\right)\Psi; \ \Psi(0, x) = f(x) \quad (7)$$

where Δ is the Laplace operator in $\mathbb{R}^d$, V is a real Borel function on $\mathbb{R}^d$, (the potential). From the point of white noise analysis it is preferable to start with the heat equation

$$\frac{\partial u}{\partial t} = \left(\frac{\lambda}{2}\Delta - iV\right)u; \ u(0, x) = f(x) \text{where} \lambda > 0. \quad (8)$$

The solution of (8) is given by the Feynman-Kac formula as

$$u(t, x) = E\left[f(\sqrt{\lambda}B_t + x)\exp\{-i\int_0^t V(\sqrt{\lambda}B_s + x)ds\}\right]$$

In terms of the Hida distribution $F(\sqrt{\lambda})$ (see Sect. 5.11) this can be rewritten as

$$u(t, x) = \int f(B_t + x)e^{-i\int_0^t V(B_s+x)ds}d\mu^{(\sqrt{\lambda})} =<< F((\sqrt{\lambda})f(B_t + x)e^{-i\int_0^t V(B_s+x)ds}, 1 >> .$$

Let $f(x) = \delta_y(x)$. Then the fundamental solution of (8) becomes

$$u^\lambda(t, x, y) =<< F(\sqrt{\lambda})\,\delta(B_t - y + x)e^{-i\int_0^t V(B_s+x)ds}, 1 >> \quad (9)$$

Suppose u^λ has an analytic continuation in λ, (the conditions for validity of this analytic continuation are given in Corollary 4.4, [Yan]), then the fundamental solution of (7) can be asserted to be

$$\Psi(t, x, y)^\lambda =<< F(\sqrt{i})\,\delta(B_t - y + x)\,e^{-i\int_0^t V(B_s+x)ds}, 1 >>, F(\lambda) \in (E)_{\mathbb{C}}^* \quad (10)$$

The two factors $F(\sqrt{i})$ and $\delta(B_t - y + x)$ are already known to be in $\mathcal{H}_{s,\mathbb{C}}$ by Sect. 5.11 and the above discussions. Therefore it remains to discuss the meaning of the exponential factor.

$V(B_s + x)$ can be expressed via Bochner integral as

$$V(B_s + x) = \int_{\mathbb{R}^d} V(z)\delta(B_s + x - z)dx. \quad (11)$$

For $\Delta_n = \{(t_1, \cdots, t_n) : 0 < t_1 < \cdots < t_n < t\}$

$$\exp\{-i \int_0^t V(B_s + x)ds = \sum_{n=0}^{\infty} (-i)^n \int_{\Delta_n} \prod_{j=1}^{n} V(B_{t_j} + x) \, d^n t$$

$$= \sum_{n=0}^{\infty} (-i)^n \int_{\Delta_n} d^n t \prod_{j=1}^{n} \int_{\mathbb{R}^d} V(z_j) \, \delta(B_{t_j} + x - z_j) dz_j \quad (12)$$

so that by (10) the full Feynman integrand for the propagator becomes

$$\sum_{n=0}^{\infty} (-i)^n \int_{\Delta_n} d^n t \prod_{j=1}^{n} \int_{\mathbb{R}^d} V(z_j) \, \delta(B_{t_j} + x - z_j) \, dz_j \, F(\sqrt{i}) \delta(B_t - y + x)$$

$$= \sum_{n=0}^{\infty} (-i)^n \int_{\Delta_n} d^n t \prod_{j=1}^{n} \int_{\mathbb{R}^d} V(z_j) \, dz_j \, F(\sqrt{i}) \, \delta(B_t - y + x) \prod_{j=1}^{n} \delta(B_{t_j} + x - z_j). \quad (13)$$

In [Khan-Str] and [Yan] considerable effort is spent to give a reasonable meaning to the factor $F(\sqrt{i})\delta(\ )\prod(\ )$ as a Hida distribution. But clearly this factor in the algebra $\mathcal{H}_{s,\mathbb{C}}$. Furthermore in [Khan-Str] it is shown that under the assumption that $d = 1$, and $V(y)dy$ is a compactly supported signed measure, the product of the three terms in the right-hand side of (10) is a Hida distribution.

To perform the t_i integration in (12) and convergence of the series involved, the assumption $d = 1$ is essential. For in the evaluation, the integral $M_n = \int_{\Delta_n} = d^n t \prod_{i=1}^{n} |t_i - t_{i-1}|^{-d/2}$ is needed and it exists only for $d = 1$. In this case M_n rapidly decreases and $\sum_{n=1}^{\infty} M_n < \infty$.

However in our case when $d = 1$, without needing the introduction of the net Δ_n and the elaborate evaluation of (12) and (13) we can show the following:

Theorem *If $d = 1$ and $V(z)dz$ is a bounded signed measure, then the product of the three terms in (10) is in $\mathcal{H}_{s,\mathbb{C}}$.*

Proof Firstly let us show that $V(B_s + x) = \int_{\mathbb{R}} V(z) \delta(B_s + x - z)dz$ given by (11) is in $\mathcal{H}_s$. It will have a representative,

$$\Gamma_\epsilon^{\phi,f}(psi) = \frac{1}{\epsilon} \int_{\mathbb{R}} V(z) \phi \left(\frac{W_{\sqrt{s}f}(\psi) - (z - x)}{\epsilon} \right) dz; \ f \in E, \ |f| = 1, \ \phi \in \mathcal{D}(\mathbb{R}), \ \psi \in E^*.$$

By bounded convergence theorem the operators D_h and D can be inserted into the integral and an approach parallel to those in previous sections yields

$$D\Gamma_\epsilon^{\phi,f}(\psi) = \frac{1}{\epsilon^2} \left[\int_{\mathbb{R}} \phi' \left(\frac{\sqrt{s} <\psi, f> -(z - x)}{\epsilon} \right) \sqrt{s} V(z) \, dz \right] f$$

$$D^k \Gamma_\epsilon^{\phi,f}(\psi) = \frac{1}{\epsilon^{k+1}} \left[\int_{\mathbb{R}} \phi^{(k)} \left(\frac{\sqrt{s} <\psi, f> -(z - x)}{\epsilon} \right) s^{k/2} V(z) dz \right] f^{\otimes k}$$

$$|D^k \Gamma_\epsilon^{\phi,f}|_{H^{\otimes k}}^p = \frac{1}{\epsilon^{k+1}} \Big| \int_{supp\phi^{(k)}} \phi^{(k)} \left(\frac{\sqrt{s} <\psi, f> -(z - x)}{\epsilon} \right) s^{k/2} V(z)dz \Big|^p < \infty$$

since $|f^{\otimes k}|_{H^{\otimes k}} = 1$. As $V(z)^+ dz - V(z)^- dz$ is a bounded signed measure we find that $\|D^k \Gamma_\epsilon^{\phi,f}\|_{L^p(\mu,H^{\otimes k})} < \infty$. Since $n \in \mathbb{N}$ and $, 1 < p < \infty$ are arbitrary it follows that $\Gamma_\epsilon^{\phi,f} \in \mathbb{D}^\infty$, $(0 < \epsilon < 1)$. Furthermore $\|\Gamma_\epsilon^{\phi,f}\|_{k,p} = O(\epsilon^{-k-1})$, hence $(\Gamma_\epsilon^{\phi,f})_\epsilon \in \mathcal{M}_s$. Then its class in $\mathcal{H}_s$ represents $V(B_s + x)$.

$\int_0^t V(B_s + x)ds$ regarded as the limit of finite sums over the partitions of $(0, 1]$ is in $\mathcal{H}_s$ due to its completion under its topology.

Now $e^{-i \int_0^1 V(B_s+x)ds} = \cos \int_0^1 V(B_s + x)ds - i \sin \int_0^1 V(B_s + x)ds$ and as a complex linear combination of smooth bounded functions of members of $\mathcal{M}_s$, this expression is included in $\mathcal{M}_{s,\mathbb{C}}$. Its class in $\mathcal{H}_{s,\mathbb{C}}$ represents $e^{-i \int_0^1 V(B_s+x)}ds$. As a consequence each factor of the product in (10) is in $\mathcal{H}_{s,\mathbb{C}}$, so is their product. $\square$

GLOSSARY

Absorbent set.   In a vector space the set A is said to absorb the set B if there is some $\alpha > 0$ such that $B \leq \lambda A$ for all $\lambda \geq \alpha$.

Absolutely convex.   A subset A of a vector space E is called absolutely convex if for all $x \in A$ and $y \in A$, $\lambda x + \mu y \in A$ whenever $|\lambda| + |\mu| \leq 1$. This is equivalent to saying that A is both convex and balanced.

Balanced.   A subset A of a vector space E is called balanced if for all $x \in A$, $\lambda x \in A$ whenever $|\lambda| \leq 1$.

Barrel.   In a locally convex topological vector space (shortly a convex space) a subset is called a barrel if it is absolutely convex, absorbent and closed. Every convex space has a neighborhood base consisting of barrels.

Barreled space.   A convex space is called barreled if every barrel is a neighborhood.

Bochner theorem.   A complex-valued function f is a characteristic function of a probability measure if and ony if:
(i) It is positive-definite, i.e. for any finite set of real numbers t_j and complex numbers z_k (with complex conjugate $\bar{z}_k$), $1 \leq j \leq n$ $\sum_{j=1}^{n} \sum_{k=1}^{n} f(t_j - t_k) z_j \bar{z}_k \geq 0$,
(ii) it is continuous, (iii) $f(0) = 1$.

Bornological space.   It is a convex space E with the property that every bounded linear mapping on it is continuous, (sometimes called a Mackey space).

Cauchy filter.   In a locally convex topological vector space the filter $\mathcal{F}$ is called a Cauchy filter if, for each neighborhood U, $\mathcal{F}$ contains a set A small of order U, i.e. $x - y \in U$ for all $x, y \in A$, (see also filter).

Closed and closable operators.   Let X, Y be linear topological vector spaces. Then a linear operator $T : X \to Y$ is called a closed linear operator when its graph $G(T)$ constitutes a closed linear subspace of $X \times Y$, $(G(T) = \{(x, Tx) : x \in D(T)\}$, ($D$ is the domain).
T is closed iff the following condition is satisfied : $\{x_n\} \subset D(T)$, $\lim_{n \to \infty} T x_n = y$ imply that $x \in D(T)$ and $Tx = y$.
T is called closable if the closure in $X \times Y$ of the graph $G(T)$ is the graph of a linear operator, say $S : X \to Y$.

© The Editor(s) (if applicable) and The Author(s), under exclusive license to Springer Nature Switzerland AG 2026
U. Çapar, *A Guide to Generalized Functions*,
https://doi.org/10.1007/978-3-032-09184-0

Compact operator.   Let $A \in \mathcal{L}(H, K)$, i.e. a linear bounded operator between Hilbert spaces H and K. If A maps the unit ball (or any bounded set) of H to a relatively compact subset of K, then A is called a compact operator (or completely continuous operator).

Convex spaces.   See locally convex topological vector spaces.

Countably Hilbertian space.   A complete countably Hilbert normed space is called countably Hilbertian space. We may assume that the countable norms are non-decreasing : $\| \cdot \|_1 \leq \| \cdot \|_2 \leq \cdots \leq \| \cdot \|_n \leq \cdots$.

A countably Hilbertian space H is called countably Hilbertian nuclear if moreover $\forall n \in \mathbb{N}$, $\exists m \geq n$ so that the imbedding mapping $I_{mn} : X_m \hookrightarrow X_n$ is nuclear. (X_m, X_n) are Hilbert spaces with norms $\| : \|_m, \| \cdot \|_n$ respectively.

(See also nuclear (trace class) operator).

Diffeomorphism.   A $\mathcal{C}^\infty$ mapping $F : M \to N$ between $\mathcal{C}^\infty$ manifolds is a diffeomorphism if it is a homomorphism and F^{-1} is $\mathcal{C}^\infty$.

M and N are diffeomorphic if there exists a diffeomorphism $F : M \to N$.

Dominated convergence theorem.   (Lebesgue). Let $\{f_n\}$ be a sequence of complex-valued measurable functions on a measurable space (S, Σ, μ). Suppose that the sequence converges pointwise to a function f, i.e. $\lim_{n\to\infty} f_n(x) = f(x)$ exists for every $x \in S$. Assume further that the sequence f_n is dominated by some integrable function g in the sense that $|f_n(x)| \leq g(x)$ for all points $x \in S$ and all n in the index set. Then f_n, f are integrable (in the Lebesgue sense)and $\lim_{n\to\infty} \int_S f_n d\mu = \int_S \lim f_n d\mu = \int_S f d\mu..$

In fact we have the stronger statement $\lim_{n\to\infty} \int_S |f_n - f| d\mu = 0$.

Dual pair.   Let E and E' be any two vector spaces over the same scalar field and let $< x, x' >$ be a bilinear form between E and E' satisfying the conditions:

D: for each $x \neq 0$ in E, there is some $x' \in E'$ with $< x, x' > \neq 0$,

D': For each $x' \neq 0$ in E', there is some $x \in E$ with $< x, x' > \neq 0$.

Then there is a natural linear mapping of E' into E^* (the algebraic dual of E), in which the image of $x' \in E'$ is the linear form f on E with $f(x) = < x, x' >$.This mapping is $(1, 1)$ because of D' and so E' is isomorphic to a vector subspace of E^*. Similarly, D ensures that E is isomorphic to a vector subspace of $(E')^*$. Then (E, E') is called a dual pair.

Filter.   Many topological properties of metric spaces can be suitably defined in terms of sequences, such as a mapping f is continuous at a if $f(x_n) \to f(a)$ whenever $x_n \to a$. These characterizations are in default for a general topological space unless the notion of a sequence is also generalized. One way of doing this is to use filters.

Let E be any set. A non-empty set $\mathcal{F}$ of non-empty subsets of E is called a filter if it satisfies:

(F_1) $A \in \mathcal{F}, B \in \mathcal{F} \implies A \cap B \in \mathcal{F}$,

(F_2) $A \in \mathcal{F}, A \subset B \implies B \in \mathcal{F}$. For example if A is a fixed subset of E, then the set of all subsets containing A is a filter.

Filter base. A non-empty set $\mathcal{B}$ of non-empty subsets of the set E is called a filter base if:
$$A \in \mathcal{B}, B \in \mathcal{B} \implies \exists C \in \mathcal{B}, C \subset A \cap B.$$
The set $\mathcal{F}$ all subsets containing a set of $\mathcal{B}$ is then a filter, called the filter generated by $\mathcal{B}$.

Filter convergence. A filter $\mathcal{F}$ is said to converge to a if every neighborhood of a contains some set of $\mathcal{F}$. The convergence of a filter base is defined the same way. The definitions clearly indicate that a filter base $\mathcal{B}$ converges to a if and only if the filter $\mathcal{F}$ generated by $\mathcal{B}$ converges to a.

The filter convergence is related to sequence convergence in the following way: Let $X_n = \{x_i, i \geq n\}$. Then $\{X_n\}$ is a filter base called the 'elementary filter'. Then x_n converges to a if and only if the associated elementary filter converges to a.

Fock space. Let H be a Hilbert space. Then $\mathcal{F}(H) = \bigoplus_{n=0}^{\infty} H^{\otimes n}$ (by convention $H^{\otimes 0} = K$ the scalar field) is called the Fock space over H.

If $\Gamma(H) = \bigoplus_{n=0}^{\infty} H^{\hat{\otimes} n}$ then $\Gamma(H)$ is called the symmetric Fock space, (for $\hat{\otimes}$ see 'symmetrized tensor product').

Fréchet space. A locally convex topological vector space X is called a Fréchet space if its topology τ is induced by a complete invariant metric d. Every Fréchet space is barreled.

Free Sheaf. Let $(X, \mathcal{O}_X)$ be a ringed space. Consider a sheaf with topological space X and for $\mathscr{S}$, the set of spaces $\mathcal{O}_X$-modules. Then this sheaf is free if it is isomorphic to a direct sum of copies of $\mathcal{O}_X$.

(Ringed space is a family of (commutative) rings parametrized by open subsets of a topological space together with ring homomorphisms).

Gevrey class. Gevrey class functions on a domain $\Omega \subset \mathbb{R}^n$ are between the space of analytic functions $\mathcal{A}(\Omega)$ and the space of smooth functions $C^{\infty}(\Omega)$. In particular, for $\sigma \geq 1$, the Gevrey class $G^{\sigma}(\Omega)$ consists of $g \in C^{\infty}(\Omega)$ such that for every compact subset $K \subset \Omega$ there exists a constant C, depending on g, K such that

$$\sup_{x \in K} |D^{\alpha} g(x)| \leq C^{|\alpha!|+1} |\alpha!|^{\sigma} \quad \forall \alpha \in \mathbb{Z}_+$$

When $\sigma = 1$, $G^{\sigma}(\Omega)$ coincides with the class of analytic functions $\mathcal{A}(\Omega)$. But for $\sigma > 1$ there are compactly supported functions in the class that are not identically zero. It is in this sense that they interpolate between $\mathcal{A}(\Omega)$ and $C^{\infty}(\Omega)$.
The homogeneous heat equation has solutions in $G^2(\Omega)$.

Hilbert-Schmidt operator. Let H and K be Hilbert spaces and let $A \in \mathcal{L}(H, K)$, i.e. a bounded linear operator. If for some base $\{e_n\}$ of H $\sum \|Ae_n\|^2 < \infty$, then A is called a Hilbert-Schmidt operator. Put $\|A\|_2 = (\sum \|Ae_n\|^2)^{1/2}$. $\|A\|_2$ is called the Hilbert-Schmidt norm of A.

Inductive limits and inductive topologies. For each $\gamma \in \Gamma$, let E_γ be a convex t.v.s., u_γ a linear mapping of E_γ into a vector space E, so that $\bigcup_{\gamma \in \Gamma} u_\gamma(E_\gamma)$ spans E. Then there is a finest convex topology on E under which all the u_γ are continuous. A base of neighborhoods for this topology is formed by the set $\mathcal{U}$ of all absolutely convex subsets U of E, such that for each γ, $u_\gamma^{-1}(U)$ is a neighborhood in E_γ.

The convex space E with this topology is called the inductive limit of the convex spaces E_γ by the mappings u_γ.

Infinitesimal generator. Let X be a Banach space and $\{T_t\}_{t \geq 0}$ be a semigroup of continuous linear operators $T_t : X \to X$ such that (i) $T_0 = I$ and (ii) $T_s \circ T_t = T_{t+s}$ for all $t, s \geq 0$. The infinitesimal operator is the operator $\mathcal{L} : D(\mathcal{L}) \to X$: $\lim_{t \to 0} \frac{T_t x - x}{t}$ exists and $\mathcal{L}(x) = \lim_{t \to 0+} \frac{T_t x - x}{t}$ for all $x \in D(\mathcal{L})$.

Ito stochastic differential equation. Consider a differential equation determined in the differential form $d(x)t = a(t, x(t))dt$ with initial condition $x(0) = x_0$. The stochasticity is introduced if the initial value is a stochastic process $X_0(\omega) = Y(\omega)$. Then we have a random differential equation $dX_t = a(x, X_t)dt$, $X_0(\omega) = Y(\omega)$. In many natural and social phenomena there is also an additional perturbation of the solution due to a white noise created by the Brownian motion W_t. Thus we have a so-called Ito stochastic differential equation $dX_t = a(t, X_t)dt + b(x, X_t)dW_t$, $X_0(\omega) = Y(\omega)$ where we have $a(t, x)$ and $b(, x)$ are deterministic functions. This equation can be interpreted as the stochastic integral equation $X_t = X_0 + \int_0^t a(s, X_s)ds + \int_0^t b(t, X_t)dW_t, 0 \leq t \leq T$ where the first integral on the right-hand side is a Riemann integral and the second one is an Ito stochastic integral.

Ito stochastic integral. It is an integral of Riemann-Stieltjes type $\int_0^T X_t(\omega)dW_t(\omega)$, where W_t is a Brownian motion and X_t, the integrand process is adapted to Brownian motion on $[0, T]$, i.e. X_t is a function W_s, $s \leq t$.
In general we can not define the Ito stochastic integral as a pathwise limit of Riemann-Stieltjes sums. Instead the integral is defined as the mean square limit of suitable Riemann-Stieltjes sums. For this purpose we start by simple inegrand processes X_t, i.e. processes assuming constant random values on a finite number sub-intervals of $[0, T]$. When the limit exists in the mean square sense it is the Ito stochastic integral which will be a random variable. If the integral is on $[0, t)$ it will be a stochastic process.

Locally convex topological vector spaces. Let E be a vector space over real or complex field K. Let ξ be a topology on E. If ξ is compatible with the algebraic structure of $x + y$ and λx, i.e. these functions are continuous under ξ, then E is said to be a topological vector space.
If $\mathcal{U}$ is a base of neighborhoods (of the origin) then for each $U \in \mathcal{U}$, i) U is absorbent, ii) $\exists V \in \mathcal{U}$ with $V + V \subset U$, iii) there is a balanced neighborhood $W \subset U$, (c.f. [Rob-R], Proposition 3.3)
In the most important and useful topological vector spaces there is also a base of convex neighborhoods of the origin. Such a space, usually abbreviated as locally convex spaces even simply 'convex spaces'.

L^p **Martingale convergence theorem.** A stochastic process $\{M_t\}_{t \in I}$ is called a martingale if it satisfies $E(M_t | \mathcal{F}_s) = M_s$ whenever $s \leq t$. $E(M_t | \mathcal{F}_s)$ is the conditional expectation given the events up to time s. Martingale property roughly states that at any time s process forgets its past and given the events up to time s, any future expectation is equal to present the value M_s.

If $(M_t)_{t \in T}$ is L^p-bounded, then there exists a random variable M_∞, such that $M_t \longrightarrow M_\infty$ a.s.

Microlocal analysis.   Comprises techniques developed from the 1950 onwards, based on Fourier transforms related to the study of variable coefficients linear or nonlinear partial differential equations. This includes generalized functions, pseudo-differential operators, wave front sets and Fourier integral operators.

The term microlocal implies localization not only with respect to localization in the space, but also with respect to the cotangent space at a given point. This gains importance on manifolds of dimension greater than one.

Montel space.   A separated barreled space with the further property that its closed and barreled subsets are compact is called a Montel space. A montel soace is reflexive. (c.f. [Rob-R]).

Nuclear operator.   (see Trace class operator).

Nuclear space.   Suppose that the topology of a locally convex space X is generated by a family Γ of Hilbertian seminorms. If $\forall p \in \Gamma, \exists q \in \Gamma$ with $p \prec q$, (i.e. p is topologically weaker) and such that $I_{pq} : X_q \to X_p$ is nuclear (is of trace class), then X is called a nuclear space.

Paracompact space.   A topological space is paracompact if every open covering has a refinement which is locally finite.

Polar set.   See 'strong topology'.

Projective limits and projective topologies.   This is a method of topologizing a vector space which in a sense dual to that of taking inductive limits. This time we topologize a vector space F by means of linear mappings v_γ of F into locally convex spaces F_γ such that if $x \in F$ and $x \neq 0$, there is some γ for which $v_\gamma \neq 0$. If $\bigcap_{\gamma \in \Gamma} v_\gamma^{-1}(0) = \{0\}$, then there is a coarsest topology on F compatible with the algebraic structure under which all the v_γ are continuous. Under this topology F is a convex space. If $\mathcal{V}_\gamma$ is a base of absolutely convex neighborhoods in F_γ, the finite intersections of the sets $v_\gamma^{-1}(V_\gamma) (V_\gamma \in \mathcal{V}_\gamma, \gamma \in \Gamma)$, form a base $\mathcal{V}$ of absolutely convex neighborhoods for F with this topology.

The convex space F with this topology is called the projective limit of the convex spaces F_γ by the mappings v_γ.

Quasi-invariance of Wiener measures.   This means the translation invariance of Wiener measures along the Cameron-Martin (C-M) space elements, (c.f. Sect. 5.2.2 Def. 3 for C-M space and also Sect. 5.3, Theorem 1).

Reflexive topological vector spaces.   Suppose that (E, E') is a dual pair and the bidual of E is just E itself. Then we call the dual pair (E, E') reflexive.

The dual pair (E, E') is reflexive if and only if every bounded set in E is contained in a weakly compact set ([Rob-R], iv, Proposition 4).

Riemann-Lebesgue Lemma.   If $f \in L^1(\mathbb{R}^n)$, then $\hat{f} \in \mathcal{C}_0(\mathbb{R}^n)$ and $\| \hat{f} \|_\infty \leq \| f \|_1$. Here $\mathcal{C}_0(\mathbb{R}^n)$ is the supremum normed Banach space of all complex continuous functions on $\mathbb{R}^n$ that vanish at infinity.

In one dimension if $f(x) \in L^1(-\infty, \infty)$, then $\int_{-\infty}^{\infty} f(x)e^{i\lambda x}dx \to 0$ as λ (real) $\to +\infty$ or $-\infty$.

Riesz representation theorem.   Let H be a Hilbert space over $\mathbb{R}$ or $\mathbb{C}$ and T be a bounded linear functional on H. Then there exists some $g \in H$ such that for every $f \in H$ we have $T(f) =< f, g >$. Moreover $\|T\| = \|g\|$. ($\|T\|$ denotes the operator norm of T).

Rotational invariance of Gaussian measures. Consider a standard Gaussian measure with the density $f(x) = \frac{1}{2\pi^{n/2}} e^{-\frac{\|x\|^2}{2}}$, $x \in \mathbb{R}^n$. Let Q be an orthogonal matrix of n dimensions. Then we have

$$\int_{Q^{-1}(B)} f(x)dx = \int_B f(x)dx, \; B \in \mathcal{B}_n.$$

Self-adjoint and essentially self-adjoint operators. Let H be a Hilbert space and $L(H)$ all linear operators $H \to H$. Let $T \in L(H)$ be densely defined. Put $D(T^*) = \{y \in H : \exists c_y > 0, \; \text{such that} \; \forall x \in D(T), | < Tx, y > | \leq \|x\|\}$. ($D(T)$ denotes the domain of T). Then by the Riesz representation theorem $\forall y \in D(T)$, there exists a unique element of H, denoted by T^*y, such that

$$< xiT^*y >=< Tx, y >, \forall x \in D(T).$$

Clearly $T^* \in L(H), T^*$ is called the adjoint of T. If $< Tx, y >=< x, Ty >, \forall x, y \in D(T)$, then T is said to be symmetric if furthermore $T = T^*$, then T is said to be self-adjoint.

Let $T \in L(H)$ be symmetric. Then T is closable, T^{**} is the closure of T and T^{**} is symmetric, ([Hu-Yan], Theorem 1.6, (1)).

If T is symmetric and its closure, i.e. T^{**}, is self-adjoint, then T is said to be essentially self-adjoint.

Seminorms. Let E be a vector space real or complex vector space. A non-negative (finite) real valued function p defined on E is called a seminorm if it satisfies:

(i) $p(x) \geq 0$,

(ii) $p(\lambda x) = |\lambda| P(x)$,

(iii) $p(x + y) \leq p(x) + p(y)$

for all $x, y \in E$ and all $\lambda \in K$, (real or complex scalar field).

By (ii) $p(0) = 0$, but it may happen that $p(x) = 0$ for some $x \neq 0$. In fact $p^{-1}(0)$ is a vector subspace of E.

Given any set Q of seminorms on a vector space E, there is a coarsest topology on E compatible with the algebraic structure in which every seminorm is continuous. Under this topology E is a convex topological vector space (or simply a convex space) and a base of closed neighborhoods is formed by the sets $\{x : \sup_{1 \leq i \leq n} p_i(x) \leq \epsilon\}$, $(\epsilon > 0, \; p_\epsilon \in Q)$.

Soft sheaf. It is a sheaf of spaces $\mathfrak{S}$ on a topological space X, any section of which over some closed subset in X can be extended to a section of $\mathfrak{S}$ over all of X.

Stopping time. Suppose that $(\Omega, \mathcal{F})$ is a measurable space an $I \subset [0, \infty]$ is an index set. A filtration is a family $\{\mathcal{F}_t\}_{t \in I}$ of sub σ-algebras of $\mathcal{F}$ such that $\mathcal{F}_s \subset \mathcal{F}_t$ whenever $s \leq t$. Most of the time the index set is either $\mathbb{R}_+$ or $\mathbb{N}$. In these cases it is simply written $t \geq 0$ and $n \in \mathbb{N}$.

In a stochastic process filtration members $\mathcal{F}_t$ may represent the events occurring up to time t, i.e. a filtration mirrors the increase of information as time increases.

Suppose that $(\Omega, \mathcal{F})$ is a measurable space with a filtration $\mathcal{F}_{t \geq 0}$. A map $T : \Omega \to [0, \infty]$ is called a stopping time if $\{T \leq t\} \in \mathcal{F}_t^+$ where $\mathcal{F}_t^+ = \cap_{\epsilon > 0} \mathcal{F}_{t+\epsilon}$.

Strong Markov property. Let B be a d-dimensional $\mathcal{F}_t$-Brownian motion (see 'stopping time' for $\mathcal{F}_t$) and T be a.s. finite stopping time. Let further $f : \mathcal{C}(\mathbb{R}_+, \mathbb{R}^d) \to \mathbb{R}_+$ be a Borel function. Then the equality $E(f(B_{T+s})_{s \geq 0}|\mathcal{F}_T^+) = E^{B_T}(f)$ is called the strong Markov property, (c.f. [Weiz-Wink], Corollary 2.1.11).

Strong topology. Let (E, E') be a dual pair. If A is a subset of E, the subset of E' consisting of those x',for which $\sup\{| < x, x' > | : x \in A \leq 1\}$ is called the polar of A in E' and is denoted by A^0.

Let $\mathcal{A}$ be any set of weakly bounded subsets of E, (see the weak topology). Then the sets $A^0 (A \in \mathcal{A})$ are absolutely convex and absorbent, (c.f [Rob-R], I.4, Proposition 9) and so there is a coarsest topology ξ' on E' in which they are neighborhoods, a base of neighborhoods in ξ' is formed by the sets $\epsilon \cap_{1 \leq i \leq n} A_i^0 = (\epsilon^{-1} \cup_{1 \leq i \leq n} A_i)^0$, $(\epsilon > 0, A_i \in \mathcal{A})$.

It turns out that convergence in ξ' means uniform convergence on each $A_i \in \mathcal{A}$ and so the topology of uniform convergence on the sets of $\mathcal{A}$ or the topology of $\mathcal{A}$ convergence. Any topology defined in this way is called a polar topology.

The finest polar topology is obtained by taking $\mathcal{A}$ to be the set of all weakly bounded subsets of E. This topology is denoted by $\beta(E', E)$ and is called the strong topology.

Supple sheaf. Let U be open in X; if for Z_1, Z_2 closed in U and $Z = Z_1 \cup Z_2$. Then any $f \in \sigma_Z(U)$, (a section with support in Z) can be split as a sum $f_1 + f_2$ with $f_i \in \sigma_{Z_i}(U)$, $i = 1, 2,$

A flabby sheaf is supple.

Symmetrized tensor product. Let H be a separable Hilbert space and $\mathscr{S}$ be the permutation group on $\{1, 2, \cdots n\}$. For $\sigma \in \mathscr{S}$ define $\pi_\sigma(\phi_1 \otimes \cdots \otimes \phi_n) \equiv \phi_{\sigma(1)} \otimes \cdots \otimes \phi_{\sigma_n})$. Then π_σ extends to an automorphism of the n-fold tensor product $H^{\otimes n}$ such that for $\sigma, \tau \in \mathscr{S}$, $\pi_{\sigma,\tau} = \pi_{\tau,\sigma}$. Thus $\pi_n \equiv \frac{1}{n!}\Sigma_{\sigma \in \mathscr{S}} \pi_\sigma$ is an orthogonal projection on $H^{\otimes n}$. The range of the orthogonal projection π_n which is a closed subspace of n-fold symmetric tensor product and is denoted by $H^{\hat{\otimes}}$. For $\phi_1, \cdots, \phi_n \in H$ the projection of $\otimes_{j=1}^n \phi_j$ is

$$\hat{\otimes}_{j=1}^n \phi_j \equiv \pi_n(\otimes_{j=1}^n \phi_j) = \frac{1}{n}\Sigma_{\sigma \in \mathfrak{S}} \otimes_{j=1}^n \phi_{\sigma(j)}$$

is called the symmetric tensor product of $\phi_1, \cdots, \phi_n$.

Tensor product of Hilbert spaces. Let H_1 and H_2 be Hilbert spaces with inner products $(., .)_1$ and $(., .)_2$ For $\phi_1 \in H_1$ and $\phi_2 \in H_2$, we define their tensor product as a conjugate bilinear form on $H_1 \times H_2 : \phi_1 \otimes \phi_2(\xi_1, \xi_2) \equiv (\phi_1, \xi_1)_1(\phi_2, \xi_2)$, $\xi_1 \in H_1, \xi_2 \in H_2$. Denote by $\mathfrak{E}$ the linear span of $\{\phi_1 \otimes \phi_2 : \phi_1 \in H_1, \phi_2 \in H_2\}$. For $\phi \otimes \phi_2, \psi_1 \otimes \psi_2 \in \mathfrak{E}$ we define the inner product by $b(\phi_1 \otimes \phi_2, \psi_1 \otimes \psi_2) \equiv (\phi_1, \psi_1)_1(\phi_2, \psi_2)_2$ and linearly extend it to $\mathfrak{E}$.

The Hilbert space obtained by the completion of the inner product space $\mathfrak{E}, b)$ is called the Hilbertian tensor product of H_1 and H_2 and denoted by $H_1 \otimes H_2$, (c.f. [Hu-Yan], I.2, Proposition 2.1).

Tensor product of linear operators. Let H_i, K_i, $(i = 1, 2)$ be Hilbert spaces, A_i be a densely defined linear operator from H_i to K_i, $D(A_i) \subset H_i$ be its domain. Denote by $D(A_i) \otimes D(A_2)$ the linear span of $\{\phi_1 \otimes \phi_2 : \phi_i \in D(A_i), i = 1, 2\}$. Then $D(A_1) \otimes D(A_2)$ is dense in $H_1 \otimes H_2$. Define $A_1 \otimes A_2(\phi_1 \otimes \phi_2) = A_1\phi_1 \otimes A_2\phi_2$, $\phi. \in D(A_i)$, $i = 1, 2$ and extend it to be a linear operator on $D(A_1) \otimes D(A_2)$. This defines the tensor product $A_1 \otimes A_2$ of linear operators A_1 and A_2.

Trace class operator. (Nuclear operator). Let A be a densely defined closed operator from Hilbert space H into another Hilbert space K. Put $T = (A^*A)^{1/2}$. There exists a unique linear isometry U from $\mathcal{R}(T)$ (range of T) to K such that $A = UT$ (Polar decomposition). Further let T have the spectral resolution $Tx = \sum_n \lambda_n < x, e_n > e_n(\lambda \neq 0)$. If $\sum_n \lambda_n < \infty$, then A is called a trace class (or nuclear) operator. Put $\|A\|_1 = \sum_n \lambda_n$, (trace norm of A).

Ultrametric space. An ultrametric on a set M is a real-valued function $d : M \times M \to \mathbb{R}$ such that for all $x, y, z \in M$

1. $d(x, y) \geq 0$
2. $d(x, y) = d(y.x)$, symmetry
3. $d(x, x) = 0$
4. If $d(x, y) = 0$ then $x = y$
5. $d(x, z) \leq \max\{d(x, y), d(y, z)\}$.

An ultrametric space is a pair $M; d)$ consisting of a set M together with an ultrametric d on M.

Ultra pseudo-seminorm and ultra pseudo-norm. An ultra pseudo-seminorm is obtained by means of a valuation (see Sects. 3.4.10 and 3.5.1) on $\mathcal{G}$ an ultra pseudo-norm is an ultra pseudo seminorm P such that $P(u) = 0$ implies $u = 0$.

Weak topology. Let (E, E') be a dual pair. To each $x' \in E'$ corresponds a seminorm p on E by $p(x) = | < x, x' > |$. The coarsest topology on E making all these seminorms continuous is called the weak topology on E determined by E' and denoted by $\sigma(E, E')$. It is clearly the coarsest topology on E under which all the linear forms in E' are continuous.

In $\sigma(E, E')$ the sets $\{x : \sup_{1 \leq i \leq n} | < x, x'_i > | \leq 1\} (x'_i \in E')$ form a basis of (closed) neighborhoods. Clearly $| < x, x' < | \leq \epsilon$ if and only if $| < x, \epsilon^{-1}y > | \leq 1$.

The topology $\sigma(E, E')$ is convex and separated. The dual of E under $\sigma(E, E')$ certainly contains E', In fact it is exactly equal to E', (c.f. [Rob-R], Ch. II, Lemma 5.).

Wiener stochastic integral. It is an integral of the form $\int f \, dX$ where f is a deterministic function and X is a stochastic process. More precisely:

Let $(\Omega, \mathcal{F}, \mu)$ be a measure space and $\{W_t\}$, $t \in I$ be a Brownian motion starting at $t = 0$. Let $f \in L^2(I, , \mathcal{B}(I), \lambda)$, λ the Lebesgue measure. First for the step functions of the form $f(t) = \sum_{i=0}^{n-1} f_i I_{(t_i, t_{i+1}]}$, $0 \leq t_0 < t_1 < \cdots t_n t_i \in I$. The integral is given pathwise as

$$\int_I f(s)\,dW_s(\omega) = \sum_{i=0}^{n-1} f_i(W_{t_{i+1}}(\omega) - W_{t_i}(\omega)).$$

The map $f \longrightarrow \int f\,dB$ is an isometry between the space of step functions E as a subspace of $L^2(I, \mathcal{B}(I), \lambda)$ and its image in $L^2(\Omega, \mathcal{F}, \mu)$ where μ is the Wiener measure (c.f. [Weiz-Wink], 1.2 [C])
We may extend the integral to the closure of E in $L^2(\lambda)$ to all of $L^2(\lambda)$. In this way we establish a linear isomorphism between $L^2(I, \mathcal{B}(I), \lambda)$ and $Ł^2(\mathcal{C}(I), \mathcal{B}(\mathcal{C}(I)), \mu|_I)$ which is called the Wiener integral.

References

[Allen]   Allendoerfer, C.B.: Calculus of several variables and differentiable manifolds. MacMillan Publishing Co. (1974)

[Amb]   Ambrose, W.: Products of distributions with values in distributions. J. Reine u. Angew. Math. **315**, 73–91 (1980)

[Bad]   Badrikian, A.: Mesure cylindrique, processus de Wiener et processus aleatoires lineaire. Lecture Notes no. 379. Springer (1971)

[Ban-Cr]   Ban-Crainic: Distributions on manifolds, lecture 2 (pdf). www.staff.sxcience.uu.nl/ (2008)

[Bau]   Bauer, H.: Wahrscheinlichkeitstheorie und Grundzüge der Masstheorie. Walter de Gruyter (1974)

[Bern]   Bernstein, J.N.: Modules over a ring of differential operators. Study of the fundamental solutions of equations with constant coefficients. Funct. Anal. Appl. **5**(2), 1–10 (1971)

[Bia]   Biagioni, A Nonlinear theory of generalized functions. Lecture Notes in Mathematics, no. 1421. Springer (1980)

[Booth]   Boothby, W.M.: An Introduction to Differentiable Manifolds and Riemannian Geometry, 2nd edn. Academic Press (1986)

[Bour]   Bourbaki, N.: Eléments de mathematique, topologie general. Herman, Paris (1971)

[Br-Dn-He]   Brouder, C., Viet Dang, N., Hélein, F.: A smooth introduction to the wavefront set. arXiv:1404.1778 vi [math-phy] 1-29 (2014)

[Carl]   Carlson, L.: On convergence and growth of partial sums of Fourier series. Acta Math. **116**, 135–157 (1966)

[Cart]   Cartier, P.: Processus aléatoire généralisés; Séminaire Bourbaki, 16e anné, no. 272 (1963–64)

[Chow]   Chow, P.L.: Stochastic Partial Differential Equations. Chapman & Hill (2007)

[Col-1]   Colombeau, J.F.: New Generalized Functions and Multiplication of Distributions. North-Holland Math. Studies 113 (1985)

[Col-2]   Colombeau, J.F.: Differential calculus and holomorphy, real and complex analysis in locally convex spaces. North-Holland Math. Studies 64 (1982)

[Col-Sp1]   Colombini, F., Spagnoli, S.: An example of weakly hyperbolic Cauchy problem not well-posed in C^∞. Acta Math. **148**, 243–253 (1982)

© The Editor(s) (if applicable) and The Author(s), under exclusive license to Springer Nature Switzerland AG 2026   251
U. Çapar, *A Guide to Generalized Functions*,
https://doi.org/10.1007/978-3-032-09184-0

[Col-Sp2]   Colombini, F., Spagnolii S.: Some examples of hyperbolic equations without local solvability. Annales Scientifique de l'E.N.S. 4ê series tome 22, no. 1, 109–125 (1989)

[Col-Ober]   Colombeau, J.F., Oberguggenberger, M.: On a hyperbolic system with a compatible quadratic term, generalized solutions, delta waves and multiplication of distributions. J. Commun. Partial Differ. Equ. **15**, 905–938 (1990)

[Çap-Ak 1]   Çapar, U., Aktuğlu, H.: An overall view of stochastics in Colombeau related algebras. In: Çapar, U., Üstünel, A.S. (Eds.) Stochastic Analysis and Related Topics VIII, Progress in Probability, vol. 53, pp. 67–90. Birkhauser (2003)

[Çap-Ak 2]   Çapar, U., Aktuğlu, H.: A new construction of random Colombeau distributions. Stat. Probab. Lett. **54**, 291–299 (2001)

[Çap-Ak 3]   Çapar, U., Aktuğlu, H.: On the extension problems in processes generalized in Colombeau sense. In: Proceedings of the 13th Prague Conference in Probability Theory and Stochastic Processes, vol. 1, pp. 71–81. Academia Prague (1998)

[Çap-1]   Çapar, U.: Imbedding of infinite dimensional distributions into simplified Colombeau type algebras. Monatshefte für Mathematik, vol. 198, no. 4, December, pp. 755–770 (2022)

[Çap-2]   Çapar, U.: Imbedding of Meyer-Watanabe generalized functions via Donsker's delta function. Trends in Pure and Applied Mathematics, Alba Iulia, Romania; abs. pp. 20–22 (2017)

[Çap-3]   Çapar, U.: Colombeau solutions of a nonlinear stochastic predator-prey equation. Turk. J. Math. **37**(6), 1048–1060 (2013)

[Çap-4]   Çapar, U.: On the solution of a nonlinear stochastic predator-prey equation. Ukranian Mathematical Congress (dedicated to the centennial of N.N. Bogoliubov. absç. pp. 8 (2009)

[Çap-5]   Çapar, U.: An excursion in Wiener and whitenoise space distributions and the Korezlioğlu - Ustunel distributions; (invited paper). In: International Workshop on Recent Developments in Mathematical Finance and Stochastic Calculus (in memory of Professor H. Korezlioğlu, 4–12 METU (2008)

[Çap-6]   Çapar, U.: Imbedding of Hida distributions in Colombeau algebras. In: 7th Bernouilli Society, IMS World Congress on Probability and Statistics, abs. no. 47, Singapore (2008)

[Çap-7]   Çapar, U.: White noise functionals as infinite dimensional random Colombeau distributions. In: ICM (International Conference of Mathematicians, Madrid, abs. no. 34 (2006)

[Çap-8]   Çapar, U.: Introduction to random nonlinear distributions. In: Proceedings of 22nd Conference on Stochastic Processes, and its Applications, 25–27, Vrije Universiteit, Amsterdam (1993)

[Deb-Ver-Vin]   Debrouwere, A., Vernavere, H., Vindas, J.: Optimal embedding of ultradistributions into differential algebras. Monatsh. Math. 407–438 (2018)

[Del-Scar1]   Delcroix, A., Scarpalezos, D.: Asymptotic scales, asymptotic algebras. Integr. Transform. Spec. Funct. **60**, 157–166 (2000)

[Del-Scar2]   Delcroix, A., Scarpalezos, D.: Topology on asymptotic algebras of generalized functions and applications. Monatshefte für Mathematics **29**, 1–14 (2000)

[Ego]   Egorov, Yu.V.: A contribution to the theory of generalized functions. Russ. Math. Surv. **45**(5), 1–49 (1990)

[Ehr]   Ehrenpreis, L.: On the theorem of kernels of Schwartz. Proc. Am. Math. Soc. **7**, 713–718 (1956)

[Fri-Jo]   Friendlander, F.G., Joshi, M.: Introduction to the Theory of Distributions, 2nd edn. Cambridge University Press (1998)

[Gar-1]   Garetto, C.: Topological structures in Colombeau algebras: topological $\tilde{\mathbb{C}}$-modules and duality theory. Acta Appl. Math. **88**, 81–123 (2005)

[Gar-2]   Garetto, C.: Pseudo-differential operators in algebras of generalized functions and global hypoellipticity. Acta Appl. Math. **80**, 123–174 (2004)

[Gar-3]   Garetto, C.: Generalized Fourier integral operators on spaces of Colombeau type. arXiv: 0803.0284vI [math AP] 3 Mar 2008

[Gar-4]   Garetto, C.: On duality theory and pseudodifferential techniques for Colombeau algebras: generalized delta functionals, kernels an wave front sets. arXiv: math/0507222v2 [math.FA] (2005)

[Gel-Vil]   Gel'fand, I.M., Vilenkin, N.Ya.: Generalized Functions, vol. 4. Academic Press (1964)

[Gross-et-al] Grosser, M., Kunzinger, M., Steinbauer, P.: A global theory of generalized functions. Adv. Math. **166**, 50–72 (2003)

[Groth]   Grothendieck, A.: Local cohomology. Lecture Notes in Math, vol. 41. Springer, Berlin (1967)

[Hicks]   Hicks, N.J.: Notes on differential geometry. Van Nostrand Mathematical Studies, vol. 3 (1965)

[Hida-1]   Hida, T.: Stationary Stochastic Processes. Princeton University Press (1970)

[Hida-2]   Hida, T.: Brownian Motion. Springer (1998)

[Hi-Kuo]   Hida, T., Kuo, H.H., Pothoff, J., Streit, L.: White noise—An infinite dimensional calculus. Kluwer Academic Publishers (1993)

[Hir-Og]   Hirata, V., Ogato, H.: On the exchange formula for distributions. J. Sci Hiroshima Univ. Ser. A **22**, 147–152 (1958)

[Hor]   Horváth, J.: Topological Vector Spaces and Distributions, vol. 1. Addison-Vesley Publishing Co. (1966)

[Hörm-1]   Hörmander, L.: On the theory of general partial differential operators. Acta Math. **94**(1), 161–246 (1953)

[Hörm-2]   Hörmander, L.: The Analysis of Linear Partial Differential Operators, I, II. Springer (1983)

[Hörm-3]   Hörmander, L.: Fourier integral operators. Acta Math. **1207**, 79–183 (1971)

[Hug]   Hugoniot, J.Y.: Memoire sur la propagation des mouvements dans les corps et spécialement dans les gaz parfaits. J. de l'École Poytechnique **57**, 3–97

[Hu-Yan]   Huang, Z.Y., Yan, J.A.: Introduction to Infinite Dimensional Stochastic Analysis. Kluwer Academic Publishers and Science Press (2000)

[Ike-Wat]   Ikeda, N., Watanabe, S.: Stochastic Differential Equations and Diffusion Processes, 2nd edn. North Holland (1989)

[Kal]   Kallianpur, G.: Abstract Wiener spaces and their reproducing kernel Hilbert spaces. Z. Wahr. verw. Gebiete (17), 113–123 (1971)

[Kal-Xi]   Kallianpur, G., Xiong, J.: Stochastic differential equations in infinite dimensional spaces. Lecture Notes- Monograph Series, vol. 26

[Kam]   Kaminski, A.: Convolution, product and Fourier transform of distributions. Stud. Math. **74**, 83–96 (1982)

[Kanek]   Kaneko, A.: Introduction to Hyperfunctions. Kluwer Academic Publishers (1989)

[Kar-Shr]   Karatzas, I., Shreve, E.S.: Brownian Motion and Stochastic Calculus. Springer (1988)

[Kel]   Keller, J.L.: General Topology. Springer (1955)

[Khan-Str]   Khandekar, D.C., Streit, L.: Constructing the Feynman integral. An. Physik 49–55 (1992)

[Komat-1]   Komatsu, H.: Ultradistributions I, structure theorems and a characterization. J. Fac. Sci. Univ. Tokyo Sect IA Math **20**, 25–105 (1973)

[Komat-2]   Komatsu, H.: An elementary theory of hyperfunctions and microfunctions. P.D.E.'s, Banach center publications, vol. 27, Institute of Mathematics, Polish Academy of Sciences, Warszawa (1992)

[Komat-3]   Komatsu, H.: Ultradistributions, hyperfunctions and linear differential equations. Asterisque, tome 2–3, 252–271 (1973)

[Kre-Kre]   Krée, M., Krée, P.: Continuite de la divergence dans les espaces de Sobolev relatif á l'espace de Wiener. C.R. Acad. Sci. **296**, 833–836 (1983)

[Kuo-1]   Kuo, H.H.: White Noise Distribution Theory. CRC Press, Newyork (1996)

[Kuo-2]   Kuo, H.H.: Gaussian measures in Banach spaces. Lecture Notes in Mathematics, no. 463. Springer (1975)

[Kuo-3]   Kuo, H.H.: Lectures on white noise analysis. Soochow J. Math. **18**, 222–300 (1992)

[Kuo-Pot]   Kuo, H.H., Pothoff, J., Streit, L.: A characterization of white noise test functionals. Nagoya Math. J. 185–194 (1991)

[Lee]   Lee, Y.J.: Generalized functions on infinite dimensional spaces and its application to white noise calculus. J. Funct. Anal. **82**, 429–464 (1989)

[Lax]   Lax, P.D.: The formation and decay of shock waves. Am. Math. Mon. 227–241 (1972)

[Lewy]   Lewy, H.: An example of a smooth linear partial differential equation without solution. Ann. Math. **56**(2) (1956)

[Li-1]   Li, C.K.: An approach for distributional products on $\mathbb{R}^n$. Integr. Transform., Spec. Funct. **16**, 139–151 (2005)

[Li-2]   Li, C.K.: A review of the products of distributions. Mathematical Methods in Engineering, pp. 71–96. Springer (2007)

[Martin]   Martin, D.: Manifold theory, an introduction for mathematical physicists. Ellis Horwood (1991)

[Mey-1]   Meyer, P.A.: Transformation de Riesz pour les lois Gaussien. Sém. Probab., XVII, Lecture Notes in Mathematics, vol. 1059, pp. 179–193. Springer (1983)

[Mey-2]   Meyer, P.A.: Quelques resultats analytique sur le semigroupe d'Ornstein - Uhlenbeck de dimension infinie. Lecture Notes in and Information, vol. 49, pp. 201–214. Springer (1983)

[Miko]   Mikosch, T.: Elementary Stochastic Calculus. World Scientific (1998)

[Miku-1]   Mikusinski, J.: Irregular operations on distributions. Stud. Math. **20**, 163–169 (1960)

[Miku-2]   Mikusinski, J.: Criteria for the existence and associativity of the product of distributions. Stud. Math. **21**, 253–259 (1962)

[Morim]   Morimoto, M.: An Introduction to Sato's Hyperfunctions. AMS, Providence R.I. (1993)

[Munk]   Munkres, J.R.: Elementary Differential Topology. Princeton University Press (1963)

[Nel]   Nelson, E.: The free Markov field. J. Funct. Anal. **12**, 117–227 (1973)

[Nig]   Nigsch, E.: Colombeau generalized functions on manifolds; pdf, Diplomarbeit, Vienna University of Technology (2002)

[Nua]   Nualart, D.: The Malliavin Calculus and Related Topics. Springer (1995)

[Oba]   Obata, N.: White noise calculus and Fock space. Lecture Notes in Mathematics no. 1577, Springer (1994)

[Ober-1]   Oberguggenberger, M.: Products of distributions, non-standard methods. Z. Anal. Anwendungen **7**(4), 347–365 (1988)

[Ober-2]   Oberguggenberger, M.: Generalized solutions to semilinear hyperbolic systems. Monatsh. Math. **103**, 133–144 (1987)

[Ober-3]   Oberguggenberger, M.: Generalized functions and stochastic processes. Progress in Probability, vol. 36. Birkhauser (1995)

[Ober-Rus]   Oberguggenberger, M. Russo, F.: Nonlinear SPDEs, Colombeau solutions and pathwise limits. In: Proceedings of 6th Oslo-Silivri Workshop on Stochastic Analysis and Related Topics, pp. 325–339. Birkhauser (1998)

[Pot-St]   Pothoff, J., Streit, L.: A characterization of Hida distributions. J. Funct. Analy. **101**, 212–229 (1990)

[Prot]   Protter, P.: Stochastic Integration and Differential Equations (A New Approach). Springer (1990)

[Reed-Sim]   Reed, M., Simon, B.: Methods of Modern Mathematical Physics, vol. II. Academic Press (1975)

[Rich-Youn]   Richards, I., Youn, H.: Theory of Distributions (A Non-technical Introduction). Cambridge University Press (1990)

[Rob-R] Robertson, A.P., Robertson, W.J.: Topological Vector Spaces. Cambridge University Press (1966)

[Rod] Roddier, F.: Distribution et Transformation de Fourier. Ediscience (1971)

[Ros] Rosinger, E.E.: Generalized solutions of nonlinear partial differential equations. North-Holland Mathematics Studies, 146 (1987)

[Rosay] Rosay, J.-P.: A very elementary proof of the Malgrange-Ehrenpreis theorem. Am. Math. Mon. **98**(6), 518–523 (1991)

[Royden] Royden, H.L.: Real Analysis, 2nd edn. Macmillan, Newyork (1968)

[Rud] Rudin, W.: Functional Analysis. Tata McGraw-Hill Publishing Co. (1974)

[Sato] Sato, M.: Theory of hyperfunctions I. J. Fac. Sci. Univ. Tokyo Sect **I**(8), 139–193 (1959)

[Schf] Schaefer, H.H.: Topological Vector Spaces. Graduate Texts in Mathematics, 2nd edn. Springer (1999)

[Schw-1] Schwartz, L.: Théorie des distributions I, II. Hermann, Paris (1950, 1951)

[Schw-2] Schwartz, L.: Sur l'impossibilité de la multiplication des distributions. C. R. Acad. Sci. Paris **239**, 847–848 (1954)

[Schw-3] Schwartz, L.: Espaces de fonctions differentiables á valeurs vectorielles. J. Analyse Math. **4**, 88–148 (1954)

[Shi] Shigekawa, I.: Stochastic Analysis. Amazon (1987)

[Sob] Sobolev, S.L.: Certaines applications de l'analyse fonctionelle $à$ la physique mathématique. Leningrad (1945)

[Sobc] Sobczyki, K.: Stochastic Differential Equations (with Applications to Physics and Engineering). Kluwer Academic Publications (1991)

[Sug] Sugita, H.: Sobolev Spaces of Wiener Functionals and Malliavin Calculus. J. Math., Kyoto Univ. 25, 31–48 (1985)

[Sze] Szegö, G.: Orthogonal Polynomials, vol. 23. AMS Colloqium Publication (1939)

[Takig] Takiguchii, T.: On the structure of hyperfunctions and ultradistributions. W. kurims, Kyoto-u .ac.jp, 71–82 (2013)

[Tor] Torstadius, D.: Pseudo-differential operators and their properties. U.U.D.M. Project Report 2021: 38, Uppsala Universitet (2021)

[Üst-Zak] Üstünel, A.S., Zakai, M.: Transformation of measures on Wiener space. Springer Monographs in Mathematics (2000)

[Üst-1] Üstünel, A.S.: Representation of distributions on Wiener space and stochastic calculus of variations. J. Funct. Anal. **70**, 126–139 (1987)

[Üst-2] Üstünel, A.S.: An introduction to analysis on Wiener space. Springer Lecture Notes in Mathematics, no. 1610 (1996)

[Vak] Vakhania, N.N.: Probability Distributions in Linear Spaces. Elsevier, North Holland (1981)

[Vlad] Vladimirov, V.S.: Obobshchennye funktsii; ikh preminenya. Generalized functions and their applications, Znanie, Mosom (1990)

[Wal] Walsh, J.B.: An introduction to stochastic partial differential equations, Ecole d'Ete de probabilités de Saint Flour XIV - 1994, pp. 265–439. Springer (1986)

[Wat-1] Watanabe, S.: Lectures on stochastic differential equations and Malliavin calculus. Springer, Tata Inst. Fund. Research (1984)

[Wat-2] Watanabe, S.: Analysis of Wiener functionals (Malliavin calculus) and its applications to heat kernels. Ann. Prob. **15**, 1–39 (1988)

[Wag-1] Wagner, P.: Parameter integration zur berechnung von fundamental lösungen. Rozprawy Matematyczne Panstwowe Wydawnictwo Naukowe (1984)

[Wag-2] Wagner, P.: A new constructive proof of the Malgrange-Ehrenpreise theorem. Am. Math. Mon. **116**(5), 457–462 (2009)

[Weiz-Wink] Weizsacker, H.W., Winkler, G.: Stochastic Integrals. Friedr. Vieg & Son. Braunschweig/Wiesbaden

[Yan] Yan, J.A.: From Feynman-Kac formula to Feynman integrals via analytic continuation. Stoch. Proc. Appl. **54**, 215–232 (1994)

[Yos] Yosida, K.: Functional Analysis. Springer (1980)

The manufacturer's authorised representative in the EU is Springer
Nature Customer Service Centre GmbH, Europaplatz 3, 69115 Heidelberg,
Germany. If you have any concerns regarding our products, please
contact ProductSafety@springernature.com

Printed and bound by CPI Group (UK) Ltd, Croydon, CR0 4YY
02/07/2026
02154962-0001